Current Research Progress of Optical Networks

Maode Ma
Editor

Current Research Progress of Optical Networks

Springer

Editor
Dr. Maode Ma
Nanyang Technological University
School of Electrical & Electronic Engineering
50 Nanyang Avenue
Singapore 639798
Singapore
Maode_Ma@pmail.ntu.edu.sg

ISBN 978-90-481-8218-3 e-ISBN 978-1-4020-9889-5

DOI 10.1007/978-1-4020-9889-5

Printed on acid-free paper

9 8 7 6 5 4 3 2 1

springer.com

Preface

Optical communication networks have played and will continue to play a prominent role in the development and deployment of network infrastructures. The optical systems and networks with protocols are capable to meet diverse requirements from wide range of applications and services. Nowadays, optical networks have evolved dramatically to more flexible, intelligent and reliable network infrastructures with new optical switching architectures and technologies as well as advanced control and management protocols to offer core backbone, metro and access network services. The widespread deployment of optical communication networks will certainly produce many significant changes to our life in the new century.

This edited book is a product from great contributions made by many experienced researchers who have much knowledge and rich teaching and/or research experience in the optical communications and networking. This edited book is intended to be a comprehensive reference book to address the recent research and technology developments of optical communications and networking for both academia and industry. It can serve as an introduction book for beginners to get the fundamental knowledge of various aspects of optical communication systems. It is also expected to be a rich reference for researchers and engineers to understand the recent developments of the technology in order to promote further development of optical communications and services.

This book consists of 12 chapters on the optical system and network design, MAC and higher layer protocols, traffic modeling and routing, network control and management, etc. Each of the chapters is either a technical overview/literature survey on a particular topic or a proposed solution to a research issue in optical communications and networking. The 12 chapters can be roughly classified into 3 parts. The first part is major on optical burst switching (OBS) and optical packet switching (OPS) networks, consisting of Chapters 1, 2, 3 and 4. Chapter 1 addresses the problem of quality of service (QoS) provisioning in the OBS networks. Several QoS scenarios which are based on the most referenced QoS mechanisms have been presented. Chapter 2 presents a proposal of a novel scheme to provide end-to-end proportional differentiated services to an arbitrary number of traffic classes in the OBS networks. A mathematical model to evaluate the loss probabilities in multiservice OBS networks has been derived. Chapter 3 is to study the switch architectures applicable to synchronous fixed-length OPS networks. Several analytical models on these switches have been proposed to describe them as discrete time Markov chains. Chapter 4 presents a novel approach for the performance analysis of OPS bus-based networks with unslotted carrier sense multiple access with collision avoidance (CSMA/CA) protocol. The network has been modeled as a multiple-priority M/G/1 queuing system with preemptive-repeat-identical (PRI) service discipline. The second part of the book is major on resource allocation, traffic scheduling, and performance evaluation of Ethernet Passive Optical Networks (EPON) and Wavelength Division Multiplexing (WDM) optical networks, consisting of Chapters 5, 6, 7 and 8. Chapter 5 presents a novel Early-Dynamic Bandwidth Allocation (E-DBA) mechanism incorporated with a Prediction-Based Fair Excessive Bandwidth Allocation (PFEBA) scheme in EPONs. Chapter 6 aims to present a comprehensive survey on the up-to-date DBA schemes for EPON networks. Numerous DBA schemes have been classified into different categories with an introduction on common features as well as the merits and shortcomings. Besides, the descriptions and comments on each individual DBA scheme have also been presented to show their comparisons. Chapter 7 is a proposal to employ a WDM passive optical network as an optical access network to show that it is much more attractive than the traditional access networks due to its huge bandwidth provisioning. The issue of QoS service offered to video traffic over a passive WDM optical network, served as an access network, has been studied. Chapter 8 is an introduction on single-hop passive-star coupled WDM optical networks, followed by a comprehensive survey of the-state-the-art MAC protocols for WDM optical networks. The third part of the book is mainly to address the issues related to robust routing, wavelength assignment, and dynamic traffic grooming in WDM optical networks, consisting of Chapters 9, 10, 11 and 12. Chapter 9 is a design to develop a logical topology and a routing scheme over the topology that minimizes the congestions of the network. Chapter 10 is a proposal of a novel solution for high-speed optical networks which reconciles packet switching and optical transparency requirements while avoiding actual technology bottlenecks. A new concept of traffic aggregation has been introduced in optical mesh networks with aims to eliminate both the bandwidth underutilization and scalability problems. Chapter 11 proposes a guaranteed quality of recovery (GQoR) mechanism for WDM mesh networks. Four GQoR levels are used to support customized services. Each of them is mapped to the adaptive recovery methodology. Once a failure occurs, the control system activates the recovery mechanism in compliance with the GQoR level. Chapter 12 studies the reactions of different versions of TCP protocols to a failure in a continental-scale network with aim to find out the particular failure duration to cause file transfer times increase markedly. The resilience behavior of SACK, NewReno, and Reno TCP are studied in the case of a single TCP session and multiple TCP flows.

It is obvious that without the great contributions and profound, excellent knowledge on optical communications and networking from the authors of each chapter, this book could not be published to serve as a reference book to the world. I wish to thank each contributor of the book for his/her time, huge efforts, and great enthusiasm to the publication of the book. I would also thank the publisher of the book and the representatives, Mr. Mark de Jongh, Mrs. Cindy Zitter, and Mr. Rajasekar Subramanian, Integra for their patience and great helps in the publication process.

Singapore Maode Ma

Contents

Contributors

Y. Aneja University of Windsor, Windsor, Ontario Canada N9B 3P4

Pablo Jesús Argibay-Losada Departamento de Enxeñería Telemática, Universidade de Vigo, Campus Universitario s/n, E-36310 Vigo, Spain

Tülin Atmaca Institut National des Télécommunications, 9, Rue Charles Fourier, 91011, Evry, France

S. Bandyopadhyay University of Windsor, Windsor, Ontario Canada N9B 3P4

Nizar Bouabdallah INRIA, Campus de Beaulieu, F-35042 Rennes, France

Alexandre Brandwajn University of California at Santa Cruz, Baskin School of Engineering Santa Cruz, CA 95064, USA

Davide Careglio Universitat Politècnica de Catalunya, C. Jordi Girona 1–3, 08034 Barcelona, Spain

Chun-Che Chang Department of Computer Engineering and Science, Yuan-Ze University, Chung-Li, Taiwan, 32026

Manuel Fernández-Veiga Departamento de Enxeñería Telemática, Universidade de Vigo, Campus Universitario s/n, E-36310 Vigo, Spain

I-Feng Huang National Taiwan College of Performing Arts, Taipei, Taiwan

Xiaohong Huang Network Technology Research Institute, Beijing University of Posts and Telecommunications, Beijing, China

I-Shyan Hwang Department of Computer Engineering and Science, Yuan-Ze University, Chung-Li, Taiwan, 32026

A. Jaekel University of Windsor, Windsor, Ontario Canada N9B 3P4

Liang-Yu Ke Department of Computer Engineering and Science, Yuan-Ze University, Chung-Li, Taiwan, 32026

Mirosław Klinkowski Universitat Politècnica de Catalunya, C. Jordi Girona 1–3, 08034 Barcelona, Spain; National Institute of Telecommunications, 1 Szachowa Street, 04-894 Warsaw, Poland

Ivan Kotuliak Slovak University of Technology, Ilkovicova 3, 812 19 Bratislava, Slovakia

Cándido López-García Departamento de Enxeñería Telemática, Universidade de Vigo, Campus Universitario s/n, E-36310 Vigo, Spain

Y. Lu University of Windsor, Windsor, Ontario Canada N9B 3P4

Maode Ma School of Electrical and Electronic Engineering, Nanyang Technological University, Nanyang Avenue, Singapore 639798

Mike H. MacGregor Department of Computing Science, University of Alberta, Edmonton, Alberta, Canada T6G 2E8

Marian Marciniak National Institute of Telecommunications, 1 Szachowa Street, 04-894 Warsaw, Poland

Viet Hung Nguyen Institut National des Télécommunications, 9, Rue Charles Fourier, 91011, Evry, France

Yang Qin School of Electrical and Electronic Engineering, Nanyang Technological University, Nanyang Avenue, Singapore 639798

Hung-Jing Shie Department of Computer Science and Engineering, Yuan-Ze University, Chung-Li, Taiwan, 32026

Zen-Der Shyu Department of General Studies, Army Academy, Chung-Li, Taiwan, 32092

Josep Solé-Pareta Universitat Politècnica de Catalunya, C. Jordi Girona 1–3, 08034 Barcelona, Spain

Andrés Suárez-González Departamento de Enxeñería Telemática, Universidade de Vigo, Campus Universitario s/n, E-36310 Vigo, Spain

Joanna Tomasik SUPELEC, Plateau de Moulon, 91 192 Gif-sur-Yvette Cedex, France

Qiang Ye Department of Computer Science and Information Technology, UPEI, Charlottetown, PE, Canada C1A 4P3

Yongqing Zhu Data Storage Institute, A*STAR, Singapore

About the Editor

Dr. Maode Ma received his BE degree in computer engineering from Tsinghua University in 1982, ME degree in computer engineering from Tianjin University in 1991 and Ph.D. degree in computer science from Hong Kong University of Science and Technology in 1999. Dr. Ma is an Associate Professor in the School of Electrical and Electronic Engineering at Nanyang Technological University in Singapore. He has extensive research interests including optical networking, wireless networking, and so forth. He has been a member of the technical program committee for more than 80 international conferences. He has been a technical track chair, tutorial chair, publication chair, and session chair for more than 40 international conferences. Dr. Ma has published more than 130 international academic research papers on optical networks and wireless networks. He currently serves as an Associate Editor for *IEEE Communications Letters*, an Editor for *IEEE Communications Surveys and Tutorials*, an Associate Editor for *International Journal of Wireless Communications and Mobile Computing*, an Associate Editor for *International Journal of Security and Communication Networks*, an Associate Editor for *International Journal of Vehicular Technology* and an Associate Editor for *International Journal of Network and Computer Applications*.

Chapter 1
A Performance Overview of Quality of Service Mechanisms in Optical Burst Switching Networks

Mirosław Klinkowski, Davide Careglio, Josep Solé-Pareta and Marian Marciniak

Abstract This Chapter addresses the problem of quality of service (QoS) provisioning in optical burst switching (OBS) networks. OBS is a photonic network technology aiming at efficient transport of IP traffic. The lack of optical memories, however, makes the operation in such networks quite complicated, especially if one wants to guarantee a certain level of service quality. Indeed quality demanding applications such as, for instance, real-time voice and video transmissions, need for additional mechanisms so that to preserve them from low priority data traffic. In this context the burst blocking probability metric is perhaps of the highest importance in OBS networks. In this Chapter we present a general classification of QoS provisioning methods considered for OBS networks. We study several QoS scenarios that are based on the most referenced QoS mechanisms and we confront their performance in the same evaluation scenario consisting of a single isolated node. Among all the mechanisms analysed, the best overall performance is achieved with a burst preemptive mechanism. Since the preemptive mechanism produces the problem of resources overbooking in the network we address this issue as well.

Keywords Brust preemption · Offset time differentiation · Preemption window · QoS mechanisms · Optical burst switching · Performance evaluation · Quality of service

1.1 Introduction

Optical switching (OBS) [1] is a promising solution for reducing the gap between switching and transmission speeds in future networks. The client packets are aggregated and assembled into optical data *bursts* in the edge nodes of an OBS network. A *burst control packet* is transmitted in a dedicated control channel and delivered with a small *offset time* prior to the data burst. In this way the electronic controller of an intermediate (core) node has enough time both to reserve a wavelength on its

M. Klinkowski (✉)
Universitat Politècnica de Catalunya, C. Jordi Girona 1–3, 08034 Barcelona, Spain;
National Institute of Telecommunications, 1 Szachowa Street, 04-894 Warsaw, Poland

M. Ma. (ed.), *Current Research Progress of Optical Networks*,

DOI 10.1007/978-1-4020-9889-5_1,

output link, usually for the duration time of the incoming burst, and to reconfigure dynamically the switching matrix. When the burst transmission is finished in a node the output wavelength is released for other connections. Such a temporary usage of wavelengths allows for higher resource utilization as well as for better adaptation to highly variable input traffic in comparison to optical circuit-switching networks. Moreover the aggregation of data packets helps to overcome the fast processing and switching requirements of optical packet switching (OPS) technology.

There are two distinct signalling architectures considered for OBS networks. The first one is based on a connection-oriented signalling protocol which performs end-to-end resources reservation with acknowledgment in so called *two-way reservation* mode [2]. The other exploits a connection-less signalling protocol which allocates the resources on-the-fly, a while before the burst arrival, in a *one-way reservation* mode [1]. Since the problem of the two-way reservation signalling concerns the latency due to the connection establishment process [3, 4] such architectures are considered mostly for short-distant metropolitan networks.

The one-way reservation signalling that can operate effectively in large distance OBS networks performs according to a statistical multiplexing paradigm; hence it encounters the problem of burst contention inside the network. Indeed, when a burst control packet enters a node in order to perform the wavelength reservation for its data burst, it may happen that the requested resources are not available at the output link and the burst has to be dropped. The lack of optical random access memories complicates the resolution of burst contention in optical networks. To alleviate this problem several mechanisms based on wavelength conversion, deflection routing and fibre delay line (FDL) buffering [5] together with dedicated burst scheduling algorithms [6] have been proposed.

A similar difficulty appears when we try to preserve *high priority* (HP) loss/delay sensitive traffic from *low priority* (LP) regular data traffic. For non-real-time applications, such as data file transfers or e-mails, the loss of data burst is not so critical issue since adequate packet level protocols can provide retransmission capability to recover the dropped packets. However, in the transmission of real-time information, for instance in voice, video, telemedicine applications, packets must arrive within a relatively narrow time window to be useful to reconstruct the multimedia signal. Retransmission in this case would add extensive delay to the reconstruction and would cause clipping or unintelligible speech as well as discontinuous picture. Here the loss of data burst means an unrecoverable loss of some information. Taking into account the foregoing, the burst loss probability is considered as the primary metric of interest in the context of quality of service (QoS) provisioning in OBS networks.

There are several techniques that enable QoS differentiation in OBS networks. The most addressed are based on *offset differentiation*, [7] *preemptive dropping*, [8, 9] *threshold-based dropping*, [10, 11] and *intentional dropping* [10] principle. All these techniques try to resolve the burst contention problem with an assumption that the bursts belonging to HP class are treated somehow better than LP bursts. As long as each QoS mechanism achieves it in a different way each one may offer different performance.

There can be found several works in the literature that provide a comparative performance analysis of selected QoS mechanisms. For instance Zhang [10] studies different QoS scenarios built on a wavelength threshold-based principle and an intentional dropping principle with the purpose of absolute quality guarantees. Vokkarane [9] compares the performance of different QoS schemes with a burst segmentation approach applied. Also, a comparative performance study of different optical packet-dropping techniques evaluated in an OPS network scenario is presented in [11].

In this Chapter we make an extension to these studies. In particular, we confront the performance of a frequently referenced offset time differentiation mechanism with two burst-dropping techniques, namely, with a preemptive dropping and a wavelength threshold-based dropping. All these mechanisms aim at the differentiation of burst loss probabilities in a connection-less OBS network.

The rest of the Chapter is organized as follows. In Section 1.2 we discuss some basic concepts of QoS provisioning in OBS networks. In Section 1.3 we present a general classification of QoS schemes considered for OBS networks. In Section 1.4 we study the performance of selected QoS mechanisms and highlight their pros and cons. In Section 1.5 we discuss the problem of resources overbooking that is inherent to a burst preemptive mechanism. Finally Section 1.6 concludes the Chapter.

1.2 Basic Concepts of QoS in OBS Networks

1.2.1 QoS Metrics

An effective QoS provisioning in OBS should engage both the definition of specific QoS classes to be given for higher level applications and the dedicated mechanisms for providing such classes. In general, each class can be characterized by a specific statistical traffic profile and has to satisfy distinct QoS requirements. In particular, the requirements concern to ensure a certain upper bounds on end-to-end delay, delay variation (also called the jitter) and burst loss probability.

The end-to-end delay arises mostly due to the propagation delay in fibre links, the introduced offset time, edge node processing (i.e., burst assembly) and optical FDL buffering. The first two values can be easily bounded by properly setting up the maximum hop distance allowed by a routing algorithm. Also, the delay produced in the edge node can be controlled by a proper setup of a timer-based burst assembly algorithm. Finally the optical buffering, which in fact has limited application in OBS, introduces relatively small delays. As long as there are many factors that have impact on the end-to-end data delay in an OBS network the problem of jitter is more complicated and needs for a special treatment. This topic, however, is out of the scope of this Chapter.

In a well-designed OBS network the data loss should arise only due to the resources (wavelength) unavailability in fibre links. The probability of burst blocking in a link depends on several factors, among others on implemented contention resolution mechanisms, burst traffic characteristics, network routing, traffic load offered to the network, and relative class load. Since the relation between these factors is usually very complex the control of burst losses may be quite awkward in a buffer-less OBS network.

1.2.2 Absolute vs. Relative QoS Guarantees

There can be distinguished two basic models of QoS provisioning in OBS networks, namely, a *relative* QoS model and an *absolute* QoS model. In the former the performance of a class is defined with respect to other classes; for instance, it is guaranteed that the loss probability of bursts belonging to HP class is lower than the loss probability of bursts belonging to LP class. In the latter an absolute performance metric of quality such as, for example, an acceptable level of burst losses is defined for a class. The performance of a given class in the relative QoS model usually depends on traffic characteristics of the other classes, whilst the absolute QoS model aims at irrelative quality provisioning. On the other hand the absolute QoS model requires more complex implementations in order to achieve desired levels of quality under a wide range of traffic conditions whilst at the same time to preserve high network link utilisation.

Providing the absolute QoS guarantees is desired by upper level applications. The lack of optical memories, however, makes the implementation of absolute QoS model very complicated in OBS networks, for instance, comparing to electrical data networks. For this reason most of QoS mechanisms considered for OBS networks basically offer relative QoS guarantees.

1.2.3 QoS in Connection-Oriented and Connection-Less OBS

The problem of QoS guarantees in connection-oriented OBS networks is similar to the one existing in dynamic wavelength-switched networks. In particular it concerns providing low establishment delays and low connection blocking probabilities, especially for HP connection requests. The establishment delay is a particularly critical problem in such networks. The reason is that the burst has to wait in an electrical buffer at the edge node until the connection establishment process terminates. This may produce the buffer overflow and, as a consequence, the data loss. After the connection is established, there is no data loss inside the network and the transmission delay is only due to the optical signal propagation delay. Notice, that in this case the connection-oriented OBS operation can provide absolute quality guarantees for the end-to-end connection.

On the contrary, the one-way reservation model needs for additional support in QoS provisioning in order to preserve HP traffic from LP traffic during both the resource reservation process and the burst transmission.

1.3 Categories of QoS Mechanisms

In this Section we provide a general classification of QoS mechanisms considered for OBS networks. In most cases, the same contention resolution-based QoS mechanisms can be applied in both OBS and OPS networks. Nevertheless, OBS possesses

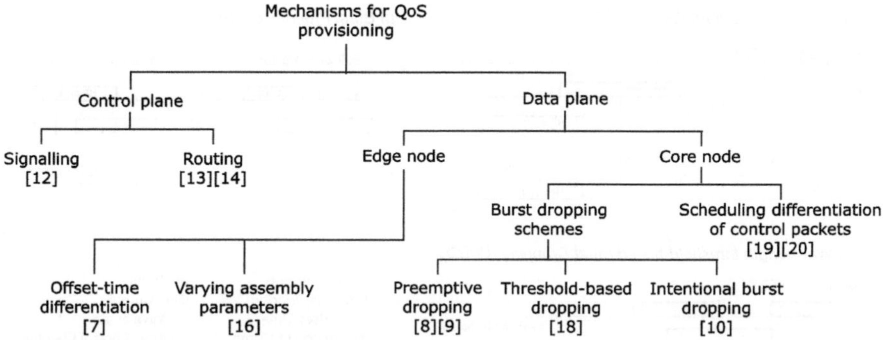

Fig. 1.1 Categories of QoS mechanisms in OBS networks

some additional features such as, for instance, the introduction of pre-retransmission offsets and the ability to operate with different signalling modes. These capabilities enable the implementation of other QoS schemes, which are proper only to OBS networks.

In general several components can contribute to QoS provisioning in one-way reservation OBS networks (see Fig. 1.1). They are related with the control plane through signalling and routing functions and with the data plane through the functions performed in both edge and core nodes.

Two mechanisms involving the control plane operation can provide service differentiation. On one hand a hybrid signalling protocol that consists of a co-operation of two-way and one-way resources reservation modes [12] can support absolute QoS guarantees. In such a scenario the established end-to-end connections can provide the guarantees inside the network such as, no losses and negligible delays, whilst the unreserved resources can be used to transmit the best-effort data burst traffic.

On the other hand, QoS provisioning can be supported by the routing function in a similar way as in OPS networks [13, 14]. In particular, a properly designed routing protocol may both minimize the length of routing path for delay-sensitive applications and preserve the selection of overloaded links for loss-sensitive applications, for instance, thanks to a deflection routing mechanism.

Regarding the data plane, at first, the edge node is responsible for the burst assembly process, when the incoming client packets are aggregated into data bursts in electronic buffers according to their class and destination. The solutions where bursts are unaware class assembled [9] present more drawbacks than benefits and they are not considered here. Then QoS can be achieved in the following ways:

- *Offset Time Differentiation*, [7] which is probably the most addressed QoS technique for OBS networks. The idea here is to assign an *extra* offset-time to high priority bursts, what results in an earlier reservation, in order to favour them while the resources reservation is performed (see Fig. 1.2a). The offset time differentiation mechanism allows achieving an absolute HP and LP class isolation, i.e., (almost) none HP class burst is blocked by an LP class burst. To

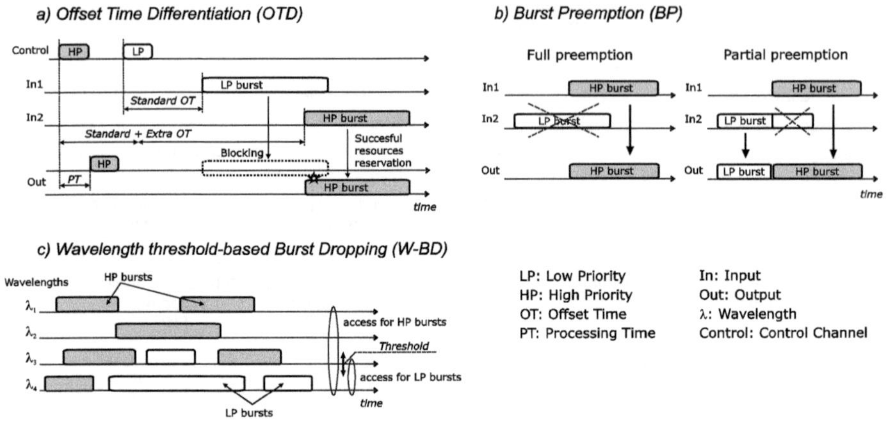

Fig. 1.2 Selected QoS mechanisms

have such a feature, however, the length of the extra offset time has to surpass several times the average LP burst duration [7]. The main advantage of this technique is its simplicity; it makes use only of the postponed transmission of HP bursts in the edge node and it does not require any differentiation mechanism in core nodes. The disadvantages are both the sensitivity of HP class to burst length characteristics [15] and extended pre-transmission delay, which may not be tolerated by some time-constrained applications. Another problem in conventional OBS networks is multiplication of effective burst classes due to the offset variation [15]. In order to limit its impact on QoS performance the transmission offset, which gives the margin for processing and switching operation in core nodes, should be small enough comparing to the extra offset.

- *Varying burst assembly parameters* such as, preset timers and burst lengths. In particular, the packets belonging to an HP class can be aggregated with shorter burst assembly periods than LP packets [16]. In this way the latency experienced by the HP traffic can be minimized. The designing of a burst assembly function is a delicate task since the resulting traffic characteristics may affect the overall network performance.

Another function of the edge node is traffic classification with assignation of specific attributes to the bursts such as, e.g., labels and priorities. These attributes are carried by the burst control packets with the purpose of their further discrimination and processing in core nodes.

First of all, QoS provisioning in core nodes takes place when resolving the burst contention problem and is achieved with an adequate burst drooping technique. The contention resolution usually is assisted by some mechanism(s) such as, wavelength conversion, FDL buffering, and deflection routing [5]. The following burst dropping techniques have been proposed for QoS differentiation in OBS networks:

1.4 Performance Comparison of QoS Mechanisms

In this Section we evaluate the performance of selected QoS mechanisms that aim at the provisioning of relative QoS guarantees. We focus on the mechanisms that implement a one-way reservation signalling protocol and are frequently mentioned in the literature (see Section 1.3 for more details), in particular:

1. Offset Time Differentiation (OTD),
2. Burst Preemption (BP), and
3. Wavelength threshold-based Burst Dropping (W-BD).

1.4.1 QoS Scenario Details

The QoS mechanisms are studied in a unified network scenario with a number of edge nodes and a single core node (see Fig. 1.3). Two classes of traffic are considered, namely, a high priority (HP) class and a low priority (LP) class. The edge nodes generate some HP class and LP class burst traffic pattern. The traffic is handled in the core node according to a given resources reservation and burst drooping policy. At the node output link we evaluate:

- the *burst loss probability* (BLP), for both HP class (BLP_{HP}) and LP class (BLP_{LP}) as well as for overall traffic, that corresponds to the amount of data burst traffic lost as a fraction of the data burst traffic offered, and
- the *throughput*, which represents the percentage of data traffic served with respect to overall data traffic offered to the core node.

We focus on a (nowadays) technologically feasible OBS core node [21, 22] of relatively low number of input ports and wavelengths, but with fast, sub-microsecond switching operation and short burst durations.

The burst scheduler implements a *latest available unused channel with void filling* (LAUC-VF) algorithm [6]. The algorithm searches for a wavelength that minimizes the time gap between currently and previously scheduled bursts. We assume that the searching procedure is performed according to a round-robin rule, i.e., it starts from the less-indexed wavelength each time.

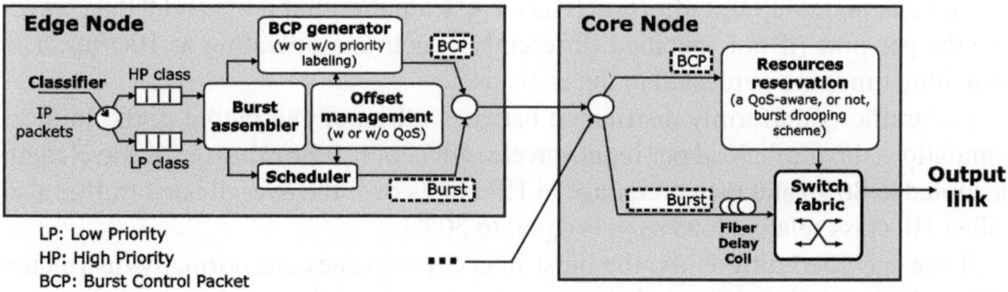

Fig. 1.3 The QoS scenario under study

The core node implements an *offset time-emulated* OBS architecture, [23] i.e., it comprises an additional fibre delay coil component which is responsible for the introduction of processing offset time. On the contrary to conventional OBS architectures, there is no additional offset, except an optional extra offset time for QoS purposes, introduced in the edge node between the burst control packet and the data burst. Thanks to this architecture we avoid the impact of variable offsets on scheduling performance [24] and thus we can gain a deeper insight into the mechanisms behaviour. Nonetheless, since the scheduling operation affects all the mechanisms equally we can expect that their relative performance will be also preserved in the conventional OBS.

The implementation of QoS mechanisms is the following:

- The duration of extra offset time assigned to HP bursts in the offset time differentiation mechanism is 4 times longer than an average LP burst duration. Such a setup allows achieving *quasi*-absolute class isolation [9].
- The burst preemption mechanism applies a simple full-preemptive scheme where each HP burst is allowed to preempt at most one LP burst if there are no free wavelengths available. The preemption concerns an LP burst that, when dropped, minimizes the gap produced between the preempting HP burst and the other burst reservations.
- The wavelength threshold-based burst dropping mechanism operates according to a *restricted* approach [11]. In particular, the threshold value specifies the maximum number of wavelengths that can be occupied simultaneously by LP bursts. On the contrary, HP bursts are allowed to access the whole pool of wavelengths. The threshold selection problem is discussed in Section 1.4.3.1.

If either the burst preemption mechanism or the wavelength threshold-based burst dropping mechanism is applied, the edge node implements a traffic classification function that assigns appropriate priorities to the bursts.

1.4.2 Simulation Scenario

The performance of QoS mechanisms is evaluated in an ad-hoc event-driven simulator. The simulator imitates an OBS core node with full connectivity, full wavelength conversion, and no FDL buffering. It has 4×4 input/output ports and 8 data wavelengths per port (if not specified differently), each one operating at $10 Gbps$. The switching times are neglected in the analysis.

The traffic is uniformly distributed between all input and output ports. In most simulations the traffic load per input wavelength is $\rho = 0.8 Erlang$ (each wavelength occupied in 80%) and the percentage of HP bursts over the overall burst traffic, also called HP class relative load α_{HP}, is equal to 30%.

If not specified differently, the burst inter-arrival times are normally distributed [25] with the mean that depends on the offered load and the standard deviation $\sigma = 5 \cdot 10^{-6}$. The burst durations are normally distributed [25] with the mean $L = 32 \mu s$

and the standard deviation $\sigma = 2 \cdot 10^{-6}$. In further discussion we express the burst length in *bytes* and we neglect the guard bands; thus, the mean burst duration L corresponds to $40kbytes$ of data (at $10Gbps$ rate).

All the simulation results have 99% level of confidence.

1.4.3 Results and Discussion

1.4.3.1 Threshold Selection in W-BD Mechanism

A critical designing issue for all threshold-based mechanisms is the setup of threshold parameter. If we assume independent exponentially distributed (i.e.d.) burst inter-arrival times and lengths, [27] the W-BD mechanism can be modelled as a queuing system [11]. We use such an analysis to assist the threshold selection process. In the discussion, we will also make use of the Erlang B-loss formula, which was shown to approximate well the link-level burst loss probabilities in OBS networks [26]:

$$Erl(A, c) = \frac{A^c}{c!}\left[\sum_{i=0}^{c}\frac{A^i}{i!}\right]^{-1}, \tag{1.1}$$

where A is an offered traffic load and c is a number of wavelengths.

We consider the link has $c = 16$ wavelengths, the overall traffic load ρ is equal to 0.8, and T denotes the threshold parameter, i.e., the number of wavelength accessible to LP class bursts.

In Fig. 1.4a–c we present some analytical results of HP and LP class burst loss probabilities and the throughput. We can see that the performance of W-BD mechanism depends both on HP class relative load α_{HP} and on threshold T value. For given α_{HP}, the $\mathrm{BLP_{HP}}$ can be controlled by a proper selection of the threshold, however, at the cost of effective throughput. The lower bound on $\mathrm{BLP_{HP}}$ is obtained when $T = 0$ (i.e., the LP class traffic is not served) and equal to $b_1 = Erl(\alpha_{HP}\rho,\ c)$. The upper bound on $\mathrm{BLP_{HP}}$ is obtained for $T = c$ (i.e., no class differentiation) and equal to $b_2 = Erl(\rho,\ c)$.

Assume, there is some level of burst loss probability, denote it as $\mathrm{BLP_{HP}}^*$, to be guaranteed for HP class. Then, if $\mathrm{BLP_{HP}}^*$ is higher than b_1, we can find threshold T^* such that complies $\mathrm{BLP_{HP}}(T^*) \leq \mathrm{BLP_{HP}}^*$ and, at the same time, maximizes the throughput. In Fig. 1.4d we present the threshold values obtained for $\mathrm{BLP_{HP}}^* = 10^{-4}$ and $c = 8$, as a function of offered traffic load.

1.4.3.2 Burst Loss Probability and Throughput

In our implementation of QoS mechanisms, both OTD and BP mechanism can achieve absolute class isolation. In other words, the extra offset time we assign to the HP class in the OTD assures that the contention of an HP burst is only due to other

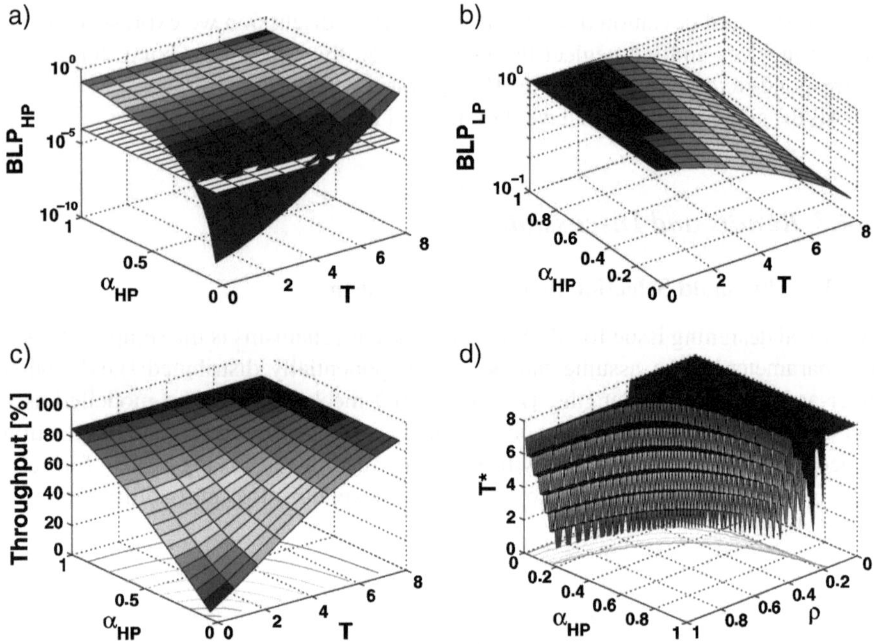

Fig. 1.4 Performance of the wavelength threshold-based burst dropping mechanism ($c = 8$), (**a**) HP class BLP, (**b**) LP class BLP, (**c**) throughput, (**d**) threshold value guaranteeing $BLP_{HP} \le 10^{-4}$

HP burst reservations. If we assume i.e.d. burst inter-arrival times and independent and identically distributed (i.i.d.) burst lengths, [27] the burst loss probability of HP traffic class can be modelled with the Erlang loss formula and it equals to $Erl(\alpha_{HP}\rho,\ c)$. Similarly, the BP mechanism allows preempting any LP reservation by an HP burst and an HP burst is lost only if all the wavelength resources are occupied by other HP reservations. Thus again the loss probability of HP bursts is equal to $Erl(\alpha_{HP}\rho,\ c)$. Note that LP bursts are successfully transmitted either if there are free wavelength resources, not occupied by any earlier HP reservations (in case of the OTD), or the LP burst are not preempted by HP bursts (in case of the BP).

As we have already discussed, the W-BD mechanism achieves its topmost HP class performance if there is no threshold established ($T = 0$), i.e., only HP bursts are transmitted at the output port. In this case, the W-BD mechanism offers the same burst loss performance with respect to the HP class of traffic as the other two QoS mechanisms we study. However, the throughput of the W-BD mechanism is seriously deteriorated as long as none LP burst is served. In Fig. 1.4 we can see that by increasing the threshold value we can improve the throughput but still we achieve it at the cost of HP class performance.

In Fig. 1.5 we provide comparative performance results obtained in the simulation scenario (see Section 1.4.2 for more details). The evaluation is performed for $\rho = 0.8$ and $\alpha_{HP} = 30\%$, and different number of data wavelengths (c). We setup

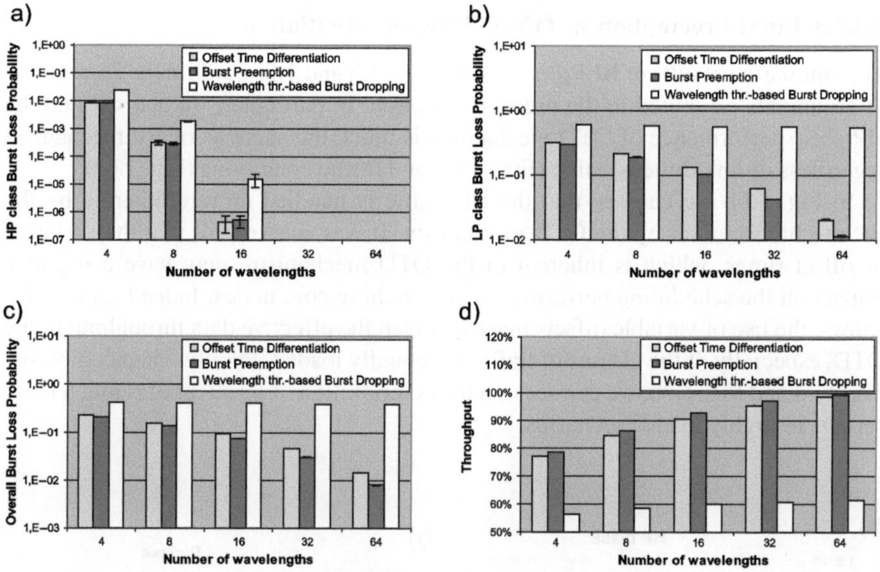

Fig. 1.5 Performance of QoS mechanism vs. link dimensioning ($\rho = 0.8$, $\alpha_{HP} = 30\%$), (**a**) HP class BLP, (**b**) LP class BLP, (**c**) overall BLP, (**d**) effective data throughput

T, the threshold in W-BD mechanism, to be equal to 50% of c, so that the LP class bursts can access at most half of all the available wavelengths at the same time.

In Fig. 1.5a we can see that by increasing the number of wavelengths in the output link we improve the effectiveness of QoS differentiation. The improvement of BLP_{HP} in both OTD and BP mechanism can be really high, for instance, as of *three* orders of magnitude when having 16 instead of 8 wavelengths. Also, we can see that W-BD mechanism offers the poorest HP class performance.

In Fig. 1.5b–d we present the results of BLP_{LP}, overall BLP, and the effective throughput. Although, the performance of both OTD mechanism and BP mechanism is very similar with respect to these metrics, still, the results are in the favour of BP mechanism; in the next section we discuss this issue in more details.

We can also observe that the W-BD mechanism once again achieves very poor performance that hardly depends on available link resources. The reason is that this mechanism has effectively fewer wavelengths available at the output link than the other two mechanisms. Indeed, it provides only 50% of wavelengths for LP class, while it attempts to serve the same amount of input traffic. As a result, both the LP class performance and the throughput are seriously deteriorated.

Although, the FDL buffering is not suitable for conventional OBS networks that operate with long data bursts, still, in OBS networks with short data burst transmission it may significantly help in the contention resolution and QoS provisioning problem. The application of FDL buffers should improve the utilization of link resources, and thus the node throughput, as well as it should decrease the loss probabilities of bursts belonging to each priority class.

1.4.3.3 Burst Preemption vs. Offset-Time Differentiation

The simulation results of BLP_{HP} shown in Figs. 1.5 and 1.6a confirm the correctness of arguments presented in the preceding section. In particular, we can see that the HP class performance of OTD mechanism is much the same as of BP mechanism regardless of link dimensioning (Fig. 1.5a) and traffic conditions (Fig. 1.6a).

In Fig. 1.6b we can see that the LP traffic is handled more efficiently by the BP mechanism than by the OTD mechanism. It was shown [24] that the variation of offset times, which is inherent in the OTD mechanism, may have a negative impact on the scheduling performance in switching core nodes. Indeed, as Fig. 1.7 shows, the use of variable offsets makes worsen the effective data throughput in the OTD, especially, if the classes of traffic are equally loaded. Finally when comparing Fig. 1.7a and Fig. 1.7b, we can see that the deterioration of throughput is much more serious in highly loaded scenarios.

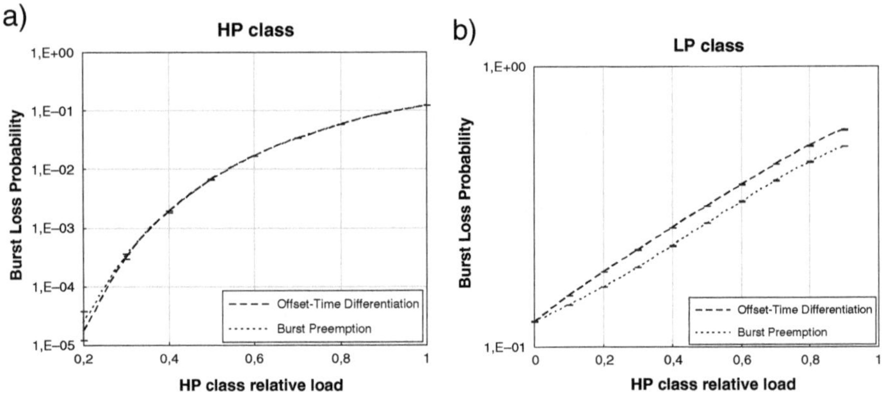

Fig. 1.6 Burst loss probabilities vs. HP class relative load in OTD and BP mechanism ($\rho = 0.8,\ c = 8$), (**a**) HP class, (**b**) LP class

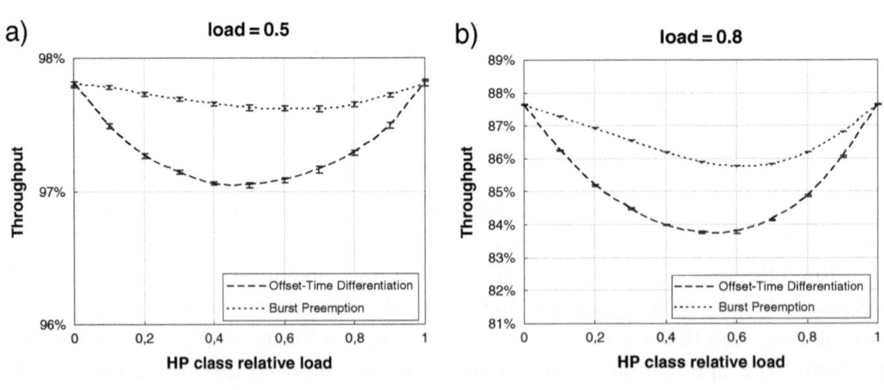

Fig. 1.7 Effective throughout vs. HP class relative load in OTD and BP mechanism, with overall traffic load: (**a**) $\rho = 0.5$, (**b**) $\rho = 0.8$

We can also observe some deterioration of throughput in the BP mechanism. It results from the preemptive operation which allows dropping an LP burst even if it has been partially transmitted at the output link. In such a case, the actual traffic load offered to the output link is increased and it comprises both entirely transmitted data bursts and the parts of preempted LP burst reservations. Since the probability of burst blocking increases accordingly the throughput decreases.

1.5 Effective Burst Preemption

As previously mentioned, the general drawback of burst preemptive mechanisms is possible waste of resources on the ongoing path in case of successful burst preemption. In conventional OBS networks, the burst control packet which belongs to a preempted LP data burst does not have any knowledge about the preemption. On the contrary, it continues its trip towards the destination node and consumes unnecessarily both the control-plane resources, when being processed in the node controllers, and data-plane resources, when reserving the wavelengths for its (preempted) data burst.

In order to assess such an overhead, we develop an approximate estimation of the preemption effect that is produced in a single node. In particular, we introduce a *preemption rate* (R) metric that expresses the number of preempted bursts over all the bursts (successfully) transmitted at the node output link.

If we assume i.e.d. burst inter-arrival times and i.i.d. burst lengths, the preemption rate of a full burst preemption scheme can be calculated as (see Appendix A for a derivation):

$$R = \frac{\alpha_{HP}\,[Erl(\rho, c) - Erl(\alpha_{HP}\rho, c)]}{1 - Erl(\rho, c)}, \tag{1.2}$$

where ρ, α_{HP}, c are, respectively, the overall load, HP class relative load, the number of wavelengths in the link, and $Erl(\cdot)$ is given by (1.1).

The formula can be interpreted as following: the numerator represents the reduction of burst losses of the HP class after the application of the preemption mechanism whilst the denominator conditions it on those bursts that have been successfully transmitted.

In Fig. 1.8 we present analytical and simulation results of the preemption rate. As we can see, R increases if either the traffic load increases or the number of wavelengths in the link decreases. A small disparity between the analytical and the simulation results comes from the fact that the simulated bursts as stream-like arranged in the data channel (bursts do not overlap each other) and their arrivals are not more exponentially distributed (as we assumed in the analytical model).

R corresponds to the percentage of additional burst control packets that have to be processed at each node on their outgoing routing paths. These burst control packets are responsible for the wastage of both processing and transmission resources as long as their data bursts are not going to be transmitted anymore (they have been

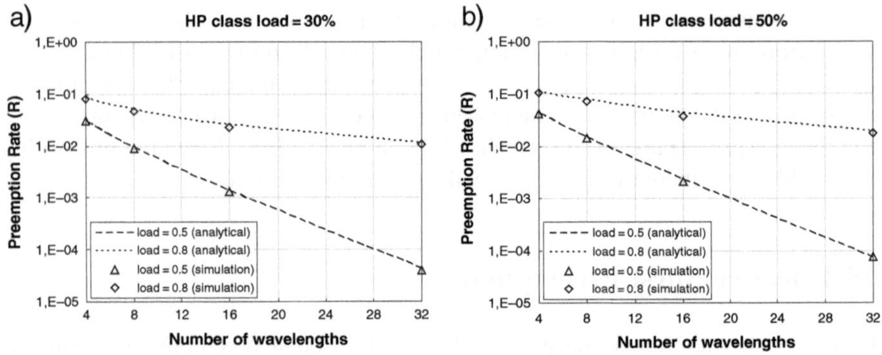

Fig. 1.8 Preemption rate in an OBS node, with HP class relative load: (a) $\alpha_{HP} = 30\%$, (b) $\alpha_{HP} = 50\%$

preempted). In large networks, of high number of nodes, the problem might be intensified since all the nodes undergo a similar effect. Such a study, however, is out of the scope of this work.

A particular attention should be paid to preemption-based routing mechanisms [28, 29]. Such mechanisms assume that the bursts carried over alternative (duplicate) paths can be preempted by the bursts carried over primary paths. In such scenarios, the amount of preempted bursts might be really high as long as both ρ and α_{HP} are assumed to be high. As a consequence the useless burst reservations may decrease the effectiveness of preemption-based routing mechanisms.

The problem of the preemption-related overhead can be effectively avoided in OBS networks with a *preemption window* control mechanism [30] applied (see Fig. 1.9). The mechanism assumes that the offset time is enlarged by additional offset which defines a preemption window period. The preemption of an LP burst is

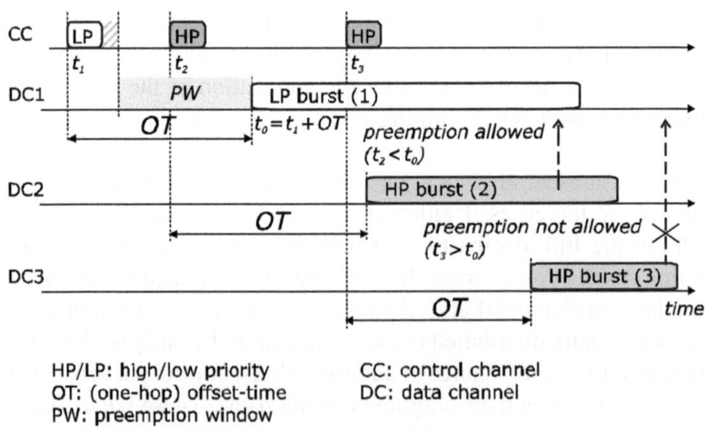

Fig. 1.9 Preemption Window mechanism

allowed only during this period. A burst control packet, after its processing, has to wait in the switch controller until the preemption window expires. Then it is either sent towards the next node (if its data burst has not been preempted) or dropped (in case of successful preemption). After the burst control packet is sent the preemption of its burst is not allowed in the node. Thanks to these rules, there are no burst reservations in the ongoing nodes that belong to the preempted bursts.

1.6 Conclusions

In this Chapter we study the performance of the most addressed mechanisms providing relative QoS differentiation in OBS networks. We show that the burst preemptive mechanism can efficiently utilize transmission resources and, at the same time, it can offer highly effective QoS differentiation. The offset time differentiation mechanism is characterized by high HP class performance as well. Nevertheless, its scheduling efficiency, and thus the throughput, is deteriorated by the variation of offset-times. Finally, the wavelength threshold-based mechanism can be characterised by the poorest overall performance, which significantly depends on its wavelength threshold value. The application of this mechanism may be reasonable only for the links of a large number of wavelengths so that the threshold would be relatively high (in order to serve efficiently the LP traffic) and could adapt to traffic changes. Although, the evaluation of the performance of QoS mechanisms is obtained in a single node scenario, still, we can expect the mechanisms will behave similarly in a network scenario.

The high performance of burst preemption mechanism designates it to be a suitable mechanism for QoS differentiation in OBS. Although, in this study we concern on relative quality guarantees, still, the preemption mechanism can support absolute QoS guarantees [17] as well. A drawback of the preemption mechanism in conventional OBS networks is the waste of resources if the preemption occurs. Nonetheless, such a problem can be avoided in OBS networks with a preemption window mechanism applied.

Acknowledgments The authors would like to thank to Dr Christoph Gauger of the University of Stuttgart for his helpful comments. This work has been partially funded by the COST 291 action, the e-Photon/ONe+ project (FP6-IST-027497) and the MEC (Spanish Ministry of Education and Science) under the CATARO project (TEC2005-08051-C03-01/TCM).

1.7 Appendix A: The Preemption Rate in a Buffer-Less OBS Node

Here we show how we derive the expression (1.2).

Let $n_{preempt}$ be the number of successful preemptions, $n_{lost\_HP}{}^{(np)}$ and $n_{lost\_HP}{}^{(p)}$ be, respectively, the number of HP bursts that are lost in a non-preemptive (without burst preemption) and a preemptive (with full burst preemption) scenario, n_{in_HP} be the number of incoming HP bursts, n_{in} be the total number of incoming bursts and n_{out} be the total number of bursts transmitted at the output link in a given time period.

Since each preemption means the acceptance of an HP burst instead of an LP burst, $n_{preempt}$ can be also interpreted as a difference between all the HP bursts that are lost in the non-preemptive scenario and the HP bursts that are lost in the preemptive scenario:

$$n_{preempt} = n_{lost\_HP}{}^{(np)} - n_{lost\_HP}{}^{(p)} \tag{1.3}$$

Obviously:

$$n_{lost\_HP}{}^{(np)} = n_{in\_HP} \cdot B_{HP}{}^{(np)} \tag{1.4}$$

$$n_{lost\_HP}{}^{(p)} = n_{in\_HP} \cdot B_{HP}{}^{(p)} \tag{1.5}$$

where $B_{HP}{}^{(np)}$ and $B_{HP}{}^{(p)}$ are the HP burst loss probabilities in the non-preemptive and the preemptive scenario, respectively.

From 1.3, 1.4, and 1.5 we have:

$$n_{preempt} = n_{in\_HP} \cdot (B_{HP}{}^{(np)} - B_{HP}{}^{(p)}) = \alpha_{HP} \cdot n_{in} \cdot (B_{HP}{}^{(np)} - B_{HP}{}^{(p)}) \tag{1.6}$$

where α_{HP} is the HP class load ratio.

Than the preemption rate is equal to:

$$R = \frac{n_{preempt}}{n_{out}} = \frac{\alpha_{HP} \cdot n_{in} \cdot (B_{HP}{}^{(np)} - B_{HP}{}^{(p)})}{n_{in} \cdot (1 - B^{(p)})} \tag{1.7}$$

Note, that the overall burst loss probability in the preemptive scenario ($B^{(p)}$) and the HP burst loss probabilities in the non-preemptive scenario ($B_{HP}{}^{(np)}$) are the same. Moreover $B_{HP}{}^{(p)}$ depends only on the HP class relative load (α_{HP}) due to absolute class isolation. Finally, assuming the exponentially distributed burst arrivals and lengths, we use (1.1) to calculate burst loss probabilities. Therefore, by the proper substitution in (1.7) we obtain (1.2).

References

1. C. Qiao and M. Yoo, Optical Burst Switching (OBS) – a New Paradigm for an Optical Internet, Journal of High Speed Networks, vol. 8, no. 1, 1999, pp. 69–84.
2. M. Duser, E. Kozlovski, R.I. Killey, and P.Bayvel, Design Trade-Offs in Optical Burst Switched Networks with Dynamic Wavelength Allocation, in Proceedings of ECOC, Munich (Germany), Sep. 2000.
3. E. Kozlovski and P. Bayvel, QoS Performance of WR-OBS Network Architecture with Request Scheduling, in Proceedings of IFIP ONDM, Turin (Italy), Feb. 2002.

4. J. Wan, Y. Zhou, X. Sun, and M. Zhang, Guaranteeing Quality of Service in Optical Burst Switching Networks Based on Dynamic Wavelength Routing, Optics Communications, vol. 220, no. 1–3, May 2003, pp. 85–95.
5. C. Gauger, Trends in Optical Burst Switching, in Proceedings of SPIE/ITCOM, Orlando (FL), Sep. 2003, vol. 5247, pp. 115–125.
6. Y. Xiong, M. Vanderhoute, and C. Cankaya, Control Architecture in Optical Burst-Switched WDM Networks, IEEE Journal of Selected Areas in Communications, vol. 18, no. 10, Oct. 2000, pp. 1838–1851.
7. M. Yoo, C. Qiao, and S. Dixit, Optical Burst Switching for Service Differentiation in the Next-Generation Optical Internet, IEEE Communications Magazine, vol. 39, no. 2, Feb. 2001, pp. 98–104.
8. A. Kaheel and H. Alnuweiri, A Strict Priority Scheme for Quality-of Service Provisioning in Optical Burst Switching Networks, in Proceedings of ISCC, Turkey, Jun. 2003.
9. V. M. Vokkarane and J. P. Jue, Prioritized Burst Segmentation and Composite Burst-Assembly Techniques for QoS Support in Optical Burst-Switched Networks, IEEE Journal on Selected Areas in Communications, vol. 21, no. 7, Sep. 2003, pp. 1198–1209.
10. Q. Zhang, V. M. Vokkarane, J. P. Jue, and B. Chen, Absolute QoS Differentiation in Optical Burst-Switched Networks, IEEE Journal on Selected Areas in Communications, vol. 22, no. 9, Nov. 2004, pp. 1781–1795.
11. H. Øverby and N. Stol, QoS Differentiation in Asynchronous Bufferless Optical Packet Switched Networks, Wireless Networks, vol. 12, no. 3, Jun. 2006.
12. I. De Miguel, J. C. Gonzalez, T. Koonen, R. Duran, P. Fernandez, and I. T. Monroy, Polymorphic Architectures for Optical Networks and their Seamless Evolution towards Next Generation Networks, Photonic Network Communications vol. 8, no. 2, 2004, pp. 177–189.
13. P. Zaffoni, F. Callegati, W. Cerroni, G. Muretto, and C. Raffaelli, QoS Routing in DWDM Optical Packet Networks, in Proceedings of WQoSR (Co-Located with QoFIS), Barcelona (Spain), Sep. 2004.
14. S. Yao, B. Mukherjee, and S. J. B. Yoo, A Comparison Study Between Slotted and Unslotted All-Optical Packet-Switched Network with Priority-Based Routing, in Proceedings of OFC, Anaheim (CA), Mar. 2001.
15. K. Dolzer and C.M. Gauger, On Burst Assembly in Optical Burst Switching Networks – a Performance Evaluation of Just-Enough-Time, in Proceedings of ITC 17, Salvador (Brazil), December 2001.
16. N. Barakat and E.H. Sargent, On Optimal Ingress Treatment of Delay-Sensitive Traffic in Multi-Class OBS Systems, in Proceedings of WOBS (Co-Located with BroadNets), San Jose (CA), Oct. 2004.
17. L. Yang, Y. Jiang, and S. Jiang, A Probabilistic Preemptive Scheme for Providing Service Differentiation in OBS Networks, in Proceedings of IEEE Globecom, Singapore, Dec. 2003.
18. F. Callegati, W. Cerroni, C. Raffaelli, and P. Zaffoni, Wavelength and Time Domain Exploitation for QoS Management in Optical Packet Switches, Computer Networks, vol. 44, no. 1, Jan. 2004, pp. 569–582.
19. Y. Wang and B. Ramamurthy, CPQ: A Control Packet Queuing Optical Burst Switching Protocol for Supporting QoS, in Proceedings of WOBS (co-located with BroadNets), San Jose (CA), Oct. 2004.
20. A. Kaheel and H. Alnuweiri, Quantitative QoS Guarantees in Labeled Optical Burst Switching Networks, in Proceedings of IEEE Globecom, Dallas (TX), Nov. 2004.
21. H. Guo, J. Wu, X. Liu, J. Lin, and Y. Ji, Multi-QoS Traffic Transmission Experiments on OBS Network Testbed, in Proceedings of ECOC, Glasgow (Scotland), Sep. 2005.
22. A. Al Amin et al., 40/10 Gbps Bit-rate Transparent Burst Switching and Contention Resolving Wavelength Conversion in an Optical Router Prototype, in Proceedings of ECOC, Cannes (France), Oct. 2006.
23. M. Klinkowski, D. Careglio, and J. Solé-Pareta, Offset-Time Emulated OBS Control Architecture, in Proceedings of ECOC, Cannes (France), Oct. 2006.

24. J. Li, C. Qiao, J. Xu, and D. Xu, Maximizing Throughput for Optical Burst Switching Networks, in Proceedings of IEEE INFOCOM, Hong Kong (China), Mar. 2004.
25. X. Yu, J. Li, X. Cao, Y. Chen, and C. Qiao, Traffic Statistics and Performance Evaluation in Optical Burst Switched Networks, IEEE Journal of Lightwave Technology, 22(12), Dec. 2004, pp. 2722–2738.
26. Z. Rosberg, H. L. Vu, M. Zukerman, and J. White, Performance Analyses of Optical Burst Switching Networks, IEEE Journal on Selected Areas in Communications, vol. 21, no. 7, Sep. 2003, pp. 1187–1197.
27. M. Izal and J. Aracil, On the Influence of Self-Similarity on Optical Burst Switching Traffic, in Proceedings of IEEE Globecom, Taipei (Taiwan), Nov. 2002.
28. C. Cameron, A. Zalesky, and M. Zukerman, Shortest Path Prioritized Random Deflection Routing (SP-PRDR) in Optical Burst Switched Networks, in Proceedings of ICST WOBS, San Jose (CA), Oct. 2004.
29. J. Li and K. L. Yeung, Burst Cloning with Load Balancing, In Proceedings of OFC, Anaheim (CL), Mar. 2006.
30. M. Klinkowski, D. Careglio, D. Morató, and J. Solé-Pareta, Effective Burst Preemption in OBS Network, in Proceedings of IEEE HPSR, Poznan (Poland), Jun. 2006.

Chapter 2
End-to-End Proportional Differentiation Over OBS Networks

Pablo Jesús Argibay-Losada, Andrés Suárez-González, Manuel Fernández-Veiga and Cándido López-García

Abstract In this paper, we propose a novel scheme to provide end-to-end proportional differentiated services to an arbitrary number of traffic classes at the *packet* level. The service classes are defined in terms of the packet loss probability measured between the ingress node and the egress node of an OBS network, where each ingress node aggregates a large number of IP flows. Our solution requires only that the OBS network is able to provide relative differentiation to two types of bursts, a task which can be accomplished in very different ways. In order to demonstrate the feasibility and the performance of the proposal, we develop a mathematical model for computing the loss probabilities in multiservice OBS networks. Specifically, we use a fixed-point model and show how to use its results to derive the desired packet loss probabilities. The second contribution of this work is to study the effects of coupling the packet service classes with reactive sources, namely, sources responsive to congestion. In particular, the well-known dynamics of TCP is embodied into the analytical model. The numerical results produced by this analytical framework show that good proportional differentiation, both in packet loss and in throughput, can be effectively achieved without sacrificing bandwidth usage.

Keywords Optical burst switching · Proportional differentiation · Segmentation

2.1 Introduction

The synergy between optical wavelength division multiplexing (WDM) transmission and the optical burst switching (OBS) paradigm is currently being regarded as a candidate architecture for the next generation Internet backbone. The basic idea of optical switching is that of eliminating all unnecessary signal conversion inside the switches, thereby creating a multi-hop all-optical path, not necessarily constrained by wavelength continuity. Optical burst switching is a realization of such principle in

P. Jesús Argibay-Losada (✉)
Departamento de Enxeñería Telemática, Universidade de Vigo,
Campus Universitario s/n, E-36310 Vigo, Spain

DOI 10.1007/978-1-4020-9889-5_2,

which the atom entity to switch is the burst, an aggregation of individual IP packets, and where the optical data channel is decoupled (both temporally and spatially) from the control channel used to convey signaling messages between neighbor switches.

It is also widely accepted, partly motivated by the increased deployment of applications requiring quality of service (QoS), that the traditional best-effort service model of the Internet should be enriched. QoS support can be generally attained within two different frameworks, namely absolute or relative assurances. Absolute QoS provision imposes a hard quantitative limit on the performance metric of interest (e.g., bandwidth, delay or loss). This service model may render itself useful for intolerant applications or for tightly controlled network services. Nonetheless, it is difficult to imagine how to achieve absolute guarantees without relying on admission control, jointly with per-flow accounting and traffic monitoring. Consequently, absolute differentiation is scarcely scalable, overly rigid and, due to the need to control strictly the admission of new traffic flows, even antagonistic to the Internet's best-effort principle.

On the other hand, relative QoS differentiation defines a model consisting of a set of service classes ordered in the QoS metrics, with the qualitative guarantee of preserving the relative ordering between them but without the capability to control the quality gap from one class to the next better one. So, the network provider commits better (or at least not worse) service to class i than to class $i + 1$ in terms of single or composite performance measures, such as delay, loss or a combination of these. Due to such flexibility, relative differentiation uses less state information, scales gracefully, and it is the constituent model of the well known Differentiated Services (DiffServ) architecture developed for IP QoS solutions [4]. Though it is certainly arguable that some applications will require absolute service differentiation, there are currently many examples of other applications (e.g., VoIP, IPTV, network games) that can tolerate an elastic service with dynamic variations in quality, adapting their response to changing network conditions. In this context, relative differentiation offers advantages both to the users and to the network provider, namely, the user having the freedom to select at any time the service class best suited to his/her desired quality, and the provider carrying more traffic and using the communication resources more efficiently. Whatever the model, a common component either to the absolute or the relative service frameworks must be a proper pricing scheme that sets the right incentives (i.e., better service always costs more).

Proportional differentiation [6, 9, 10] is a refinement and a quantification of relative differentiation. In this paradigm, the service level offered to each class is controlled according to prespecified constant factors, and these quantitative law is kept stable even on a short time scale. Hence, the quality of service, as seen by the user, is consistent and predictable while, as a network operator's concern, is controllable. More importantly, it can benefit from efficient forwarding mechanisms that keep up the scalability.

A number of previous efforts have addressed the direct provision of proportional QoS in optical networks. Some of them merely propose schemes for proportional differentiation of bursts in isolated nodes (e.g., [5, 15, 23]).

In [22], however, the quality metric is the burst loss probability measured between the ingress and egress points of the OBS network. It is further presented an algorithm to assign a different time offset to each class (route) inside the optical core in such a way that the losses in distinct classes are proportional. Not only requires this scheme the knowledge of the current network state for adapting the offsets, but it compels the nodes to exchange much state information as well. Yet the burst blocking probability is only a provider-oriented, indirect and insufficient measure of the service quality, difficult to map into a meaningful quality level for users.

In an alternative approach, Vokkarane and Jue [25] study several configurations engineered to provide different packet loss probability to an arbitrary number of traffic classes, using various combinations of burst types and schedulers in the OBS network. The authors carry out analyses of loss probability and delay for each traffic class when segmentation, deflection or preemptive discarding are in use inside the OBS area. But the whole service model is one of relative differentiation, not proportional.

In this work, we propose a novel scheme to provide end-to-end proportional differentiated services to an arbitrary number of traffic classes at the *packet* level, not merely at the burst level. The service classes are defined in terms of the packet loss probability measured between the ingress node and the egress node of an OBS network, where each ingress node aggregates a large number of IP flows. We describe the system architecture and operations when the OBS subnetwork is contained in a larger computer network like the Internet. Moreover, our solution requires only that the OBS network is able to provide relative differentiation to two types of bursts, whatever the scheduling or switching algorithms employed for this purpose. Many suitable algorithms exist to accomplish two different service levels inside the optical subnetwork (e.g., deflection, segmentation, variable offset-times, etc.), spanning from the simplest to the most intricate. In our scheme, which one to use is a largely immaterial issue, provided that it attains a strong difference in the loss probabilities of both burst classes. Additionally, the proposed architecture is remarkably simple. The unique mandatory element is a probabilistic classification of the packets received at the ingress routers to the optical domain. The classifier randomly divides the packets of all classes into two groups, where the probability of entering a group depends upon the QoS level of the class, and the packets in each group are assembled in the two internal burst classes, respectively. Incidentally, we should also emphasize that the proposed architecture can be generalized in a straightforward manner to pure optical packet switching.

In order to demonstrate the feasibility and the performance of the proposal, we develop a mathematical model for computing the loss probabilities in multiservice OBS networks. These consist of bufferless nodes and can therefore be regarded as a whole, from an ingress to an egress point, like a one-hop bufferless subnetwork characterized by a collection of loss probabilities computable with simple, approximate circuit switching models. Specifically, we use a fixed-point model and show how to use its results to derive the desired packet loss probabilities. The model is fairly general, accounting for different scheduling policies and several operating modes in the control plane. The numerical results produced by this analytical framework show that good proportional differentiation can be effectively achieved without sacrificing bandwidth usage.

The second contribution of this work is to study the effects of coupling the packet service classes with reactive sources, namely, sources responsive to congestion. In particular, the well-known dynamics of TCP is embodied into the analytical model. The goal is to understand how the packet loss probabilities can be easily converted in throughput differentiation, as well as gain insight about the performance of a realistic scenario.

The rest of the paper is structured as follows. Section 2.2 presents the probabilistic classification algorithm used in the ingress nodes to distribute the traffic among the relative differentiated internal classes. Section 2.3 describes the fixed-point approach adopted for the network model. First, the link blocking probabilities are derived (Section 2.3.1), and then the general model used in the analysis of the burst blocking probabilities for a network with fixed routing is explained (Section 2.3.2). Section 2.4 extends the basic fixed-point equations by incorporating TCP sources, so as to model a traffic load responsive to congestion. Three possible differentiation strategies that could be applied to the external classes are studied, too. With these tools, an analytical and numerical study for single-link and multi-node networks is provided at Section 2.5. A preliminary experimental validation of the scheme is addressed in Section 2.6. Finally, some concluding remarks and a summary of conclusions are given in Section 2.7.

2.2 The Packet Differentiation Algorithm

This paper analyzes the following scheme for differentiating proportionally an arbitrary number of traffic classes sharing an OBS network: suppose that we have n classes of packets —numbered $0, \ldots, n-1$, with 0 and $n-1$ the highest and lowest priority classes, respectively— and want to provide differentiated services in such a way that the packet loss probabilities for any two classes i and j satisfy

$$\frac{p_i}{p_j} = \frac{c_i}{c_j} \qquad c_i \in (0,1), \quad i \neq j$$

where p_i is the average loss probability of a class i packet, and c_i is an arbitrarily assigned coefficient that measures the quality of service desired by the packets of that class.

Assume also that the optical transport network distinguishes between two classes of bursts, say type 0 and type 1 bursts, in such a way that the loss probability of packets transported in type 0 bursts (respectively, 1) is B_0 (B_1), where $B_0 \ll B_1$. Though this assumption of two widely separated burst loss probabilities was in the past difficult to uphold, there are nowadays numerous routing and scheduling algorithms in the literature which could be used to ensure it, such as burst segmentation, deflection routing, burst preemption and others (see, for instance, [3, 17, 25, 28] and the references therein).

If packets of class x are assembled into type 0 bursts with probability h_0^x, and into type 1 bursts with probability $1 - h_0^x$, then the ratio of packet loss probability between classes $i, j \in \{1, \ldots, N\}$ is

$$\frac{p_i}{p_j} = \frac{h_0^i B_0 + (1 - h_0^i) B_1}{h_0^j B_0 + (1 - h_0^j) B_1}.$$

Under the assumption $B_0/B_1 \ll 1$ this ratio becomes

$$\frac{p_i}{p_j} \simeq \frac{1 - h_0^i}{1 - h_0^j}$$

i.e., it is approximately the ratio of two arbitrarily chosen constants, in conformance with the proportional differentiation paradigm. Consequently, by varying the fraction of packets of each class transmitted through each burst type, it is possible to provide proportional differentiation in loss probability. The range of feasible differentiations with this probabilistic scheme is bounded by the case in which one packet class goes entirely through the low priority bursts, and another class is always assigned to the high priority bursts, giving a maximum differentiation power of

$$\max\left(\frac{p_i}{p_j}\right) \simeq \frac{B_1}{B_0}.$$

It is worthy of mention that the probabilistic mapping of several external traffic classes into two internal transport levels is easy to incorporate into the network equipment, and is independent of any technology, so it could be used at the edge of any core network, either OBS or not.

From a practical viewpoint, the task of differentiating between the two burst classes should become easier when packets from the lowest external priority class (class $n - 1$) are entirely carried by the low priority bursts. Therefore, we arbitrarily fix $c_{n-1} = 1$ and $h_0^{n-1} = 0$, so the mapping from packets in class $n - 1$ to the best-effort bursts is deterministic. In that case we have

$$\frac{p_i}{p_{n-1}} = \frac{c_i}{1} \simeq \frac{1 - h_0^i}{1} \quad \Rightarrow \quad h_0^i = 1 - c_i. \tag{2.1}$$

We call this algorithm **Open-Loop Two-Class Dispatcher (OTCD)** (see Fig. 2.1).

The OTCD scheduling strategy can lead to out-of-order arrival of packets to their destinations, but the consequences of this behavior can be alleviated by a proper design of the receiving transport entity. Alternatively, the reordering problem can be tackled by means of control information indicating which ranges of packets of a route have been sent through each of the burst classes. The disassemblers can use that information to buffer, for a bounded time, the packets that seem to be out of order, waiting for the arrival of the packets belonging to the missing range. A timer can control when to declare that late packets were effectively lost (because

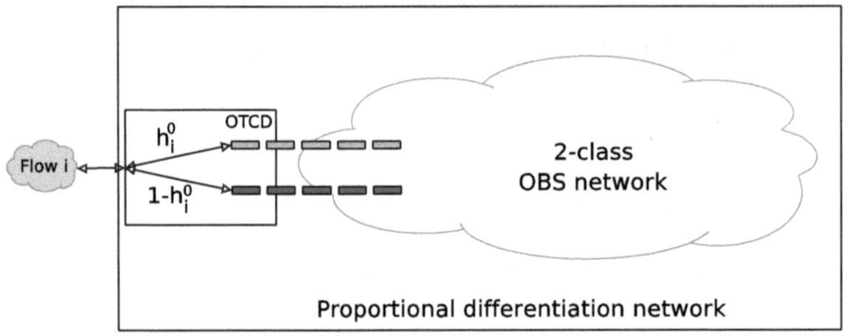

Fig. 2.1 Schematic operations of the OTCD algorithm

their bursts were blocked), and trigger the delivery of the buffered packets. This scheme could also be applied if the priority is being implemented, not by using two distinct classes of bursts, but by the reordering of packets of different priorities inside a single type of bursts [24]. In this case, the task of the reordering algorithm would be simplified, since it could be done partly by rearranging packets conveyed in the same burst. It would suffice then to add a bitmap to the control information to indicate in which order the carried packets should be forwarded to the exterior of the OBS network.

In the ensuing sections, we will use fixed h_0^i values as determined by (2.1) and aim to study analytically the algorithm accuracy in attaining proportional packet loss probabilities in several network configurations. To this end, we first formulate the mathematical framework to compute the link and burst blocking probabilities in a general OBS network with static routing. The approach is based on the common assumption of link independence, and sets a system of fixed-point equations easily solvable by iterated substitutions. Next, this system is extended to account for a new feature, the reaction of the external traffic sources upon congestion. So, the dynamics of a typical rate adaptation protocol (TCP) is incorporated into the system model. As the final step, the complete model is particularized to study analytically the performance of three simple network topologies, so as to investigate the reach of the proportional differentiation paradigm, in packet loss and in throughput too.

2.3 The Fixed Point Model for the OBS Network

We shall use a fixed point model [14] to develop a mathematical abstraction for an OBS network with service differentiation capabilities. Fixed point models have been widely used in the circuit-switching world, in packet switching environments and also in OBS to analyze the fundamental performance of those networks [11, 19, 20]. We recast the main model assumptions to the OBS framework so as to include explicitly in our formulation two fundamental elements of the network architecture: the scheduling (or contention resolution) algorithm, and the operations of the signaling plane. Specifically, the model allows the use of several burst schedulers with different policies for burst prioritization. Since our probabilistic differentiation scheme only uses two priorities, we present the model for two classes of bursts, though it is easily extensible to an arbitrary number of classes. In fact, the fixed-point model is fairly independent of the *packet* classification algorithm used at the edge nodes. A second distinctive feature is the modeling of the different signaling procedures inside the optical domain. The signaling procedure defines how an edge node attempts to reserve the necessary transmission and switching resources for a burst, and how the switching nodes are informed about this. Two common schemes that have been proposed for OBS networks are just-enough-time (JET) and just-in-time (JIT) [2, 26, 27]. We assume that the network can use any of both types in the control plane, so that the analysis is more general.

Fixed point models tend to be more accurate in densely connected networks, and also when the traffic flows can be modeled as a Markovian process. In our study, traffic with long range dependence has not been considered mainly for reasons of analytical tractability, but also because, at low or medium time scales, it can be effectively replaced by suitable Markovian models [13].

2.3.1 Link Blocking Probabilities

In this work, we consider simultaneously schedulers supporting preemption and segmentation, the former in order to substantiate the QoS capabilities, and the latter because burst segmentation offers higher utilization than other approaches to resolve contention. Recall that, in burst segmentation, a long burst involved in contention is split in several smaller segments, and only those segments overlapping another burst are dropped. Hence, segmentation allows partial transmission of bursts during contention epochs, achieving better resource utilization than alternative contention-resolution methods [7, 24] (i.e., deflection or composite switching). The performance analysis of switching architectures employing mechanism such as deflection has been carried out in other works [21], and are in some respects complementary to this paper. So, assume that bursts of high priority can preempt low priority bursts, and the burst being preempted or the one being scheduled can be segmented in order to allow the scheduling of the non-overlapping part. Assume also that, in the case of two bursts of the same priority, the head of the contending burst is dropped. This allows the control packet of the contending burst to be updated with the new burst information. If, on the contrary, the tail of the scheduled burst had been dropped, then it would be impossible to update its control packet, which had already left the node. Even worse, this inconsistent control packet would continue to reserve uselessly transmission resources in the downstream nodes of its path.

In a previous work [1], we analyzed the differentiation behavior for schedulers supporting QoS by means of preemption. The results confirm that the performance depends on the specific scheduler being used in the core nodes. As an illustration, Fig. 2.2 plots the utilization factor of a single link as a function of the number of wavelengths for several total blocking probabilities —0.001, 0.01 and 0.1—, and for

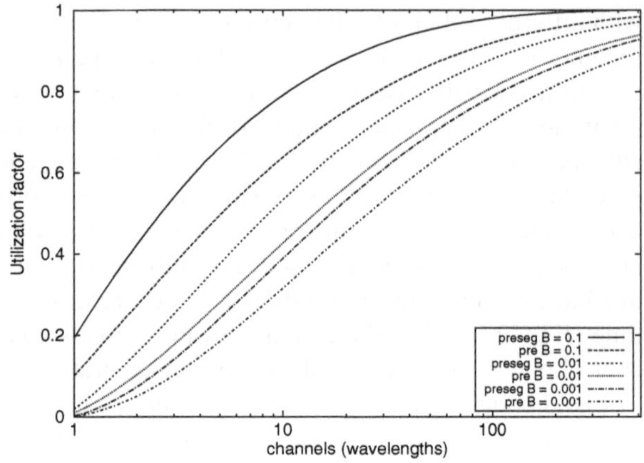

Fig. 2.2 Utilization factor for different burst schedulers

two schedulers: one implementing QoS by means of preemption (the lines labeled *pre*) and another one implementing QoS by means of preemption and segmentation (the lines labeled *preseg*). The curves are numerical examples of analytical results, not simulations.

Under the stated assumptions, a link between two OBS nodes with m wavelengths, full wavelength conversion at its head node and offered traffic of two priorities with intensities I_0 (high priority) and I_1 (low priority) can be modeled as an M/G/m queueing system with the Molina approximation [19]. This means that high priority bursts are completely unaware of the low priority traffic, and see a system with m resources all the time. Instead, the low priority bursts enter a system with the same m resources but traffic intensity $I_0 + I_1$. Then, the Molina approximation computes the blocking probability for each priority as the fraction of traffic served by servers $m+1, m+2, \ldots$ in a virtual M/G/∞ queueing system with the same offered traffics. Hence, the overall carried traffic in the link, for a given received traffic I, is (we drop the link subscript, in order to make the notation less cumbersome)

$$A_c(I) \triangleq \sum_{j=0}^{\infty} \min(j, m) \frac{I^j}{j!} e^{-I}. \tag{2.2}$$

The overall blocking probability is given by

$$B = 1 - \frac{A_c(I_0 + I_1)}{I_0 + I_1}, \tag{2.3}$$

the blocking probability for high priority bursts is

$$B_0 = 1 - \frac{A_c(I_0)}{I_0} \tag{2.4}$$

and, consequently, the low priority bursts are blocked with probability

$$B_1 = \frac{(I_0 + I_1)B - I_0 B_0}{I_1} = 1 - \frac{A_c(I_0 + I_1) - A_c(I_0)}{I_1}. \tag{2.5}$$

Note that the M/G/m and M/G/∞ models on which the above expressions are derived are insensitive to the second order statistics of the arrival process, and depend only on the mean arrival rate. For notational convenience, we regard equations (2.2–2.5) as a functional mapping $B_0 = \Lambda_0(I_0, I_1)$, $B_1 = \Lambda_1(I_0, I_1)$ giving the blocking probabilities as a function of the offered traffic.

2.3.2 Burst Blocking Probabilities

Consider an OBS network with L links, N nodes with full wavelength conversion and two internal types of bursts. Each link l is unidirectional having capacity C_l. Let $\mathcal{R}$ be the set of routes in the network, and $\alpha \in \mathcal{R}$ an origin-destination pair. The priority x burst arrivals to route α are assumed to be a Poisson process (x, α) with rate $\lambda_{(x,\alpha)}$. Each burst will attempt to reserve in each node along its path a time S dependent on the OBS discipline at use in the node. So, in JET mode, each burst will try to reserve L/C time units from its expected time of arrival, where L is the burst length and C is the capacity of each wavelength in the WDM network, whereas with JIT the reservation interval begins with the arrival of the control packet, and its length equals the offset plus the burst transmission time

$$S^{\text{JET}} = \frac{L}{C}, \qquad S^{\text{JIT}} = \text{Offset} + \frac{L}{C}.$$

Hence, each policy p , $p \in \{\text{JET}, \text{JIT}\}$ will induce an offered traffic intensity $A^p_{(x,\alpha)} = \lambda_{(x,\alpha)} \cdot S^p_{(x,\alpha)}$.

Denote by A_{xl} and B_{xl} the offered traffic intensity and blocking probability of class x bursts at link l, respectively. Then, the implicit solution of the following fixed point system

$$\begin{aligned} A_{xl} &= \sum_{r \in \mathcal{R}} A^p_{x,r} I_{r,l} \prod_{i=1}^{o(l,r)-1} \left(1 - B_{xi_r}\right) \\ B_{xl} &= \Lambda_x(A_{0l}, A_{1l}) \end{aligned} \tag{2.6}$$

gives the vector $\Phi_x = (B_{x1}, B_{x2}, \ldots, B_{xL})$ of blocking probabilities for class x at the L links in an arbitrary network. In (2.6)

$$I_{r,l} = \begin{cases} 1, & \text{if link } l \text{ belongs to route } r \\ 0, & \text{otherwise} \end{cases}$$

is the $|\mathcal{R}| \times L$ topology matrix of the network, $o(l, r)$ is the ordinal of link l in route r, i_r is the i-th link of route r, and $\Lambda_x(\cdot, \cdot)$ is a mapping giving the losses for each class as a function of the offered load, the capacity of link l and the local scheduling algorithm. The form of $\Lambda_x(\cdot, \cdot)$ has been given in the previous section.

This nonlinear equation system can be solved by means of iterated substitutions, starting with an arbitrary initial vector for the link blocking probabilities, resulting in a transformation $B_{xl}^{i+1} = F(B_{xl}^{i})$, with B_{xl}^{j} the value of the vector of blocking probabilities at iteration j. $F(\cdot)$ is a continuous mapping from the set $[0, 1]^L$ into itself, so it has a fixed point, by the Brouwer fixed point theorem [16]. In general, the uniqueness of the fixed point cannot be guaranteed, but we have not encountered convergence related problems during the solution of our models for the analyzed cases.

Note in (2.6) that the load offered to link l includes the sum from the whole set of routes traversing that link. In the system model, the traffic contributed by route α is approximated as a Poisson process with intensity $\lambda_{(x,\alpha)}$ thinned by the losses in the links preceding l along that route. This approximation is more accurate as the degree of connectivity of the network increases, i.e, as more route diversity exists. Note also that the model has been formulated only for two classes of bursts, but the generalization to an arbitrary number of traffic classes is straightforward.

After solving for the link blocking probabilities, and assuming that their blocking probabilities are independent, a burst is blocked whenever any of the links along its route is blocked, so the blocking probability of route r is

$$B_x^r = 1 - \prod_{i \in r} (1 - B_{xi})$$

and the overall network throughput

$$T = \sum_{\alpha \in \mathcal{R}} \sum_{x} \lambda_{(x,\alpha)} \left(1 - B_x^r\right).$$

2.4 Throughput Differentiation

Since in networks without admission control, as Internet, some form of congestion control is responsible for avoiding instability, inefficiency and unfairness, we extend the basic model with dynamic traffic load. This means that the sources adapt their transmission rate upon congestion (loss) in the network. The rate adaptation is that of TCP. Remarkably, this allows us to analyze also proportional differentiation in throughput to the end users. As for the differentiation strategy (the policy used to set up every class' weight), there is freedom of choice. For the sake of concreteness, we suggest here three simple possibilities, and devote some space to discuss their properties.

Let us suppose that every external class reacts upon "congestion" inside the core OBS subnetwork reducing its offered traffic when packet losses are more frequent. Consider, in particular, that each external class results from the aggregation of a similar and sufficiently large number — large enough so that the departure or the arrival of a connection does not change substantially the overall throughput of the class— of long-lived TCP Reno connections generating constant length packets. TCP is still the dominant transport traffic in Internet and its Reno implementation the most widely used. Moreover, there have been proposed TCP-friendly throughput regulators for UDP traffic [12], ensuring a fair sharing of the network capacity.

There are well established analytical models for the dynamic behavior of the Reno variant (see [18] and references therein). Under appropriate conditions, and within the range of blocking probabilities of our interest, equation (32) in [18] gives the offered rate of a single TCP session, $\lambda(p, \mathrm{RTT})$, as

$$\frac{\frac{1-p}{p} + \mathrm{E}[W] + \widehat{\mathrm{Q}}(\mathrm{E}[W]) \cdot \frac{1}{1-p}}{\mathrm{RTT} \cdot \left(\frac{\mathrm{E}[W]}{2} + 1 + \widehat{\mathrm{Q}}(\mathrm{E}[W]) \cdot \frac{\mathrm{f}(p)}{1-p}\right)} \tag{2.7}$$

where p is the packet loss probability of the TCP connection, RTT is its average round trip time, W is the unconstrained window size, $\widehat{\mathrm{Q}}(x)$ is the probability that a loss in a window of size x is a timeout, and

$$\mathrm{f}(p) = 1 + p + 2p^2 + 4p^3 + 8p^4 + 16p^5 + 32p^6.$$

For the validity of (2.7), we will assume that the maximum receiver window size is arbitrarily large, so that the congestion window is never exhausted, and also that the condition discussed in [8] holds: sources are slow, i.e., there is at most one TCP data segment from any source in a burst, and the standard deviation of the RTT is negligible.

Equation (2.7) has the following asymptotic behavior for small values of p:

$$\lambda'(p, \mathrm{RTT}) \triangleq \lim_{p \to 0} \lambda(p, \mathrm{RTT}) = \frac{\sqrt{\frac{3}{2}}}{\mathrm{RTT} \cdot \sqrt{p}}. \tag{2.8}$$

As an illustration, Fig. 2.3 shows the relative error

$$\mathrm{e}_{\lambda'} \triangleq \frac{\lambda'(p, \mathrm{RTT})}{\lambda(p, \mathrm{RTT})} - 1$$

and confirms that the approximation

$$\overline{\lambda}(p, \mathrm{RTT}) \triangleq \lambda(p, \mathrm{RTT}) \cdot (1 - p) \simeq \lambda(p, \mathrm{RTT}) \simeq \lambda'(p, \mathrm{RTT}), \tag{2.9}$$

where $\overline{\lambda}(p, \mathrm{RTT})$ refers to the throughput of a single TCP session, is in fact quite accurate for small packet loss probabilities ($p \leq 0.1$) which is the regime we precisely expect OBS networks to be designed for.

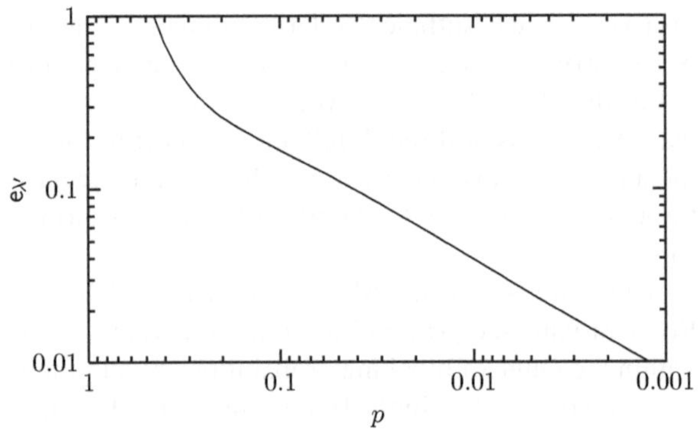

Fig. 2.3 TCP Reno behavior

Now, we will investigate if proportional packet loss differentiation can be achieved, and also whether proportional differentiation in the throughput is indirectly possible and to what extent. Thus,

$$\overline{\lambda}_i \simeq s_i \cdot \overline{\lambda}_{n-1} \tag{2.10}$$

where $s_0, s_1, \ldots, s_{n-1}$ is a strictly decreasing sequence, s_{n-1} is 1 and $\overline{\lambda}_i$ denotes the throughput of a connection of class i.

Note that, in order to enforce (2.10) acting only on the packet loss probabilities, we are implicitly making the assumption that the round trip time is approximately the same for all the TCP connections. Nevertheless, this hypothesis has nothing to do with constraining the actual RTTs to be constant, but it is rather a mathematical artifact that makes the analysis easier. Indeed, were the RTTs known they could be subsumed into the factors s_i appearing in (2.10) without losing generality.

Then, from (2.1) and (2.8), the packet loss probabilities in order to attain (2.10) must be related by

$$s_i \simeq \sqrt{\frac{p_{n-1}}{p_i}} \simeq \sqrt{\frac{1}{c_i}}. \tag{2.11}$$

Denoting by Λ_0 and Λ_1 the arrival rates of packets in an edge node to the two internal classes towards a given egress node, and by N_i the number of TCP connections of class i, the ratio between the arrivals rate r $\triangleq \Lambda_0/\Lambda_1$ can be expressed using (2.1), (2.10) and (2.11) as a function of the factors s_i and N_i. For

$$\Lambda_j = \sum_{i=0}^{n-1} h^i_j N_i \lambda_i \simeq \lambda_{n-1} \sum_{i=0}^{n-1} h^i_j N_i s_i, \qquad j = 0, 1$$

which, using $h^i_0 = 1 - c_i$, $h^i_1 = c_i$ and (2.11), for $i = 0, \ldots, n-1$, yields

$$\Lambda_0 \simeq \lambda_{n-1} \sum_{i=0}^{n-1} N_i \left(s_i - \frac{1}{s_i}\right), \qquad \Lambda_1 \simeq \lambda_{n-1} \sum_{i=0}^{n-1} \frac{N_i}{s_i}.$$

Thus, the ratio between the arrival rates is

$$\mathrm{r} \simeq \frac{\sum_{i=0}^{n-1} N_i s_i}{\sum_{i=0}^{n-1} N_i \frac{1}{s_i}} - 1. \tag{2.12}$$

We remark that the lower r is, the easier will be for the OBS network to maintain the burst loss probability differentiation. But a given value of r could be obtained with many different pairs (s_i, N_i), for $i = 0, \ldots, n-1$. For analytical simplicity, fix the same number of TCP connection per class, $N_i = N \quad \forall i \in \{0, \ldots, n-1\}$. This is equivalent to say that the *overall* per-class throughputs are also proportional, which may be seen as a worst case configuration. Doing so, (2.12) simplifies to

$$\mathrm{r} \simeq \frac{\sum_{i=0}^{n-1} s_i}{\sum_{i=0}^{n-1} \frac{1}{s_i}} - 1. \tag{2.13}$$

There are several possible choices for the scaling factors s_i, so the behavior of r is worth some closer examination. Consider these three arbitrary differentiation strategies:

- The existence of only two external classes.
- A geometric differentiation strategy: $n > 2$ and $s_i = g^{n-i-1}$, with $g > 1$.
- A linear differentiation strategy: $n > 2$ and $s_i = n - i$.

Let us examine these three alternatives in turn.

Two classes. When the number of external and internal classes is two, the rate arrival ratio (2.13) becomes

$$\mathrm{r} \simeq \frac{1 + s_0}{1 + \frac{1}{s_0}} - 1 = s_0 - 1,$$

i.e., it is approximately equal to the throughput multiplicative factor of the best external class, minus one.

Exponential. For the geometric differentiation case, the ratio is

$$\mathrm{r} \simeq \frac{\sum_{j=0}^{n-1} g^j}{\sum_{j=0}^{n-1} \frac{1}{g^j}} - 1 = g^{n-1} - 1 = s_0 - 1,$$

that is, the same approximate result as in the system with only two classes.

Linear. In the linear differentiation scheme, one gets

$$\mathrm{r} \simeq \frac{\sum_{i=0}^{n-1} n-i}{\sum_{i=0}^{n-1} \frac{1}{n-i}} - 1 = \frac{(n+1)n}{2\mathcal{H}_n} - 1$$

where $\mathcal{H}_n = \sum_{j=1}^{n} 1/j$ is the n-th harmonic number. For example, the approximate values of r for $n = 3, \ldots, 7$ are $25/11 \simeq 2.27$, $19/5 = 3.8$, $763/137 \simeq 5.57, 53/7 \simeq 7.57$ and $3557/363 \simeq 9.80$, respectively.

In summary, for each n the ratio of arrival rates is higher with the geometric scheme ($g \geq 2$) than with the linear scaling factors. Additionally, since the geometric scheme is equivalent to the reduced scenario with two classes, only the latter will be analyzed. The performance of any other strategy is expected to be bounded by that of a two-class system with a greater or equal r.

Introducing the sources' reaction to network congestion into the fixed-point model, the system equations can now be written as

$$\begin{aligned} (A^{p}_{(0,r)}, A^{p}_{(1,r)}) &= (\pi_{0,r} h^{0}_{0}, \pi_{0,r}(1-h^{0}_{0}) + \pi_{1,r}) \\ A_{xl} &= \sum_{r \in \mathcal{R}} A^{p}_{(x,r)} I_{r,l} \prod_{i=1}^{o(l,r)-1} \left(1 - B_{xi_r}\right) \\ B_{xl} &= \Lambda_x(A_{0l}, A_{1l}) \end{aligned} \tag{2.14}$$

in the same notation as Section 2.3, and where $\pi_{x,r}$ is the external class-x traffic intensity offered to route r

$$\pi_{x,r} = N_x \lambda(p_x, RTT) \cdot \overline{S}$$

with $\overline{S}$ the average packet transmission time, and p_x and RTT refer to the loss probabilities and round trip times of class-x packets traversing route r.

2.5 Performance Analysis

In this section, the mathematical framework developed so far is applied to three particular cases amenable to further analysis: a single congested link, a ring topology, and a mesh network. The general system equations (2.6) are instantiated and solved numerically. The purpose is to gain insight into the performance of the packet classification algorithm, and to that end we investigate in detail whether the traffic sources experience packet loss probabilities scaled proportionally to a given set of weights when several classes exist.

2.5.1 Congested Link

Consider now a scenario where all links in a network, except one, have very low burst loss probabilities. Thus, the congested link determines the overall performance. The vast majority of congestion episodes in large networks involve typically only one node/link, so analyzing this case has practical as well as theoretical interest.

In order to gain some insight into the behavior with a congested link, we report in the following several numerical experiments that evaluate three performance metrics for a system with two external classes exclusively and $s_0 = 2, \ldots, 6$:

- The internal loss probability ratio B_1/B_0. This is a direct measure of the differentiation power of the scheduling algorithm.
- The relative error in the loss probability differentiation
$$e_{c_i} \triangleq \frac{c_i' - c_i}{c_i}$$
(where $c_i' \triangleq p_i/p_{n-1}$ is the actual loss probability differentiation while c_i is the desired one). Hence, this quantity captures the actual impact of all the approximations and assumptions made in the previous analysis.
- The relative error in the throughput differentiation coefficients:
$$e_{s_i} \triangleq \frac{s_i' - s_i}{s_i}$$
(where $s_i' \triangleq \overline{\lambda}_i/\overline{\lambda}_{n-1}$ is the actual throughput differentiation while s_i is the expected one.

Figures 2.4 and 2.5 plot the performance of the scheduler with segmentation, setting a global loss probability equal to 0.01. Specifically, Fig. 2.4 shows the internal probability differentiation, Fig. 2.5(a) shows the relative error in the probability

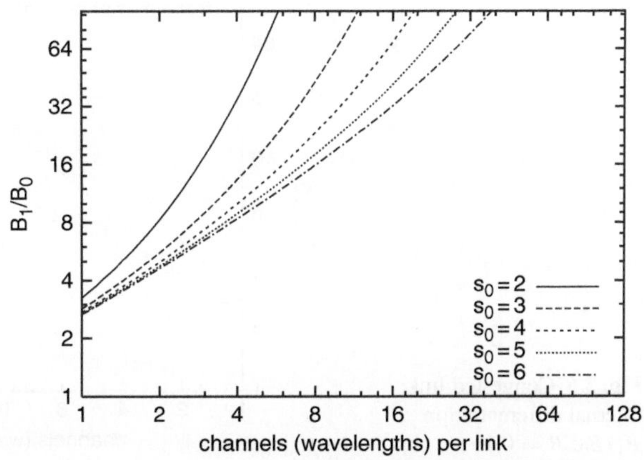

Fig. 2.4 Congested link; internal differentiation B_1/B_0; $B = 0.01$

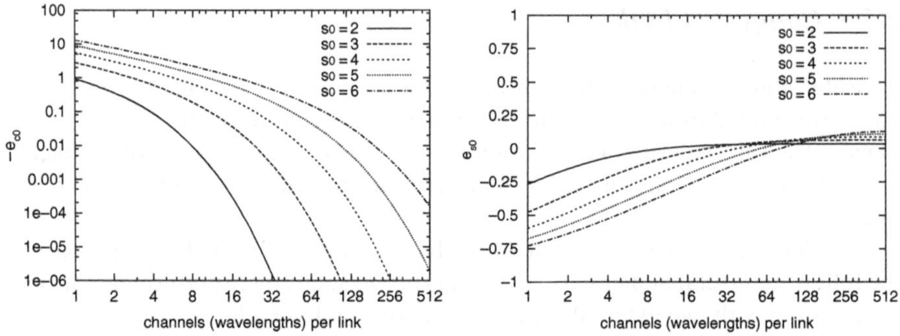

(a) Packet loss (b) Throughput

Fig. 2.5 Congested link; external differentiation relative error; $B = 0.01$

differentiation and, finally, Fig. 2.5(b) depicts the relative error in the throughput differentiation. As seen in the figures, with $B = 0.01$ the goal of proportional differentiation of the packet loss probability is achieved with only a small number of wavelengths in each link. For instance, with $m = 8$ and $s_0 = 2$, the internal differentiation ratio $B_1/B_0 \simeq 104$ produces an external probability differentiation error $e_{c_0} \simeq -2.9\%$, that is, we will have $p_1 \simeq 4.12 \cdot p_0$, $1/c'_0 \simeq 4.12$, instead of the target values $1/s_0 = c_0^2 = 4$.

One can also see in Fig. 2.5(b) the impact due to the optimistic prediction of the rate offered by TCP: the low external priority class experiences more losses than the high priority traffic. Therefore, (2.8) and (2.9) (see Fig. 2.3) give less throughput overestimation for the external high priority traffic, giving rise to $e_{s_0} > 0$ for m sufficiently large. For instance, with $m = 8$ and $s_0 = 2$ we already have an external throughput differentiation error $e_{s_0} \simeq 0.02\%$.

The behavior of the resulting configuration with global packet loss probability equal to 0.1 is depicted in Figs. 2.6, 2.7(a) and 2.7(b). The conclusion to draw from these plots is that an operating regime around a global loss probability of 0.1 is too hard in order to ensure the differentiation, except for the lower s_0. Moreover,

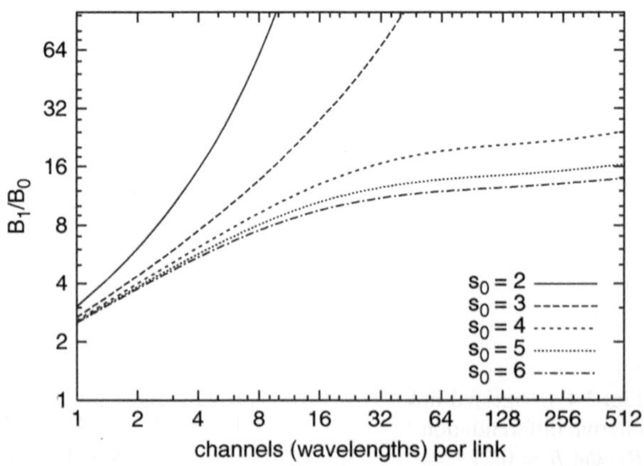

Fig. 2.6 Congested link; internal differentiation B_1/B_0; $B = 0.1$

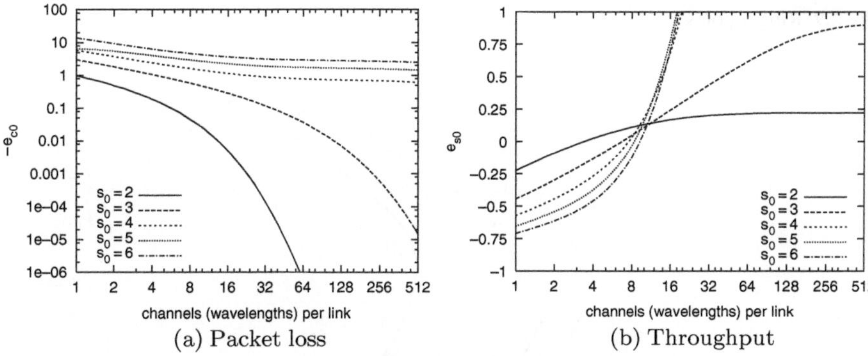

Fig. 2.7 Congested link; external differentiation relative error; $B = 0.1$

it is easy to see in Fig. 2.5(b) how the proposed mechanism penalizes more the low priority than the high priority flows (for $m \geq 64$ and $s_0 > 2$ we get $e_{s_0} \gg 50\%$) when coexist a very large number of TCP flows. Nevertheless, this undesirable performance vanishes if the arbitrary condition that all classes comprise an identical number of TCP flows is relaxed. This observation can be checked in the Fig. 2.8(a) and (b) (the internal differentiation B_1/B_0 reaches its peak earlier than in Fig. 2.6, so it is not shown here), where the curves have been computed for $N_1 = 10N_0$. Obviously, the results will improve accordingly for smaller global loss probabilities.

2.5.2 Ring Network

Another simple topology to analyze symmetrical multinode configuration shown in Fig. 2.9. Assume each node receives traffic flows of both external classes directed towards any other node from the outside, and the network uses shortest path routes. In this way, any unidirectional link between two nodes carries traffic flows from three source-destination pairs:

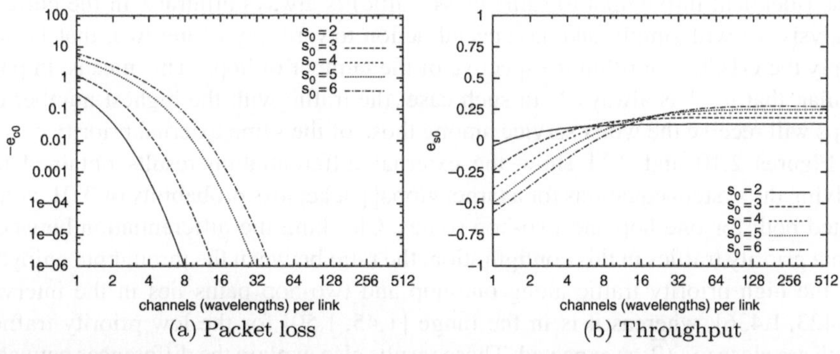

Fig. 2.8 Congested link; external differentiation relative error; $N_1/N_0 = 10$, $B = 0.1$

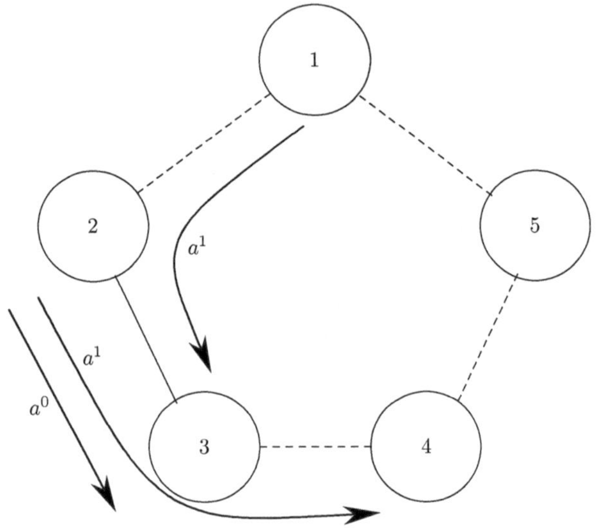

Fig. 2.9 Ring topology

- There are two traffic flows from two source-destination pairs with two hop paths. We will denote by $a^1 = a_0^1 + a_1^1$ the offered traffic intensity, composed of the internal high priority a_0^1 and the internal low priority, a_1^1.
- There is one traffic flow from one pair with a one hop path. We will denote similarly by $a^0 = a_0^0 + a_1^0$ its offered traffic intensity, sum of the intensities from the high internal priority a_0^0 and from the low priority, a_1^0.

We will also assume that the RTT of all TCP connections is approximately the same, that is, it will be mainly dominated by the queue waiting times outside the OBS network.

Apart from the external differentiation among external classes of the same source-destination pair, any actual network will also surely differentiate between traffic of different source-destination pairs pertaining to the same class. Since a trade-off between efficiency (maximum throughput to fewer hops traffic) and fairness (identical throughput to same class traffic) is always arbitrary, in the current analysis we will simply take no special action toward any of the two, that is, we apply the OTCD algorithm irrespective of the number of hops. This means, in particular, that h_1^{n-1} is always 1. In such case, the traffic with the highest number of hops will receive the worst service among those of the same external priority class.

Figures 2.10 and 2.11 show the external differentiation results obtained by solving the system equations for a target global packet loss probability of 0.01, computed both for one-hop and two-hop routes. Checking the differentiation between same priority traffics in this configuration, the ratio between the throughput enjoyed by the high priority traffic along one-hop and two-hop paths lies in the interval [1.423, 1.436], whereas it is in the range [1.45, 1.50] for the low priority traffic. Both are close to $\sqrt{2}$, as expected. These results also explain the differences between Fig. 2.11(a) and (b).

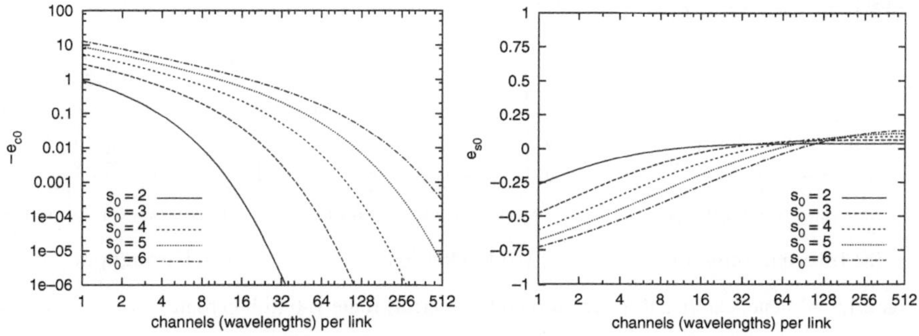

(a) Packet loss (b) Throughput

Fig. 2.10 Ring network; external differentiation relative error; 1-hop paths

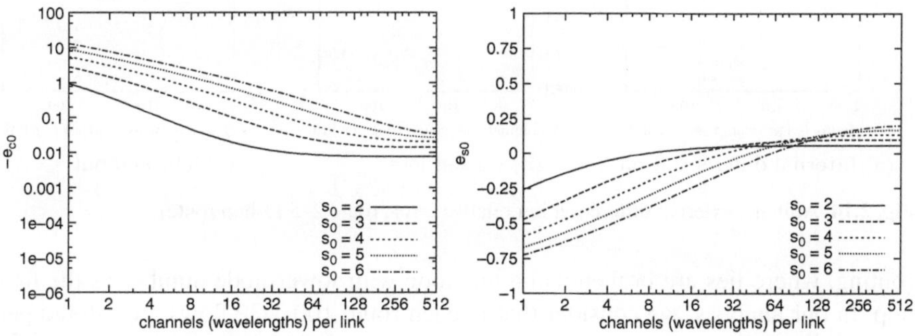

(a) Packet loss (b) Throughput

Fig. 2.11 Ring network; external differentiation relative error; 2-hop paths

2.5.3 Mesh Topology

Here we analyze what could be a typical mesh scenario, the Abilene network (Fig. 2.12). This network has a diameter of 5 hops, and we have chosen several routes with a number of hops varying between 2 and 5 in order to have some diversity of path lengths and degrees of link sharing. We consider shortest path

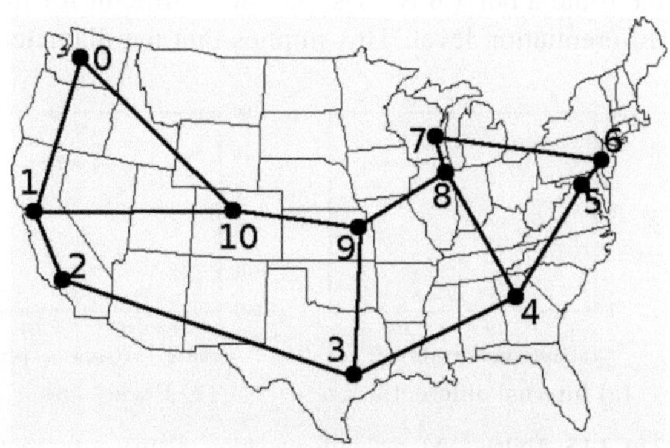

Fig. 2.12 Abilene Topology

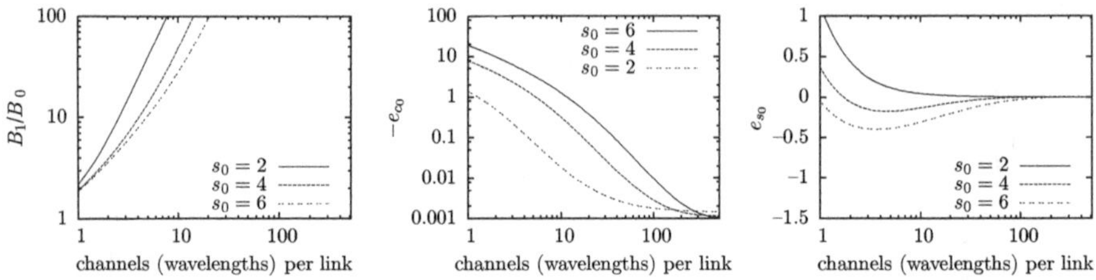

(a) Internal differentiation (b) Packet loss (c) Throughput

Fig. 2.13 Abilene; external differentiation relative error; route 1–6 (5-hop route)

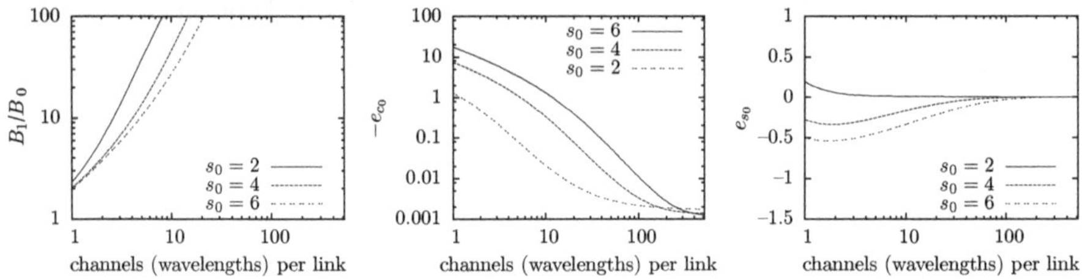

(a) Internal differentiation (b) Packet loss (c) Throughput

Fig. 2.14 Abilene; external differentiation relative error; route 2–5 (3-hop route)

routing, where ties are broken using the node with lower node number as the first hop. In our analysis, we consider that in each route 100 TCP flows are offered per lambda for the lower class, and 10 flows for the higher one. We also suppose that the maximum segment size of TCP flows is 1500 bytes—typical of bulk transfers—, a round trip time for all connections of around 10 ms, and that each wavelength is operated at 10 Gbps.

The system (2.14), customized for this topology, results in 28 equations for the 28 link blocking probabilities—14 links with 2 classes of burst per link. It can be solved to evaluate the 110 end-to-end path blocking probabilites for each kind of burst. These, in turn, can be used to draw directly the packet loss probabilities by means of the OTCD algorithm.

The results are shown in Figs. 2.13, 2.14 and 2.15. It can be seen that, the longer the route a burst traverses, the more difficult for the network is to achieve a given differentiation level. This implies that the diameter of the network will determine

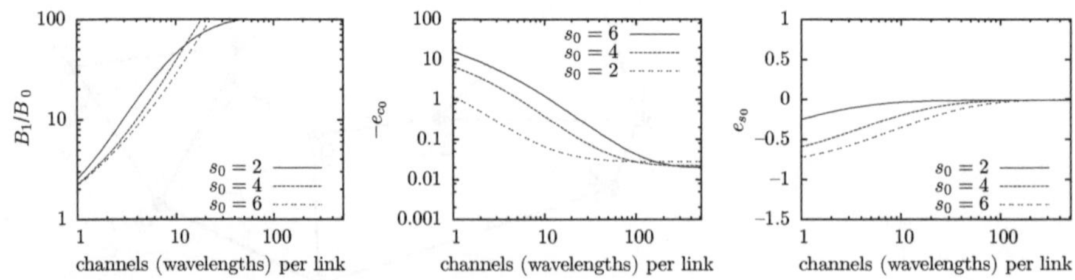

(a) Internal differentiation (b) Packet loss (c) Throughput

Fig. 2.15 Abilene; external differentiation relative error; route 4–8 (1-hop route) the feasible differentiation level in the case that all users (routes) can choose the same range of differentiations. Alternatively, the network can offer a differentiation level in a range that is dependent on the specific route a connection has to follow. This can also be influenced by the fact that the network can use methods for contention resolution such as deflection routing (affecting the path lengths), but we do not address the issue in this work.

2.6 A Simulation Study

It may seem a rather strong assumption that, in practice, the burst blocking probabilities in the OBS subnetwork may be made different by several orders of magnitude, so that the ratio of packet loss probabilities become almost insensitive to them. In order to demonstrate that the claim $B_0 \ll B_1$ holds, and that the OTCD classifier is well founded, we present at this point the results of a realistic simulated scenario in support of the scheme. We use a topology with two edge nodes and two core nodes, connected in a linear fashion. The edge nodes classify the incoming packets according to the OTCD rules before aggregating them in separate bursts types. The assembling algorithm is mixed—a burst is created after 10 milliseconds or the arrival of 10000 bytes, whichever occurs first—, the offset time is 20 μs, and the average processing time for control packets is 1 μs in each node. There are 4 control channels and 12 data channels at 1 Mbps each in every fiber, and a delay of 1 ms per link. There are two types of bursts, high-class and low-class, the difference in priority implemented through selective burst discarding: low-class bursts are discarded before trying to be scheduled with a probability of 0.1. We consider two classes of packets trying to traverse the OBS network, one is a typical best-effort class and the other has QoS constraints expressed by means of the proportional differentiation paradigm: high-class packets desire to have a loss probability K times less than the one for best-effort packets. The best-effort packets are generated by 100 TCP-Reno flows, while the high-class packets come from 10 TCP-Reno flows. Each flow has packets with a size of 1500 bytes, typical of bulk transfers. We show in Fig. 2.16(a) the evolution of a typical sample path of a simulation

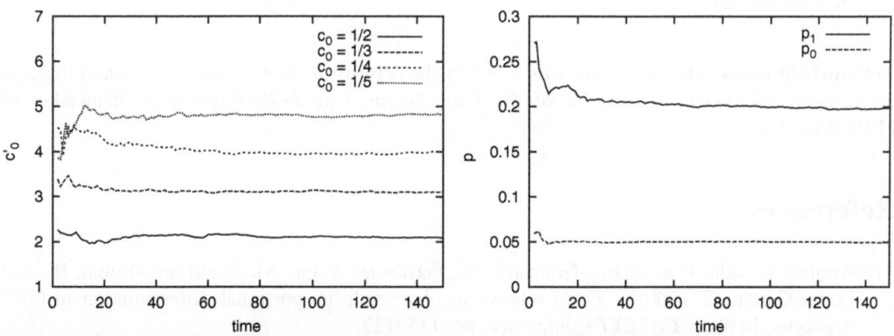

(a) Packet loss proportional differentiation for $K \in \{2, 3, 4, 5\}$. (b) Packet loss probability for the case $K = 4$.

Fig. 2.16 Performance of the OTCD classifier. The displayed traces have been produced via simulation of the described scenario. The figure plots the ratio between the loss probabilities of high and low class packets, respectively, for desired differentiations of $K \in \{2, 3, 4, 5\}$. It is clear that the simulation of each case closely approximates the desired ratios.

In addition, we have plotted in Fig. 2.16(b) the absolute measures for packet losss probability for the case of $K = 4$. The experimental results are thus in good agreement with the expected performance confirming the validity of assumptions.

2.7 Conclusions

We have devised a simple method to achieve proportional loss differentiation between packets that traverse an OBS core network. This method is solely based on an stochastic algorithm to assemble two classes of packet bursts and merely requires that the OBS network provide some form of internal relative differentiation such that one of the burst classes has a much lower loss probability than the other.

In order to gain some insight, both theoretical and numerical about the performance of the method, we have presented an analytical study, by using simple, approximate circuit switching models, of two simple but sufficiently representative scenarios: a single congested link and a multinode symmetrical network. The results have shown the validity of the proposed algorithm over a wide range of operating conditions.

Nevertheless, further work remains to be done. An important issue is to study the impact of variable RTTs on the behavior of the algorithm. The exactness of the slow source model in realistic network configurations must also be assessed and a proper burst assembly algorithm must be devised in order to reduce packet loss correlation and avoid false congestion notifications. Overall, the whole model should be validated through simulations. On the practical side, it seems interesting to extend the technique toward both ends of the communications path, including intermediate IP routers, so that genuine end-to-end proportional packet loss differentiation is actually achieved.

Acknowledgments This work was supported by the *Ministerio de Educación y Ciencia* through the project TSI2006-12507-C03-02 of the *Plan Nacional de I+D+I* (partially financed with FEDER funds).

References

1. Argibay-Losada, P. J., Suárez-González, A., Fernández-Veiga, M., Rodríguez-Rubio, R., and López-García, C. (2005). From relative to observable proportional differentiation in OBS networks. In *Proc. CoNEXT Conference*, pp. 115–123.
2. Baldine, I., Rouskas, G. N., Perros, H. G., and Stevenson, D. (2002). JumpStart: a just-in-time signaling architecture for WDM burst-switched networks. *IEEE Commun. Mag.*, 40(2): 82–89.

3. Barakat, N. and Sargent, E.H. (2005). Analytical modeling of offset-induced priority in multiclass OBS networks. *IEEE Transactions on Communications*, 53(8):1343–1352.
4. Blake, S., Black, D., Davies, M. E., Wang, Z., and Weiss, W. (1988). An architecture for differentiated services. RFC 2475.
5. Chen, Y., Hamdi, M., and Tsang, D. H. K. (2001). Proportional QoS over OBS networks. In *Proc. Globecom 2001*, pp. 1510–1514.
6. Chen, Y. and Qiao, C. (2003). Proportional differentiation: A scalable QoS approach. *IEEE Commun. Mag.*, 41(6):52–58.
7. Detti, A., Eramo, V., and Listanti, M. (2002). Performance evaluation of a new technique for IP support in a WDM optical network: Optical composite burst switching (OCBS). *J. Lightwave Tech.*, 20(2):154–165.
8. Detti, A. and Listanti, M. (2002). Impact of segments aggregation on tcp reno flows in optical burst switching networks. In *Proc. INFOCOM 2002*, pp. 1803–1812.
9. Dovrolis, C. and Ramanathan, P. (1999). A case for relative differentiated services and the proportional differentiation model. *IEEE Network Mag.*, 13(5):26–34.
10. Dovrolis, C., Stiliadis, D., and Ramanathan, P. (2002). Proportional differentiated services: delay differentiation and packet scheduling. *IEEE/ACM Trans. on Netw.*, 10(1): 12–26.
11. Gibbens, R.J., Sargood, S.K., Eijl, C. Van, Kelly, F.P., Azmoodeh, H., Macfadyen, R.N., and Macfadyen, N.W. (2000). Fixed-point models for the end-to-end performance analysis of IP networks. In *Proc. 13th ITC Specialist Seminar: IP Traffic Measurement, Modeling and Management*.
12. Handley, M., Floyd, S., Padhye, J., and Widmer, J. (2003). TCP friendly rate control (TFRC): Protocol specification. RFC 3448.
13. Karagiannis, T., Molle, M., Faloutsos, M., and Broido, A. (2004). A nonstationary poisson view of internet traffic. In *Proc. INFOCOM 2004*, pp. 1558–1569.
14. Kelly, F. P. (1986). Blocking probabilities in large circuit-switched networks. *Adv. Applied Prob.*, 18(2):473–505.
15. Liao, W. and Loi, C-H. (2004). Providing service differentiation for optical-burst-switched networks. *J. Lightwave Tech.*, 22(7):1651–1660.
16. Munkres, J. R. (1984). *Elements of Algebraic Topology*. Perseus Books.
17. Neuts, M., Rosberg, Z., Vu, H. L., White, J., and Zukerman, M. (2002). Performance analysis of optical composite burst switching. *IEEE Commun. Lett.*, 6(8):346–348.
18. Padhye, J., Firoiu, V., Towsley, D.F., and Kurose, J.F. (2000). Modeling TCP Reno performance: a simple model and its empirical validation. *IEEE/ACM Trans. on Netw.*, 8(2):133–145.
19. Rosberg, Z., Vu, H. L., Zukerman, M., and White, J. (2003). Performance analyses of optical burst-switching networks. *IEEE J. Select. Areas in Commun.*, 21(7):1187–1197.
20. Ross, K. W. (1995). *Multiservice Loss Models for Broadband Telecommunication Networks*. Springer Verlag, London, UK.
21. Sahasrabuhde, A. and Manjunath, D. (2006). Performance of optical burst switched networks: a two-moment analysis. *Computer Networks*, 50(18):3550–3563.
22. Tan, S. K., Mohan, G., and Chua K. C., (2007). Feedback-based offset time selection for end-to-end proportional QoS provisioning in WDM optical burst switching networks. *Computer Commun.*, 30(4):904–921.
23. Tan, C.-W., Gurusamy, M., and Lui, J. C.-S. (2004). Achieving proportional loss differentiation using probabilistic preemptive burst segmentation in optical burst switching WDM networks. In *Proc. Globecom 2004*, pp. 1754–1758.
24. Vokkarane, V. M., Jue, J. P., and Sitamaran, S. (2002). Burst segmentation: An approach for reducing packet loss in optical burst switched networks. In *Proc. ICC 2002*, pp. 2673–2677.
25. Vokkarane, V. M. and Jue, J. P. (2003). Prioritized burst segmentation and composite burst-assembly techniques for qos support in optical burst-switched networks. *IEEE J. Select. Areas in Commun.*, 21(7):1198–1209.

26. Wei, J. Y. and McFarland, R. I. (2000). Just-in-time signaling for wdm optical burst switching networks. *J. Lightwave Tech.*, 18(12):2019–2037.
27. Xiong, Y., Vandenhoute, M., and Conkaya, H. C. (2000). Control architecture in optical burst-switched wdm networks. *IEEE J. Select. Areas in Commun.*, 18(10):1838–1851.
28. Zalesky, A., Vu, H. L., Rosberg, Z., Wong, E. W. M., and Zukerman, M. (2004). Modelling and performance evaluation of optical burst switched networks with deflection routing and wavelength reservation. In *INFOCOM 2004*, pp. 1864–1871.

Chapter 3
Markovian Analysis of a Synchronous Optical Packet Switch

Joanna Tomasik and Ivan Kotuliak

Abstract We study switch architectures applicable to synchronous fixed-length optical packet networks in order to compare their performance in terms of *packet loss ratio* (PLR). We propose analytical models of these switches representing them as discrete time Markov chains and we solve them for incoming traffic with varying statistical properties. We compare performance measures computed using an analytical method with those obtained in an experimental way, by simulation, and we formulate conclusions on performance of considered switches. Our paper shows that Markovian models of future optical packet network architectures are efficient and that they can be applied as a tool to practical studies of network design.

Keywords Synchronous optical network · Discrete-time Markov chain · Autocorrelation coefficient · Packet loss Ratio (PLR) · Performance evaluation

3.1 Introduction

The modern technology in the area of transport networks, based on electronic packet processing and using optical technology only as the link layer, is getting to its limits. New promising technologies are based on the packet processing in the optical layer [1].

The all-optical packet switching means that packet flows incoming into studied network are converted into optical payloads at the network entries. These payloads pass through the network without being transformed into electronic form. They are converted back into their original format at the egress of the network [2].

The all-optical technology, allowing one to construct all-optical networks, is already deployed in local networks thanks to their topology (all-optical networks are limited to the ring or star topology) and to the medium-access protocol adapted to this topology [3, 4]. Such networks use either one wavelength for signalization

J. Tomasik (✉)
SUPELEC, Computer Science Departement, Plateau de Moulon,
91 192 Gif-sur-Yvette Cedex, France
e-mail: Joanna.Tomasik@supelec.fr

M. Ma. (ed.), *Current Research Progress of Optical Networks*,
DOI 10.1007/978-1-4020-9889-5_3,

or a central station, which manages a medium access policy. On the other hand, the deployment of the all-optical technology in meshed, metropolitan, and backbone networks, suffers from several problems. The major one is the lack of optical memory which makes a switching operation very difficult. Several studies have been performed in order to overcome the lack of memory, such either as applying of optical packet switching on full-meshed network [5] or as using special routing techniques, like hot-potato [6].

Two different approaches try to introduce all-optical technology into WAN environment [7]: the first one is based on the synchronous packet processing in the node and on the fixed length time slots [8]. The second one is based on the variable-length optical packets and is generally used with the burst switching [9].

We investigate first approach, a synchronous all-optical packet network whose topology is an incomplete mesh and whose routing is based on local routing tables. A synchronous optical packet network transmits packets of constant length (the length of an optical packet is usually expressed in time units, not in bytes). This approach allows one to increase the transmission speed without any change in the packet format. Details concerning the packet format can be found in [10]. The synchronous optical packet switching research has been pushed forward by KEOPS *Keys in Optical Packet Switching* project [11] in Europe. The KEOPS project proposed a technical base for the optical packet switching and described different architectures of the optical switch. Performance evaluation framework of OPS congestion resultion is given in [12]. In this article we use some achievements of the ROM project [8], which have used performance evaluation by simulation and compare these results with those obtained using discrete time Markov chains [13].

In [14], we provided an analytical study of the synchronous all-optical switch. The analysis is done using discrete-time Markov chains and the results are compared to those obtained in the ROM project by simulation. This paper extends the previous work giving more detailed analysis of the problem. The next section contains a description of the switch architecture. The proposed Markovian models are described in Section 3.3. The statistical features of traffic incoming into the node have a crucial impact on the performance of the node [15]. The characteristics of the traffic used in our study are given in Section 3.4. The obtained results, which are focused on computing of Packet Loss Ratio (PLR), using analytical models and their comparison with these obtained by simulation [16] can be found in Section 3.5. Section 3.6 provides the conclusions.

3.2 Studied Models

A switching node of the all-optical network should route the incoming packets to the outgoing ports within fixed length time slots. It should be done with strict memory constraints and it has to be very fast. It may happen that two or more packets try to take over the same outgoing port (the same wavelength on the same fiber). In

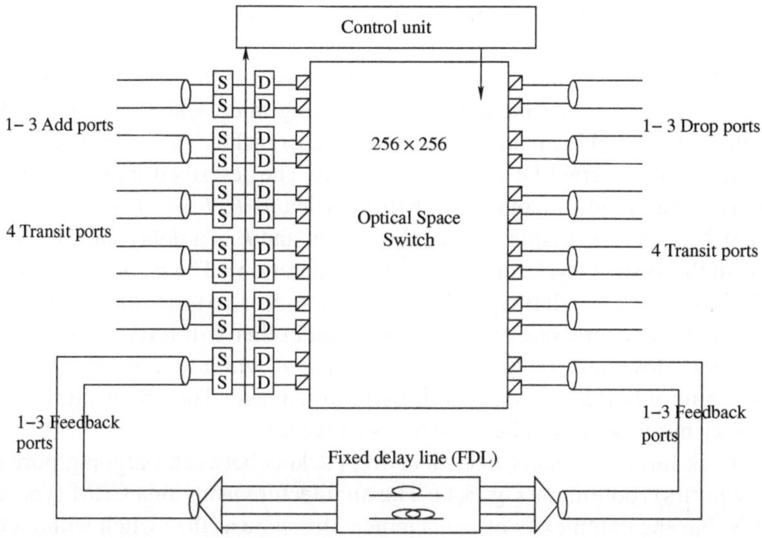

Fig. 3.1 Structure of optical packet switching node

this case, only one packet can succeed. Such a situation is called *contention* and is similar to *collision* in the Ethernet. In this case the packet, which has not succeeded, either uses a contention resolution mechanism (CRM) or is lost.

The node architecture is presented in Fig. 3.1. The functionality of the node can be described as follows: packets, entering the node on several wavelengths per fiber, are firstly synchronized (S). Then headers are separated for electronic processing. Payloads, still in the optical form, enter delay lines based on FDL (Fiber Delay Lines) [17] (D) and leaving them, they enter the non-blocking switching matrix. During the time while the payloads are being retarded in the delay lines (D), the control unit (CU) processes the routing information contained in the headers and parameterizes the switching matrix so that the payloads can pass through and leave it on the chosen port and the chosen wavelength. There are no limitations in the wavelength conversion; an incoming packet can leave the node on any wavelength and any port. Before sending the packets to the outgoing fibers, new headers indicating the routing information are generated and added in front of the packets. Depending on the used technology and the time slot duration this process does not take more time than a few hundreds nanoseconds.

It has already been mentioned above that contention existing in optical networks could cause packet losses. Packet losses can be reduced by existing congestion resolution mechanisms. Three node architectures, differing in the CRM type, will be at the center of our interest:

1. No CRM — this is the simplest architecture in terms of fabrication and management. It is exactly the same as the architecture described above. In the case of contention (two or more packets are trying to use the same outgoing port) one

succeeds and other(s) is (are) lost. This solution provides obviously the worst results.

2. Memory buffers — real memories in the optical technology do not exits, so optical buffers emulate memory behavior using fiber delay lines. This CRM is based on the switching matrix with several FDLs inside. The incoming payload is injected to the several FDLs with varied delay. The described approach allows the Control Unit to route incoming packets with delays $T, T+k, T+2k, \ldots, T+nk$, where T is the processing time of the packet and k is a delay unit of the FDLs used in the system (typically $k =$ time slot duration). The packets can be stored only for nk time, which is typically limited up to 8 or 16 time slots. Consequently, if contention occurs, one packet is routed and other will retry to go out at next time slot. However, if the packet is not routed within n time slots, it is lost. The memory buffer approach is difficult to manage. The architecture allowing memory use is usually called a MAN architecture.
3. Feedback lines — optical fiber delaying packets between outgoing port(s) and entry port(s) (bottom of Fig. 3.1). The architecture using this CRM type, called WAN, has the complexity of management the same as the switch without CRM. Therefore, the first architecture is also called WAN without CRM.

   The packets, which do not find a free destination port, are sent to a feedback line with fixed delay (an FDL). They re-appear at the switch entry once again a several time slots later. The feedback system is relatively simple to be implemented and could be deployed in very high performance switches (Tbit/s). Its drawback is a possible sequence loss that means that some packets of the same flow sent later could arrive before those sent earlier. This could cause problems for some types of traffic, namely IP and data transmissions. Consequently, this solution cannot be used for a traffic with guarantee of services, only the packets with low priority can profit from it.

The network performance in terms of packet loss (PLR) is supposed to be the best for the architecture with emulated memory buffers (MAN architecture). Feedback lines should give also good results comparing to those of WAN node without CRM. It should be emphasized that switching node can be very sensible to the traffic profile (shaped vs. self-similar traffic).

3.3 Markovian Models of Optical Switches

In this section, we present in detail Markovian models [18] of the optical switches described above, and we propose the algorithms to generate the stochastic matrices and the number of their reachable states for all of them.

Because optical packets transmitted by the network are of constant size, we choose a discrete time chain in which a time slot is equivalent to a time-length of one optical slot. Observations of the Markov chain which determine its current state are taking place just after the beginning of a time slot. An external observer of the network sees its state just after it has accomplished all possible actions at the beginning of a time slot.

3.3.1 Model Without Contention Resolution Mechanism (CRM)

We start our analysis with a Markovian model corresponding to a basic node without CRM. In Fig. 3.2 we present the model of the router with two incoming fibers (fiber (0), fiber (1)), each composed of N wavelengths ($N = 4$ in our example). In order to introduce a delay imposed by the router we use two servers with number of places equal to a number of wavelengths in the fiber, N (server (0), server (1), respectively). We assume that the service time for each packet is equal to its length.

Optical packets come into the first stage of the switch where the choice of the routes for them is determined. The probability that a packet departing from the (i) fiber is routed to the fiber (j) is equal to $r_{ij},\ i, j = 0, 1$. These probabilities form the matrix $\boldsymbol{R} = [r_{ij}]_{i,j=0,1},\ \sum_j r_{ij} = 1$. A packet has to be routed immediately and if it cannot enter the selected fiber, it is lost.

We analyze now the stochastic process describing packet arrivals. We note the probability that one wavelength on the fiber (i), $i = 0, 1$, is free (there is no packet on a given wavelength) as $q^{(i)}$ and the probability that it is occupied (there is a packet with data) as $p^{(i)} = 1 - q^{(i)}$. The probability $p^{(i)}$ is called also the mean wavelength load. At this moment we do not consider any variability of $p^{(i)}$ (the analysis with $p^{(i)}$ varying in time is presented in Section 3.4).

This assumption allows us to consider a distribution of occupancy of one wavelength as a Bernoulli distribution. We state that all the wavelengths are loaded independently, so the occupancy of the fiber seen as a group of N wavelengths is distributed binomially. If $p_n^{(i)}$ is a probability that n wavelengths are busy in the i-th fiber, it is computed as follows:

$$p_n^{(i)} = \binom{N}{n} p^{(i)^n}(1 - p^{(i)})^{N-n}. \tag{3.1}$$

Notice that the mean load for an entire fiber remains equal to the mean load of a wavelength in this fiber as the mean number of occupied wavelengths according to the binomial distribution defined by Equation (3.1) is equal to $Np^{(i)}$.

In order to build up a Markov chain describing the router behavior we have to take into consideration the current fiber occupancies and the number of busy places

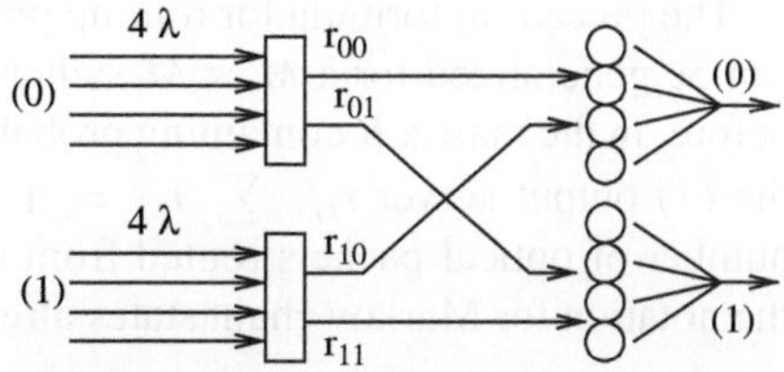

Fig. 3.2 Model of switch without storage space corresponding to WAN model

in the servers which are delaying the packets. We propose a Markov chain state made up of four components:

$$\mathbf{s} = \left(s^{(0)}, s^{(1)}, s^{(2)}, s^{(3)}\right), \qquad \text{where} \tag{3.2}$$

$s^{(0)}$ — number of occupied wavelengths in fiber (0),
$s^{(1)}$ — number of occupied wavelengths in fiber (1),
$s^{(2)}$ — number of occupied places in server (0),
$s^{(3)}$ — number of occupied places in server (1).

The notation introduced in Equation (3.2) describes a current state and $\mathbf{s}_* = (s_*^{(0)}, s_*^{(1)}, s_*^{(2)}, s_*^{(3)})$ the next Markovian state. We note also that $s^{(k)} \in S^{(k)}$, $k = 0, 1, 2, 3$, i.e. $S^{(k)}$ is a state space for the k-th state element. We put $|S^{(k)}|$ as a number of elements in the set $S^{(k)}$ and we compute the number of chain states as $|S^{(0)}| \cdot |S^{(1)}| \cdot |S^{(2)}| \cdot |S^{(3)}|$, or simply $(N+1)^4$, because all the states belonging to the Cartesian product $S^{(0)} \times S^{(1)} \times S^{(2)} \times S^{(3)}$ are reachable.

Let us establish now the conditions, which launch the transitions taking place in the Markov chain. We observe that the $s_*^{(0)}, s_*^{(1)}$ are independent of the previous state and $s^{(2)}, s^{(3)}$ do not influence the next state at all. We evaluate now the probability $p(\mathbf{s}, \mathbf{s}_*)$ of passing from $\mathbf{s}$ to $\mathbf{s}_*$. The previous remark allows us to put

$$p(\mathbf{s}, \mathbf{s}_*) = p^{(0)}_{s_*^{(0)}} p^{(1)}_{s_*^{(1)}} \mathrm{pr}(\mathbf{s}, \mathbf{s}_*), \tag{3.3}$$

where $p^{(i)}_{s_*^{(i)}}$, $i = 0, 1$, is a probability of $s_*^{(i)}$ occupied wavelengths in the (i) fiber and $\mathrm{pr}(\mathbf{s}, \mathbf{s}_*)$ is a probability of routing of incoming packets to the output. The routing probability is given by the combinatorial formula:

$$\mathrm{pr}(\mathbf{s}, \mathbf{s}_*) = \sum_{k=0}^{h} \binom{s^{(0)}}{h-k} \binom{s^{(1)}}{k} r_{00}^{h-k} r_{01}^{s^{(0)}-(h-k)} r_{10}^{k} r_{11}^{s^{(1)}-k} \tag{3.4}$$

for which we assume that if $k > n$ than $\binom{n}{k} = 0$.

To determine the upper index of the sum, h, we have to consider the packets, which had been routed but, possibly, they were lost what happens when the number of incoming packets exceeds the number of outgoing ones on the next step: $s^{(0)} + s^{(1)} > s_*^{(2)} + s_*^{(3)}$. For this reason, we write: $h = s^{(0)} + s^{(1)} - s_*^{(3)}$. Notice that in cases with no losses $h = s_*^{(2)}$.

The preceding formula for routing probability between the switch states $\mathbf{s}$ and $\mathbf{s}_*$ can be generalized for a $M \times M$ switch whose routing probabilities are stored, as before, in the matrix $\boldsymbol{R}$ containing probabilities of passing from the fiber (i) towards the (j) output server r_{ij}, $\sum_j r_{ij} = 1$, $i, j = 0, 1, \ldots, M-1$. We note $k_{(i)}^{(j)}$ a number of optical packets routed from the (i) fiber to the (j) output and we keep the notation for Markov chain states already introduced above. A state is composed of $2M$ elements, the component $s^{(i)}$ of the state indicates a number of occupied wavelengths in the (i) fiber for $i = 0, 1, \ldots M - 1$ and it indicates a number of occupied places in the (i) server for $i = M, M + 1, \ldots, 2M - 1$. The numbers of packets sent from the (i) fiber to all the destination ports have to fulfill the following condition: $\sum_{l=0}^{M-1} k_{(i)}^{(l)} = s^{(i)}$. On the other hand, numbers of packets arriving into the (j) output from all the fibers sum up to $s_*^{(M+j)}$ if there is no losses for the state s_* at the (j) output and they exceed N in a case when losses at this server occur for this state. Let us assume that a state of the considered switch is s and the fiber (i) dispatches its $s^{(i)}$ optical packets between M outputs in the following way: $\left(k_{(i)}^{(0)}, k_{(i)}^{(1)}, \ldots, k_{(i)}^{(M-1)}\right) = \left(k_{(i)}\right)$. Such a distribution can be done in $n\left(k_{(i)}\right)$ different ways:

$$n\left(k_{(i)}\right) = \binom{s^{(i)}}{k_{(i)}^{(0)}} \binom{s^{(i)} - k_{(i)}^{(0)}}{k_{(i)}^{(1)}} \cdots \binom{s^{(i)} - \sum_{l=0}^{(M-3)} k_{(i)}^{(l)}}{k_{(i)}^{(M-2)}} \binom{s^{(i)} - \sum_{l=0}^{(M-2)} k_{(i)}^{(l)}}{k_{(i)}^{(M-1)}}.$$

It should be obvious that the last factor is equal to 1 because, as we have said above, $k_{(i)}^{(M-1)} = s^{(i)} - \sum_{l=0}^{(M-2)} k_{(i)}^{(l)}$. The formula (3.4) can be written in the general case with summation performed for all possible distributions of $s^{(i)}$ packets arriving via the fiber $(i), i = 0, 1, \ldots, M - 1$ into M output servers $\left(k_{(0)}^{(0)}, k_{(0)}^{(1)}, \ldots, k_{(0)}^{(M-1)};\ k_{(1)}^{(0)}, k_{(1)}^{(1)}, \ldots,\ k_{(1)}^{(M-1)}; \ldots; k_{(M-1)}^{(0)}, k_{(M-1)}^{(1)}, \ldots, k_{(M-1)}^{(M-1)}\right) = \left(k_{(0)}, k_{(1)}, \ldots, k_{(M-1)}\right)$:

$$\text{pr}(s, s_*) = \sum_{(k_{(0)}, k_{(1)}, \ldots, k_{(M-1)})} \prod_i n\left(k_{(i)}\right) \prod_j r_{ij}^{k_{(i)}^{(j)}}, \tag{3.5}$$

where for each input fiber i we have $\sum_{l=0}^{M-1} k_{(i)}^{(l)} = s^{(i)}$ and for each output (j) we may have $\sum_{l=0}^{M-1} k_{(l)}^{(j)} = s_*^{(j)}$ if $\sum_{l=0}^{M-1} k_{(l)}^{(j)} \leq N$ (for outputs with no losses, $\sum_{i=0}^{M-1} s^{(i)} = \sum_{i=M}^{2M-1} s_*^{(i)}$) or $\sum_{l=0}^{M-1} k_{(l)}^{(j)} > s_*^{(j)} = N$ (for outputs where losses occur, $\sum_{i=0}^{M-1} s^{(i)} > \sum_{i=M}^{2M-1} s_*^{(i)}$). Assuming that a distribution of occupancy of each fiber is a Bernoulli distribution, we rewrite immediately the formula (3.4) to compute Markov stochastic matrix elements in the form:

$$p(s, s_*) = \text{pr}(s, s_*) \prod_{i=0}^{M-1} p_{s_*^{(i)}}^{(i)}. \tag{3.6}$$

The Equations (3.5) and (3.6) describe a general model of an optical switch without CRM, with M incoming/outgoing fibers and a number of wavelengths per fiber equal to N.

3.3.2 MAN Model – Model with Memory

Starting from the previous model, we introduce mėmory into the servers of second stage. In Fig. 3.3, we present a model with buffers of length $B = 4$ with FIFO schedule. If servers are occupied, routed optical packets can be stored in these buffers. In this model, we keep the binomial distribution of arrivals for the fibers and the routing probability matrix $\boldsymbol{R}$. A new factor in the model is memory so a Markov chain state has two elements more in order to take into account the optical packets waiting in the buffers:

$$s = \left(s^{(0)}, s^{(1)}, s^{(2)}, s^{(3)}, s^{(4)}, s^{(5)}\right), \qquad \text{where} \tag{3.7}$$

$s^{(0)}, s^{(1)}$ — number of occupied wavelengths on incoming fiber (0) and (1), respectively,
$s^{(2)}, s^{(3)}$ — number of packets in the memory (0) and (1), respectively,
$s^{(4)}, s^{(5)}$ — number of servers occupied on outgoing fiber (0) and (1), respectively.

We evaluate now the stochastic matrix dimension by taking into consideration classes of reachable states under condition that the number of wavelengths in a fiber is equal to N and the capacity of each buffer is B.

1. $(s^{(0)}, s^{(1)}, 0, 0, s^{(4)}, s^{(5)})$ — there are no packets stored in the buffers. This situation is similar to the one of the previous model, i.e. the elements $s^{(j)}$, $j = 0, 1, 4, 5$ can have any value which belongs to their state spaces. The number of the states of this type is $|S^{(0)}||S^{(1)}||S^{(4)}||S^{(5)}| = (N + 1)^4$

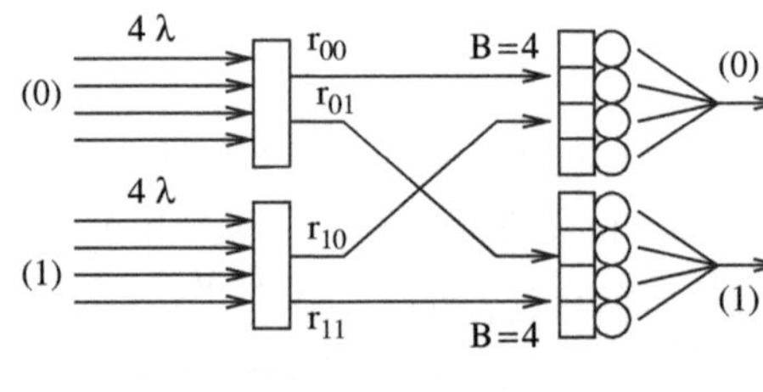

Fig. 3.3 Model with memory corresponding to the MAN model

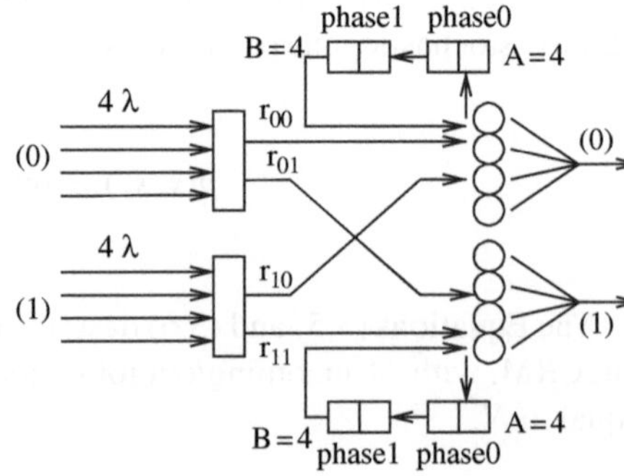

Fig. 3.4 Model with feedback lines

2. $(s^{(0)}, s^{(1)}, \{1, 2, \ldots, B\}, 0, N, s^{(5)})$ — there is at least one packet in the memory for the server (0) and no packet in the memory for the server (1). In this case element $s^{(4)}$ must be equal to its greatest value because packets are stored in the buffer only when there is no room for them in the corresponding server. The number of reachable states in this class is: $B(N+1)^3$.
3. $(s^{(0)}, s^{(1)}, 0, \{1, 2, \ldots, B\}, s^{(4)}, N)$ — the situation symmetric to the previous one. The number of states is as above: $B(N+1)^3$.
4. $(s^{(0)}, s^{(1)}, \{1, 2, \ldots, B\}, \{1, 2, \ldots, B\}, N, N)$ — there is at least one packet in each buffer. Consequently, both the servers are full. As we prove below, the total load of two buffers cannot exceed the nominal capacity of one buffer, i.e. $s^{(2)} + s^{(3)} \leq B$. This condition leads us to the conclusion, that the states belonging to the fourth group have one of the forms listed below:

$$\begin{gathered}(s^{(0)}, s^{(1)}, 1, \{1, 2, \ldots, B-1\}, N, N)\\ (s^{(0)}, s^{(1)}, 2, \{1, 2, \ldots, B-2\}, N, N)\\ \vdots\\ (s^{(0)}, s^{(1)}, B-1, 1, N, N),\end{gathered}$$

which gives the total number of states of this type equal to: $(N+1)^2 \frac{B(B-1)}{2}$.

Summing up the numbers of states computed for each group, we obtain the number of reachable states for the entire chain as:

$$(N+1)^2 \left[(N+1)^2 + 2B(N+1) + \frac{B(B-1)}{2} \right].$$

We will prove now that for the model with identical buffers the number of packets stored in them does not exceed the capacity of the single one: $s^{(2)} + s^{(3)} \leq B$.

The equation below expresses the fact that packets, which arrive or are already stored in memory during one time slot either enter a server either stay in a memory or are lost on the succeeding step:

$$s_*^{(2)} + s_*^{(3)} + s_*^{(4)} + s_*^{(5)} + \mathrm{Loss} = s^{(0)} + s^{(1)} + s^{(2)} + s^{(3)}. \tag{3.8}$$

Let us define the changes of the number of packets in the buffers (0) and (1), c_0 and c_1, respectively:

$$s_*^{(2)} = c_0 + s^{(2)}, \qquad s_*^{(3)} = c_1 + s^{(3)}. \tag{3.9}$$

We assume that the maximal number of packets arrives on each fiber, $s^{(0)} = s^{(1)} = N$, and there is no loss (Loss $= 0$) on the next step. To increase the number of packets in the memory (0), $s_*^{(2)} > s^{(2)}$, the inequality $c_0 > 0$ must be satisfied. In the case of $c_0 > 0$, there should be at least $N+1$ packets which are routed to the fiber (0) (because at most N packets are processed by a server and sent out of the router). In other case $c_0 \le 0$ and such a situation happens while the packets taken out of the memory will not be replaced by new ones.

By assembling Equations (3.8) and (3.9), with no losses, we get:

$$s_*^{(4)} + s_*^{(5)} + c_0 + c_1 = s^{(0)} + s^{(1)}, \tag{3.10}$$

which represents exactly what we have previously stated. Now if we take in Eq. (3.10) $c_0 \ge 0$ and $s_*^{(4)} = N$ (if there are packets in the memory, all the sub-servers are occupied), we can proceed with $s_*^{(4)} = s^{(0)}$:

$$s_*^{(5)} + c_1 = s^{(1)} - c_0 \tag{3.11}$$

Let us take system state $(N, N, 0, 0, 0, 0)$ and fill up memory $s^{(2)}$. Let us assume that $s^{(0)} = s^{(1)} = N$ and incoming packets are routed to the fiber (0). On each step, we can write $c_0 = N$ while $s^{(2)} < B$ and $c_1 = 0$ while $s^{(1)} - c_0 = 0$. After $\frac{B}{N}$ steps the system state will look like $(N, N, B, 0, N, 0)$. So we have completely filled up the memory $s^{(2)}$. If we try to put more packets into the memory $s^{(2)}$, losses will be observed in the system.

In this case, if we want to increase the number of packets in the memories, we should try to fill up the memory $s^{(3)}$. Eq. (3.11) says that this is possible only under the condition $c_1 = -c_0$ (notice that if we want to fill up $s^{(3)}$, the condition $s^{(1)} = s_*^{(5)} = N$ should be held). So the number of packets which we introduce into the memory $s^{(3)}$ is subtracted from the number of packets stored in the memory $s^{(2)}$ and the total number of packets in both the memories does not increase.

To conclude, the increase of the number of packets in one memory, decreases the packet number in other one. Under this condition, we are capable to fill completely one memory. Once the memory is filled, we can only hold the total number of packets but not increase it and this we wanted to prove. This proof can be easily generalized for system with M servers, each with memory capacity for N packets and the maximum packet numbers in all memories equal to $(M - 1)N$.

We build up now a probability matrix of the chain using as our scheme the formulæ (3.3) and (3.4) and taking into account the model's characteristics in order to fix the upper index of the sum, h. To begin with, we compute the number of packets accepted by the buffers and the servers (0), (1):

$$\mathrm{acc}_0 = (s_*^{(4)} - s^{(2)}) + s_*^{(2)}, \qquad \mathrm{acc}_1 = (s_*^{(5)} - s^{(3)}) + s_*^{(3)}$$

Afterward, we write $h = s^{(0)} + s^{(1)} - \mathrm{acc}_1$ when there are losses caused by memory corresponding to the fiber (0) and $h = s^{(0)} + s^{(1)} - \mathrm{acc}_0$ when there are losses caused by memory corresponding to the fiber (1).

3.3.3 Feedback Lines

Feedback lines represent an efficient CRM solution implemented by fiber connecting outgoing and incoming ports. In our analytical model, this mechanism can be described as "multilevel" memory carried out in practice as a system of delaying loops. It means that a packet which does not enter the server is stored in the memory of the first level of size A. The packet is moved in the next time-unit into the memory of the second level of size B. Packet arrival characteristics and its routing probability is the same as in the previous model.

Now a state of the Markov chain contains eight elements:

$$s = \left(s^{(0)}, s^{(1)}, s^{(2)}, s^{(3)}, s^{(4)}, s^{(5)}, s^{(6)}, s^{(7)}\right), \qquad \text{where} \tag{3.12}$$

$s^{(0)}$, $s^{(1)}$ — number of arriving packets on the fiber (0) and (1), respectively,

$s^{(2)}$, $s^{(3)}$ — number of packets in the last phase of feedback (the packets which will try to find a server free) for the outgoing fiber (0) and (1), respectively,

$s^{(4)}$, $s^{(5)}$ — number of packets in the server corresponding to the fiber (0) and (1) respectively,

$s^{(6)}$, $s^{(7)}$ — number of packets in the memory of first phase for the fiber (0) and (1).

We observe that the content of the first stage memory is moved to the second stage memory: $s_*^{(2)} = s^{(6)}$ and $s_*^{(3)} = s^{(7)}$. We observe also that the number of packets in the server is greater or equal to the number of packets in the memory of the last stage in the previous states: $s_*^{(4)} \geq s^{(2)}$ and $s_*^{(5)} \geq s^{(3)}$. The condition limiting the number of packets present in two buffers of the same degree introduced for the previous model is also valid in this case.

The number of reachable states of the Markov chain can be computed by counting them in subsets related to each class listed below:

1. Both the buffers of the second level are empty, i.e. $s^{(2)} = s^{(3)} = 0$.
   These states have to be regarded in more detailed way, considering particular cases:

$$(s^{(0)}, s^{(1)}, 0, 0, N, N, \{1, 2, \ldots, A\}, \{1, 2, \ldots, A\}) \quad \text{and} \quad s^{(6)} + s^{(7)} \leq A$$
$$(s^{(0)}, s^{(1)}, 0, 0, \{0, 1 \ldots, N-1\}, \{0, 1 \ldots, N-1\}, 0, 0)$$
$$(s^{(0)}, s^{(1)}, 0, 0, N, N, 0, 0)$$
$$(s^{(0)}, s^{(1)}, 0, 0, N, \{0, 1 \ldots, N-1\}, \{0, 1, \ldots, A\}, 0)$$
$$(s^{(0)}, s^{(1)}, 0, 0, \{0, 1 \ldots, N-1\}, N, 0, \{0, 1, \ldots, A\})$$
$$(s^{(0)}, s^{(1)}, 0, 0, N, N, \{1, 2, \ldots, A\}, 0)$$
$$(s^{(0)}, s^{(1)}, 0, 0, N, N, 0, \{1, 2, \ldots, A\})$$

The number of states in this class is:

$$n_1 = (N+1)^2 \left[\frac{A(A+1)}{2} + (N+1)(2A+N+1) \right].$$

2. Two memories of the second level contain at least one packet: $s^{(2)}, s^{(3)} \in \{1, 2, \ldots, B\}$. We can split states into seven subclasses as above taking also into account the condition which is always to be satisfied for two buffer storage: $s^{(2)} + s^{(3)} \leq B$. As the conclusion we get: $n_2 = \frac{B(B-1)}{2} \cdot n_1$.
3. In the last case exactly one memory of the second level is empty and we obtain: $n_3 = 2Bn_1$.

So the total number of state n of the Markov chain is given by a sum

$$n = \left(1 + 2B + \frac{B(B-1)}{2}\right) n_1.$$

Let us look at the routing of incoming packets. We note as before the number of packets accepted in the server as acc:

$$\mathrm{acc}_0 = s_*^{(4)} + s_*^{(6)} - s^{(2)}, \quad \mathrm{acc}_1 = s_*^{(5)} + s_*^{(7)} - s^{(3)}, \quad \mathrm{Acc} = \mathrm{acc}_0 + \mathrm{acc}_1$$

In the case with no loss at all ($\mathrm{Acc} \leq s^{(0)} + s^{(1)}$) or in the case when there are no losses at the server (0) the upper summing limit h is equal to acc_0. Otherwise $h = s^{(0)} + s^{(1)} - \mathrm{acc}_1$. The equation of the routing probability is the same as in the previous examples Equations (3.3 and 3.4).

3.4 Varying Traffic Characteristics

The traffic description in terms of binomial distribution Eq. (3.1) proposed for performance analysis of our three models of routers is valid for rather regular data streams. This remark can be explained by the fact that the autocorrelation of such a stochastic process is degrading very quickly. To slow down the deterioration of autocorrelation of a packet arrival process we introduce independent modulating Markov chains, one chain associated with a stream in one fiber [19]. A modulating chain (i), $i = 0, 1$ has $M^{(i)} + 1$ states and if it is in the state m, the probability that a wavelength in the fiber (i) transports data is equal to $p^{(i,m)}$. We change loads of all wavelengths of the fiber in the same time. For this reason the global distribution for a fiber stays binomial, despite the probabilities of occupancy are varying in time.

Incorporating modulating chains into our models requires increasing of the number of state components. We propose to append two additional elements at the end of an actual state. For example, in the model without memory (Section 3.3.1) $s^{(5)}$ will indicate modulation for the fiber (0) and $s^{(6)}$ will do the same for the fiber (1). The reader has to be aware that the stochastic matrices of these extended models require more space in computer memory because we observe here a notorious phenomenon of *state explosion*. The number of states is now equal to $n(M^{(0)}+1)(M^{(1)}+1)$, where n is a state number of a model with non-modulated stream.

To keep the notation the simplest possible for autocorrelation analysis we restrict ourselves at the beginning to the two states modulating Markov chains, which can be seen as ON/OFF switches and we apply here the notation as in [20]. Let us consider one wavelength whose probabilities of being busy (load) are equal to Ch_0 and Ch_1 depending on modulating chain state 0, 1, respectively. The autocovariance of the process $R(n,n+k)=E(\mathrm{Ch}(n)\mathrm{Ch}(n+k))$ observed on this wavelength is expressed by the formula [21]:

$$
\begin{aligned}
R(n,n+k) = {} & \mathrm{Ch}_0\pi_0(n)\mathrm{Ch}_0\pi_0(n+k|\pi_0(n)=1)\\
& +\mathrm{Ch}_0\pi_0(n)\mathrm{Ch}_1\pi_1(n+k|\pi_0(n)=1)\\
& +\mathrm{Ch}_1\pi_1(n)\mathrm{Ch}_0\pi_0(n+k|\pi_1(n)=1)\\
& +\mathrm{Ch}_1\pi_1(n)\mathrm{Ch}_1\pi_1(n+k|\pi_1(n)=1)
\end{aligned}
$$

We mark two possible initial conditions of the process, A (the state 0 is observed at $n=0$) and B (the complementary one) which are equally probable. Assuming that the condition A is satisfied at the starting point $n=0$ and taking advantage of a fact that our process is stationary, we can simplified the formula above

$$
\begin{aligned}
R^{[A]}(n,n+k) &= \mathrm{Ch}_0^2\pi_0^{[A]}(n)\pi_0^{[A]}(k)+\mathrm{Ch}_0\mathrm{Ch}_1\pi_0^{[A]}(n)\pi_1^{[A]}(k)\\
&\quad +\mathrm{Ch}_0\mathrm{Ch}_1\pi_1^{[A]}(n)\pi_0^{[B]}(k)+\mathrm{Ch}_1^2\pi_1^{[A]}(n)\pi_1^{[B]}(k)\\
&= \mathrm{Ch}_0\pi_0^{[A]}(n)E^{[A]}(k)+\mathrm{Ch}_1\pi_1^{[A]}(n)E^{[B]}(k)
\end{aligned}
$$

which yields to the autocorrelation

$$
C^{[A]}(n,n+k)=\mathrm{Ch}_1\pi_1^{[A]}(n)\left(E^{[B]}(k)-E^{[A]}(k)\right). \tag{3.13}
$$

In the stationary state, $n\longrightarrow\infty$, influence of the initial condition vanishes and we obtain: $\pi_i^{[A]}(n)=\pi_i^{[B]}(n)=\pi_i,\ i=0,1,\ E^{[A]}(n)=E^{[B]}(n)=E(\mathrm{Ch}),\ R(k)=\mathrm{Ch}_0\pi_0E^{[A]}(k)+\mathrm{Ch}_1\pi_1E^{[B]}(k)$. In this case the autocorrelation is expressed as $c(k)=R(k)-E^2(\mathrm{Ch})$.

As it may be seen from the formulæ above, the speed with which the coefficient of autocorrelation is vanishing, depends on the numerical properties of stochastic matrices describing behaviors of Markov modulating chains, i.e. their initial k transient states. If a stationary state of the modulating chains is achieved relatively late, the coefficient of correlation stays non-zero longer than in the previous, non-modulated, case.

A wavelength load being changed in time by modulating chain influences the overall fiber behavior as the mean number of occupied wavelengths is proportional to the load of a single wavelength (see page 50). Adapting the approach applied for a single wavelength we can write the statistical formulæ for a fiber expressing influence of the temporary mean on the autocovariance. We note as $[l],\ l = 0, 1, \ldots, N$ an initial condition for which l wavelengths are busy, par example the one for the autocovariance in the case in which l wavelength were transporting data at $n = 0$:

$$C^{[l]}(n, n+k) = \sum_{i=1;i\neq l}^{N} i p_i^{[l]}(n) \left[\hat{E}^{[i]}(k) - \hat{E}^{[l]}(k)\right].$$

Our other approach is based on introduction of modulation directly by defining the Markov state coordinates $s^{(0)},\ s^{(1)}$ (which indicates the numbers of wavelengths occupied in both the fibers) as states of independent Markov chains which define the number of busy wavelengths in the fibers (0) and (1). A modulating matrix $\boldsymbol{Q}$ is of size $(N+1)\times(N+1)$ and the autocovariance which it induces, is given by the formula:

$$R^{[l]}(n, n+k) = \sum_{l=0}^{N}\sum_{i=1}^{N} i q_i^{[l]}(n)\overline{E}^{[i]}(k)$$

3.5 Numerical Results

Numerical procedures applied to solve Markov chains, i.e. to find the left eigenvector corresponding to the greatest eigenvalue (equal to 1) of the stochastic matrix $\boldsymbol{P},\ \boldsymbol{\pi P} = \boldsymbol{\pi},\ \sum_i \pi_i = 1$ are iterative and they deal with a matrix $\boldsymbol{P}$ stored in the compact way. We point out here that the stochastic matrices of our models in discrete time are still sparse, however, they are denser than these, which one might expect for continuous time scale.

A performance measure characterizing efficiency of the router is a *packet loss ratio* (PLR), a ratio between the number of lost packets and the number of all packed transmitted, expressed in [%]. We use this parameter also to compare results obtained by simulation and the analytical method. In order to compute PLR out of the probability vector $\boldsymbol{\pi}$ we identify the states $\boldsymbol{s}_*$ in which losses occur and the states $\boldsymbol{s}$ from which the chain can pass towards $\boldsymbol{s}_*$. We compute the number $\Delta(\boldsymbol{s}, \boldsymbol{s}_*)$ of packets lost on this passage (in case of loss a positive value is attributed to $\Delta(\boldsymbol{s}, \boldsymbol{s}_*)$). The mean number of lost packets is equal to: $\sum_{(\boldsymbol{s},\ \boldsymbol{s}_*)} \Delta(\boldsymbol{s},\ \boldsymbol{s}_*)\boldsymbol{P}_{(\boldsymbol{s}, \boldsymbol{s}_*)}\boldsymbol{\pi}_{\boldsymbol{s}}$ and consequently

$$\text{PLR} = \frac{\sum_{(\boldsymbol{s},\ \boldsymbol{s}_*)} \Delta(\boldsymbol{s}, \boldsymbol{s}_*)\boldsymbol{P}_{(\boldsymbol{s},\ \boldsymbol{s}_*)}\boldsymbol{\pi}_{\boldsymbol{s}}}{N(\text{mean load for fiber (0)} + \text{mean load for fiber (1)})}.$$

Computation of $\Delta(\boldsymbol{s}, \boldsymbol{s}_*)$ depends on a model type. For the model without memory we put $\Delta(\boldsymbol{s}, \boldsymbol{s}_*) = s^{(0)} + s^{(1)} - (s_*^{(2)} + s_*^{(3)})$, for the model with buffers: $\Delta(\boldsymbol{s}, \boldsymbol{s}_*) = s^{(0)} + s^{(1)} + s^{(2)} + s^{(3)} - (s_*^{(2)} + s_*^{(3)} + s_*^{(4)} + s_*^{(5)})$, and for the model with feedback lines: $\Delta(\boldsymbol{s}, \boldsymbol{s}_*) = s^{(0)} + s^{(1)} + s^{(2)} + s^{(3)} - (s_*^{(4)} + s_*^{(5)} + s_*^{(6)} + s_*^{(7)})$. To adapt the formula for PLR for a model with M fibers of N wavelengths we modify its denominator which becomes $N(\sum_{i=0}^{M-1}$ mean load for fiber $(i))$. Modifications of the number of lost packets have to be made depending on the model's type. For the model of a switch without memory, a state of the Markov chain is composed of $2M$ elements: $s^{(0)}, s^{(1)}, \ldots, s^{(M-1)}$ describe input fiber states and $s^{(M)}, s^{(M+1)}, \ldots, s^{(2M-1)}$ describe server occupations. The number of lost packets is equal to $\Delta(\mathbf{s}, \mathbf{s}_*) = \sum_{i=0}^{M-1} s^{(i)} - \sum_{i=M}^{2M-1} s_*^{(i)}$. A state of the model with memory has $3M$ elements: $s^{(0)}, s^{(1)}, \ldots, s^{(M-1)}$ describe input fiber states, $s^{(M)}, s^{(M-1)}, \ldots, s^{(2M-1)}$ describe memory states, and $s^{(2M)}, s^{(2M+1)}, \ldots, s^{(3M-1)}$ describe server occupations. This implies $\Delta(\mathbf{s}, \mathbf{s}_*) = \sum_{i=0}^{2M-1} s^{(i)} - \sum_{i=M}^{3M-1} s_*^{(i)}$. The model with two phase feedback lines has a state composed of $4M$ elements: $s^{(0)}, s^{(1)}, \ldots, s^{(M-1)}$ describe input fiber states, $s^{(M)}, s^{(M-1)}, \ldots, s^{(2M-1)}$ describe last phase state, $s^{(2M)}, s^{(2M+1)}, \ldots, s^{(3M-1)}$ describe server occupations, and $s^{(3M)}, s^{(3M+1)}, \ldots, s^{(4M-1)}$ describe first phase states. For this model $\Delta(\mathbf{s}, \mathbf{s}_*) = \sum_{i=0}^{2M-1} s^{(i)} - \sum_{i=2M}^{4M-1} s_*^{(i)}$.

We state at the beginning that numerical results presented in this paper have been performed for symmetric routing, i.e. $r_{i,j} = 0.5,\ i, j = 0, 1$ and two fibers are identical. The number of wavelengths in a fiber was fixed in all the cases to $N = 4$. Varying parameters were: loads of fibers, capacities of buffers, number of retarding lines, and traffic characteristics.

To begin with, we analyze an influence of load on PLR for all three models with non-modulated traffic (Fig. 3.5). At this point, we notice that for two models representing routers with storage capacity of the same size (capacity of the buffer for the second model is equal to the capacity of the first degree retarding loops in the third model) we observe the same results (the same curve "With storage" in Fig. 3.5). The packet losses are the same although the stochastic matrices corresponding to both the models are different. For example, the number of reachable states for the model with memory is equal to 1′150 and for the one with feedback lines is equal to 6′900. The same PLR for both models is explained by the fact that a purely binomial traffic is smooth, there are no burst-idle periods. If packet arrivals are independent, the PLR results are the same for these different architectures. This conclusion has

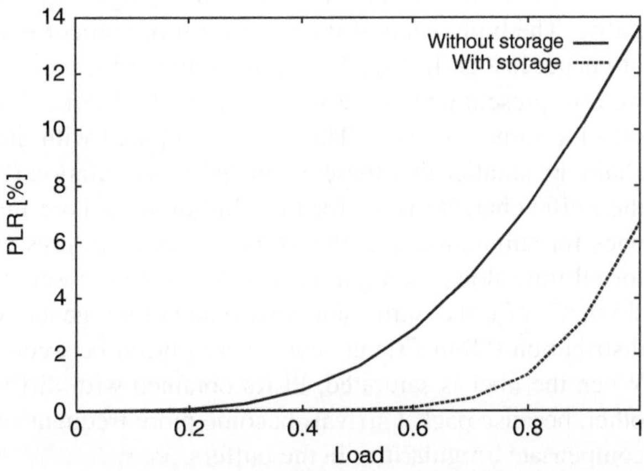

Fig. 3.5 PLR as a function of wavelength load; "Without storage" — results for the model without CRM, "With storage" — results for both the models with storage, $B = 2$ for memory, $A = 2,\ B = 2$ for feedback lines

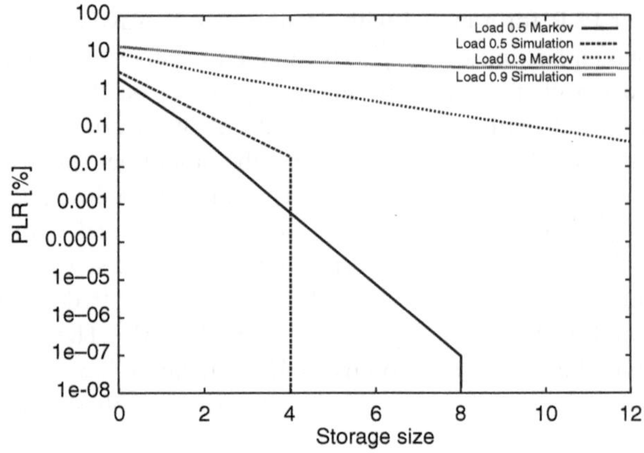

Fig. 3.6 PLR as a function of storage space for two loads: 0.5 and 0.9 expressed in logarithmic scale

also an important impact on a router design, because it proves that in situations of weak data stream fluctuations, the architecture with feedback, less expensive than the one with buffer, ensures the same PLR. The discussed property vanishes when the incoming traffic is autocorrelated, as we will show later.

Thanks to analytical models, we can estimate the capacity of storage necessary to achieve expected losses under a chosen threshold. In Fig. 3.6 we present the graphs for a weak load (0.5) and heavy load (0.9). These results are compared with the ones obtained by simulation. We may see that two methods show the same tendencies in models' behavior. As a binomial distribution generates the traffic more smooth than simulation, PLR computed with the Markov models is smaller than the one obtained by simulation. Our simulation tool requires that the memory size in models is a multiple of the number of wavelengths incoming per fiber. This feature has been implemented in the tool to meet the technological constraints of studied architectures and, therefore, we do not observe losses for storage size bigger than 4 places. Consequently, we see an advantage of the analytical solution, where the losses occur up to storage size equal to 8 places.

The performance of the models with storage is different in the case of modulated traffic. The high value of the coefficient of autocorrelation produces a higher level of packet losses. In Fig. 3.7 we show the influence of autocorrelation on PLR and we also present that the results given by the Markovian method and the simulation give the same tendency. The losses computed with the Markov ("Markov" curve) chain are smaller that these estimated by simulation ("Simulation" curve) because the traffic characteristics for the Markov model are much more regular than these ones for simulation. For the Markov chain, we preserve the mean load "locally", for all time slots. As a matter of fact, we may observe that the results obtained by simulation for the traffic smoother than before, generated with a two-phases Erlang distribution ("Sim-Erlang" curve) are placed between two ones mentioned above. When the fiber is saturated, PLRs obtained with different methods approach each other, because packet arrivals become more frequent and there is less possibility to compensate irregularities in the buffers.

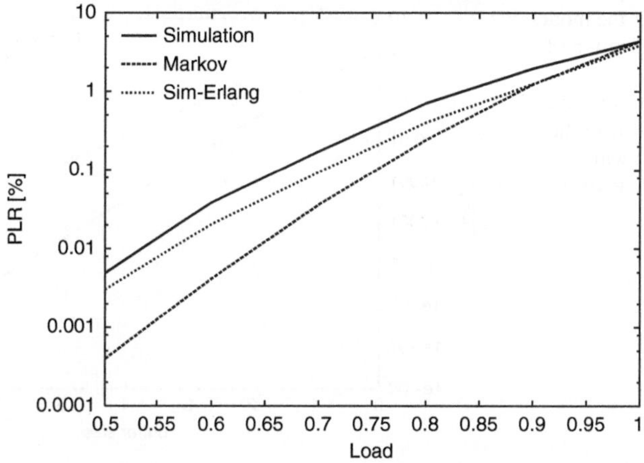

Fig. 3.7 PLR as a function of load for the Markovian method and simulation in logarithmic scale; model with buffers of size $B = 4$

To incorporate autocorrelation dependencies into the analytical models we propose a matrix $\boldsymbol{M}$ to switch between ON/OFF phases ($\mathrm{Ch}_0 = 0,\ \mathrm{Ch}_1 = C$) modulating optical packet streams,

$$\boldsymbol{M} = \begin{bmatrix} 0.25, & 0.75 \\ 0.50, & 0.50 \end{bmatrix}. \tag{3.14}$$

The differences $|\Delta(k)| = |E^{[A]}(k) - E^{[B]}(k)|$ for $C = 1$ are presented in Fig. 3.8. These differences determine the speed of degradation of process autocorrelation as we may see in the formula (3.13). We want to keep these differences as great as possible and as long as possible in order to represent in our models optical packet streams similar to those observed in real network.

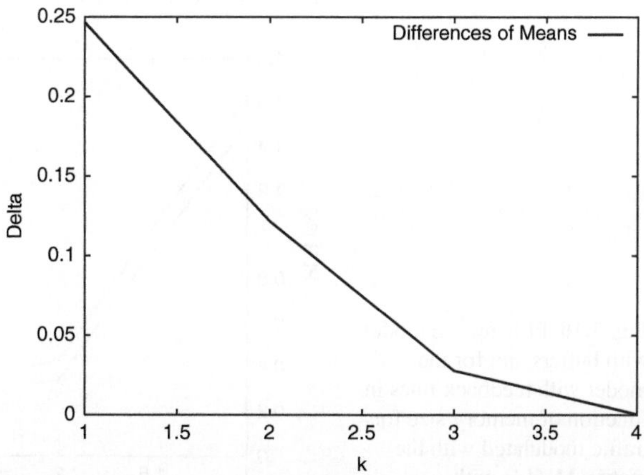

Fig. 3.8 The differences $|\Delta(k)| = |E^{[A](k)} - E^{[B](k)}|$ for the ON/OFF packet stream ($\mathrm{Ch}_0 = 0,\ \mathrm{Ch}_1 = C = 1$) modulated by the matrix $\boldsymbol{M}$, (3.14)

Fig. 3.9 PLR for the model with buffers in function of memory size for non-modulated traffic and traffic modulated with the matrix $\boldsymbol{M}$, (3.14) with preserved total load of 0.5

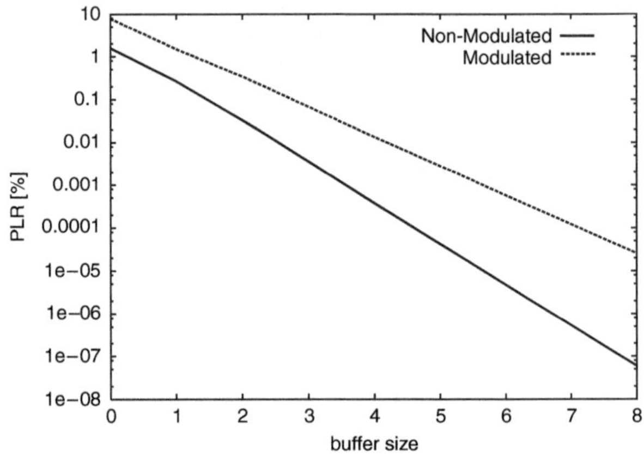

Although the modulating matrix dimension is small and, consequently, the introduced autocorrelation is short-term, we can observe its influence on PLR. In Fig. 3.9 we see PLR in function of memory size for the second model (beginning with no memory at all) for the overall fiber load equal to 0.5 ($\mathrm{Ch}_1 = C = 0.83(3)$). For instance, when there are 8 places in each buffer, the PLR for modulated traffic is more than 400 times greater than the PLR for non-modulated traffic.

The Markovian models also allow us to compare the performance of the different router architectures. Figure 3.10 shows that for the fixed total load (as before equal to 0.5) the more expensive architecture with memory performs only slightly better than the one with feedback lines.

We show now the impact induced by direct modulation using the matrix $\boldsymbol{Q}$ as follows:

$$\boldsymbol{Q} = \begin{bmatrix} 0.50,\ 0.20,\ 0.15,\ 0.10,\ 0.05 \\ 0.20,\ 0.35,\ 0.20,\ 0.15,\ 0.10 \\ 0.15,\ 0.20,\ 0.30,\ 0.20,\ 0.15 \\ 0.10,\ 0.15,\ 0.20,\ 0.35,\ 0.20 \\ 0.05,\ 0.10,\ 0.15,\ 0.20,\ 0.50 \end{bmatrix} \tag{3.15}$$

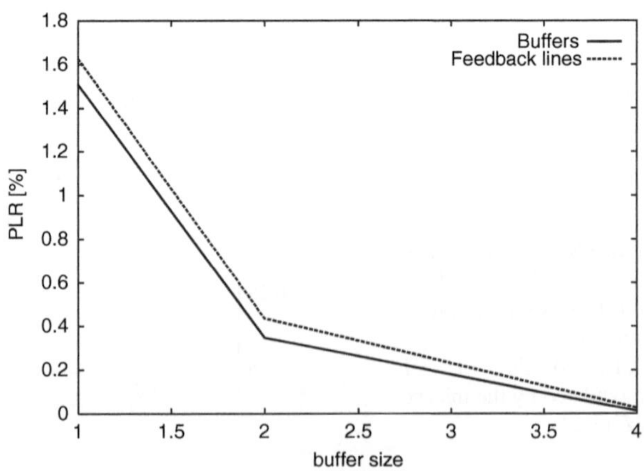

Fig. 3.10 PLR for the model with buffers and for the model with feedback lines in function of memory size for traffic modulated with the matrix $\boldsymbol{M}$, (14) with preserved total load of 0.5

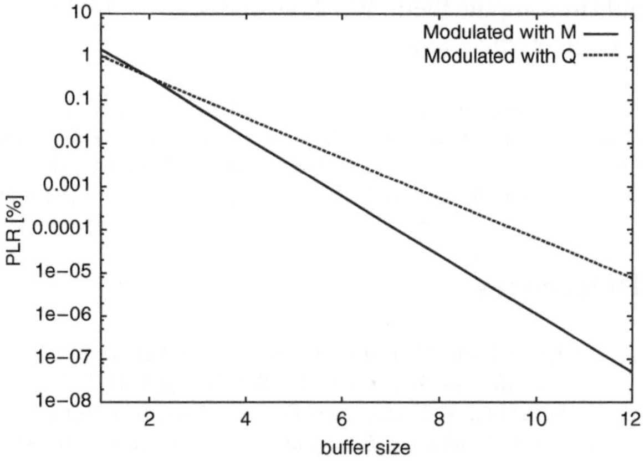

Fig. 3.11 PLR for the model with buffers in function of memory size for traffic modulated with the matrix $\boldsymbol{M}$, (3.14) and modulated with the matrix $\boldsymbol{Q}$ (3.15) and with preserved total load of 0.5

which guarantees the same mean total load as before but which generates a more regular stream then before. The solution of the matrix $\boldsymbol{Q}$ is [0.2, 0.2, 0.2, 0.2, 0.2], it means that all situation on the wavelengths (0, 1, 2, 3 or 4 wavelengths occupied) are equally probable and the mean number of occupied wavelengths is 2. On the other hand the modulation with $\boldsymbol{M}$ matrix gives with probability 0.4 all wavelengths free and with probability 0.6 the distribution of number of occupied wavelengths as $[\frac{1}{1296}, \frac{20}{1296}, \frac{150}{1296}, \frac{500}{1296}, \frac{625}{1296}]$ which gives the same mean number of occupied wavelengths equal to 2. As we see in Figure 3.11, the burstiness of the modulation with $\boldsymbol{M}$ shows up clearly for smaller buffer size (comparable with the mean). However, with the increasing storage capacity the burst traffic begins to be compensated better.

3.6 Conclusions

We presented three architectures of routers applicable to synchronized all-optical networks: without contention resolution mechanism, with emulated optical memory and with feedback implemented using delay lines. Afterwards, we proposed their analytical models based upon discrete-time Markov chains and we solved them using iterative methods. We point out that construction of these Markov chains is performed in purely combinatorial way. We do not have to analyze the model topology but we compute directly transition probabilities as functions of a current and of the next Markov chain states. This feature allows us to build up a stochastic matrix and to store it in a very efficient way. A performance measure taken as a representative one for the switch is PLR (*Packet Loss Ratio*). We present the results obtained for different types of traffic, varying autocorrelation in analytical way and compare them with those, estimated by simulation.

The conclusion is that the Markov chains define well the modeled systems' behavior and they express clearly the impact of different types of traffic on the system performance. In case of "rare simulation events" when a simulator cannot gather enough of valid data to furnish reliable results, the Markov chains are still able to compute them. We show that one can use Markovian models as a competitive tool in practical analysis of optical router performance.

Acknowledgments The work presented in this article has been performed with support of the RNRT (*Réseau National de Recherche en Télécommunications*) project RNRT ROM (*Réseau Optique Multiservice – Multiservice Optical Network*), the French national research project n° 99 S 0201 and of the Slovak Research Agency project VEGA 1/4084/07.

References

1. O. Gerstel and H. Raza. On the synergy between electrical and photonic switching. *IEEE Communication Magazine*, (4):98–104, April 2003.
2. S. Sengupta, V. Kumar, and D. Saha. Switched optical backbone for cost-effective scalable core ip networks. *IEEE Communication Magazine*, (6):60–70, June 2003.
3. K. Shrikhande et al. CSMA/CA MAC Protocols for IP-HORNET: An IP over WDM metropolitan area ring network. In *IEEE Globecom'00*, 2: 1303–1307, 2000.
4. J. Fransson et al. Design of a medium accesss control protocol for a WDMA/TDMA photonic ring network. In *IEEE Globecom'98*, 1: 307–312, 1998.
5. I. Chlamtac and A. Fumagali. An optical switch architecture for manhattan networks. *IEEE JSAC*, 11(4):550–559, May 1993.
6. A. Bononi, G. A. Castañon, and O. K. Tonguz. Analysis of hot-potato optical networks with wavelength conversion. *IEEE JLT*, 17(4):525–534, April 1999.
7. F. Callegati et al. Research on optical core networks in the e-photon/one network od excellence. In *Proceedings of IEEE INFOCOM 2006*, 1–5, 2006.
8. P. Gravey et al. Multiservice optical network: Main concepts and first achievements of the ROM Project. *IEEE Journal of Lightwave Technology*, 19:23–31, Jan. 2001.
9. C. Qiao. Labeled optical burst switching for IP-over-WDM integration. *IEEE Communications Magazine*, 38(9):104–114, Sep. 2000.
10. D. K. Hunter and I. Andonovic. Approaches to optical internet packet switching. *IEEE Communication Magazine*, (9):116–122, Sept. 2000.
11. P. Gambini et al. Network architecture and demonstration in KEOPS project. *IEEE JSAC*, 16(7):1245–1258, 1998.
12. F. Callegati, G. Muretto, C. Raffaelli, P. Zaffoni, and W. Cerroni. A framework for performance evaluation of ops congestion control resolution. In *Conference on Optical Network Design and Modelling*, 243–250, 2005.
13. H. Bruneel and B. G. Kim. *Discrete-Time Models for Communication Systems Including ATM*. Kluwer Academic Publishers, 1993.
14. J. Tomasik, I. Kotuliak, and T. Atmaca. Markovian performance analysis of a synchronous optical packet switch. In *Proceedings of IEEE/ACM MASCOTS 2003*, 254–257, 2003.
15. H. El Biaze, T. Atmaca, and G. Hébuterne. Impact of shaping on network performance with self similar traffic. In *Proceedings of TCT'2000*, May 2000.
16. I. Kotuliak and T. Atmaca. Logical performance of the optical switch in WAN and MAN networks. In *ECUMN'02*, Colmar, France, Apr. 2002.
17. L. Tancevski and L. Tamil, and F. Callegati. Nondegenerate buffers: an approach for building large optical memories, August 1999.
18. L. Kleinrock. *Queuing Systems*, volume 1: Theory. John Wiley & Sons, 1975.
19. S. Robert and J.-Y. Le Boudec. On a Markov modulated chain exhibiting self-similarities over finite timescale. *Performance Evaluation*, 27&28, 1996.
20. A. Papoulis. *Probability, Random Variables, and Stochastic Processes*. WNT, Warszawa, 1972.
21. J. Tomasik. An impact of source modulation on an autocorrelation function of a generated data stream. In *National Conference Computer Networks*, Zakopane, Poland, 1999.

Chapter 4
A Conditional Probability Approach to Performance Analysis of Optical Unslotted Bus-Based Networks

Alexandre Brandwajn, Viet Hung Nguyen and Tülin Atmaca

Abstract This paper presents a novel approach to the performance analysis of optical packet switching bus-based networks with unslotted Carrier Sense Multiple Access with Collision Avoidance (CSMA/CA) protocol. Because of the interdependence among bus nodes, an accurate performance analysis of such networks has been an open question for a long time. We model the bus as a multiple-priority M/G/1 queuing system with preemptive-repeat-identical (PRI) service discipline. To solve this model, we use a recurrent level-by-level analysis technique where the interference from higher levels (upstream nodes) is taken into account in terms of reappearance and disappearance rates for the server. The key features of this method are as follows. First, it specifically accounts for the distribution of voids seen by a node (via the number of attempts before a successful transmission) as a function of the node's position and offered network load. Second, it approximately computes the queue-length distribution at each node, and thus provides a tool for buffer size selection so as to meet a loss rate criterion. A comparison of our approximate model solution with network simulations indicates that our model generally offers good accuracy in assessing the performance of the network, including in terms of the queue-length distribution. Occasionally, the results of our model may deviate from simulation results. A discussion of the causes of such deviations, most likely to occur when nodes are close to saturation, offers additional insight into the properties of the bus-based network and its approximate solution.

Keywords Asynchronous optical bus-based network · Unslotted CSMA/CA · Performance analysis · Preemptive-repeat-identical service · Recurrent analysis

4.1 Introduction

For many years, voice service was the main traffic in Metropolitan Area Networks (MANs). Since voice service does not tolerate jitter (or delay variation), traditional MANs are based on synchronous circuit-switched network technologies

---

A. Brandwajn (✉)
University of California at Santa Cruz, Baskin School of Engineering Santa Cruz, CA 95064, USA
e-mail: alexb@soe.ucsc.edu

M. Ma. (ed.), *Current Research Progress of Optical Networks*,
DOI 10.1007/978-1-4020-9889-5_4,

(e.g., SONET/SDH) that guarantee very high quality (no jitters) for any service they transport. Recent years have witnessed a dramatic increase in the volume of new multimedia and data traffic, resulting in new service and bandwidth demands. The inefficiencies in terms of bandwidth granularity associated with traditional circuit-switched networks make the latter difficult to provision for these new demands. Therefore, a more efficient networking technology is required for modern MANs.

In this regard, the optical packet switching (OPS) technology is clearly the leading candidate for the next generation MANs. It offers bandwidth-on-demand service thanks to a high granularity of bandwidth provisioning. Additionally, in combination with new technologies such as circuit emulation [1] or GMPLS, it provides cost-effective network architectures that support both multimedia (voice, video) and data traffic. The bus topology appears as one of the best choices for next generation OPS MANs, because it inherits the reliable property of current SONET/SDH ring networks (viz., fast recovery from link failures). Furthermore, since bus topology allows many nodes to share the same transmission resource, it improves resource utilization thanks to statistical multiplexing of traffic flows. In order to efficiently share a transmission resource among nodes, OPS bus-based networks need an efficient medium access control (MAC) protocol. The Optical Unslotted Carrier Sense Multiple Access with Collision Avoidance (OU-CSMA/CA)[1] [2, 3] protocol appears an attractive solution. Its simplicity makes the network installation and management easier.

In spite of the above advantages, the OU-CSMA/CA protocol and the bus topology have several drawbacks. The bus topology may exhibit unfairness depending on the position of the nodes on the bus. For example, upstream nodes (i.e., the nodes closest to the beginning of the shared transmission resource) might monopolize the bandwidth and thus prevent downstream nodes from transmitting. Additionally, the asynchronous nature of the OU-CSMA/CA protocol may lead to an inefficient fragmentation of bandwidth, resulting in low resource utilization. Indeed, the asynchronous transmission of packets at upstream nodes may fragment the global bandwidth into small segments of bandwidth (voids) that are unusable for the transmission at downstream nodes.

Due to this interdependence among bus nodes, an accurate performance analysis of OPS bus-based networks using OU-CSMA/CA protocol has been an open question for a long time. There are performance analysis studies of packet-switched slotted-ring networks, such as [4, 5] for Fiber Distributed Data Interface, [6] for PLAYTHROUGH networks, [7, 8] for Distributed Queue Dual Bus, and [9, 10] for probabilistic p_i-persistent networks. These studies usually model a ring node as an M/G/1 vacation system with Bernoulli schedule [6, 9, 10] or with time-limited service discipline [4, 5], and obtain approximate values for the mean access delay and throughput at each node. Since the interdependence between nodes makes exact

[1] Note that the OU-CSMA/CA discussed here is unrelated to the one that is used in wireless protocols such as 802.11

analysis intractable, most studies use the assumption of independent arrival of empty slots at a tagged node. For the Distributed Queue Dual Bus protocol, an exact analysis based on Markov chain model is provided in [7], but it does not appear to scale well for a larger network size.

For the CSMA with Collision Detection (CSMA/CD) protocol, a number of performance studies have been published for both slotted and unslotted ring. An analysis treating each node as an independent M/G/1 queuing system, which services fixed-length packets, is presented in [11]. This work takes into account the interference from other nodes in terms of the service time distribution; it also considers the propagation delay between two adjacent nodes. Another approach applying complex analysis of the packet output process of unslotted CSMA/CD is used in [12], under the assumption of variable length packets. Authors of [12] derive the Laplace-Stieltjes transform (LST) of the packet inter-departure time and of the packet delay.

The aforementioned studies are not easily and directly applicable to model the network studied in this work, mainly due to the difference in access schemes (e.g., slotted versus unslotted, collision detection versus collision avoidance, etc.). More recently, several authors used priority queuing systems to analyze the performance of bus-based networks. In [13, 14, 15 and 16], authors have presented performance analysis of OPS bus-based networks employing *optical* CSMA/CA protocol (e.g., the DBORN architecture [2]). Using Jaiswal's [17] and Takagi's [18] results on M/G/1 priority queues, the authors derive mean packet delay at each node for both slotted [13, 14, 15] and unslotted [14, 15, 16] modes, which support both fixed and variable length packets. In particular, both [15] and [16] use the same approach to derive the upper and lower bounds of the packet delay at downstream nodes on the unslotted bus-based network.

The work presented in this paper focuses on the performance analysis of OPS bus-based network using OU-CSMA/CA protocol such as described in [2] or [3], supporting variable length packets (i.e., Ethernet-based network). The whole network can be modeled as a multiple-priority M/G/1 queuing system with Preemptive-Repeat-Identical (PRI) service discipline. Our approach to the analysis of this model provides an insight into the correlation between local transmission at a downstream node and transit traffic flowing from upstream nodes through the number of service interruptions before a successful transmission. In addition to the mean packet delay that was studied in [13, 14, 15, 16], we are able to compute the queue length probability at each node via simple recurrent equations. This result may help network designers select buffer sizes so as to meet a loss rate criterion.

To solve the above model, we use a recurrent level-by-level (node-by-node) analysis technique, where the interference from upstream nodes (which causes service preemptions at downstream nodes) is taken into account in terms of reappearance and disappearance rates for the server (bandwidth for transmission), as well as in terms of change of service time distribution for preempted services. The solution for each level is based on conditional probabilities and a fixed point iterative method, and requires only a limited number of iterations. The advantage of this method, compared to classical methods for solving M/G/1 priority queue, is that it provides a computationally efficient approach to the evaluation of the stationary queue-length distribution.

This paper is organized as follows. Section 4.2 describes the network architecture and our analytical model. Section 4.3 presents the outline of the recurrent solution of this model (the detailed solution is given in the Appendix) yielding approximate queue-length distributions, as well as an approximate distribution of the number of transmission attempts at each node. Section 4.4 provides numerical results, and compares the solution of the proposed analytical model with simulation and with the model proposed in [13]. Finally, Section 4.5 is devoted to conclusions and discussion of future work.

4.2 Network Architecture and Analytical Model

4.2.1 Network Architecture

The network considered consists of two unidirectional buses: a transmission (upstream) bus that provides a shared transmission medium for carrying traffic from several access nodes to a point of presence (POP) node; and a reception (downstream) bus carrying traffic from the POP node to all access nodes (e.g., the DBORN architecture [2]). Thus, an access node always "writes" to the POP node employing the transmission bus and "listens" for the POP node using the reception bus. The traffic emitted on the transmission bus by an access node is first received by the POP node, then is either switched to the reception bus to reach its destination node, or is routed to other MAN or backbone networks.

In this paper we only analyze access nodes performance on the transmission (upstream) bus where the problem of access control arises. The reception (downstream) bus is a simple broadcast or diffusion communication, which does not need to be deeply investigated. Therefore, we can describe the transmission bus under study as a unidirectional fiber connecting several nodes (stations) to a POP node (Fig. 4.1: Unidirectional OPS bus-based network). All nodes share a single wavelength (operating at 2.5 Gbps or 10 Gbps) to transmit local traffic to the POP node. For reasons of simplicity and cost-efficiency, bus nodes use passive optical technology that reduces the number of transceivers at each node. Specifically, there is no packet-drop operation at bus nodes; each node can only add/insert local traffic into the wavelength without intercepting transit traffic of upstream nodes.

Since the single wavelength is shared between many nodes, a MAC protocol is required to control the access of bus nodes to the wavelength. Thus the OU-CSMA/CA protocol (Fig. 4.1), which is based on void (i.e., free segment of bandwidth) detection, is used to ensure an asynchronous collision-free packet insertion into the transmission wavelength. This protocol has two interesting characteristics: (1) It is a fully-distributed protocol that simplifies the implementation and management of the network; (2) Its asynchronous nature offers the capability to support data traffic with variable size packets in the network, without the need for

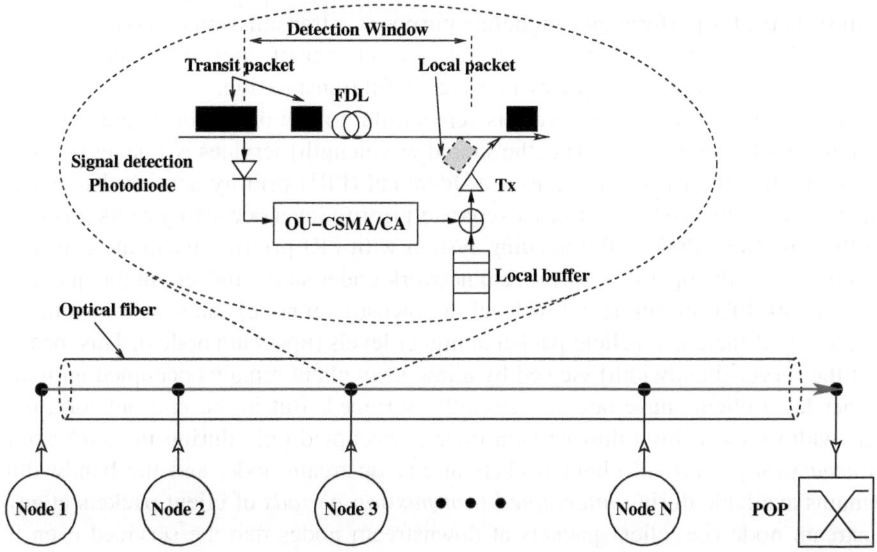

Fig. 4.1 Unidirectional OPS bus-based network

complex segmentation/assembly process. Based on the second property, a mature technology like Ethernet is used for the data plan of the network. This means that the network supports variable size optical packets, each consists of one conventional Ethernet packet encapsulated by an optical overhead [2]. It is worth noting that the maximum packet length supported by the network is limited by the Maximum Transmission Unit (MTU) of Ethernet, which is around 1500 bytes.

With OU-CSMA/CA protocol, a node begins transmitting a packet if and only if it detects a void on the transmission wavelength that is at least equal to the packet size (including an optical overhead if necessary). Generally speaking, voids seen by a node are free segments of bandwidth between transit packets coming from upstream nodes. Therefore, the most upstream node that begins the shared wavelength is always able to transmit, since it always has available bandwidth at any time (i.e., infinite void length).

4.2.2 Network Model

The transmission bus uses only one wavelength shared by the N access nodes. From modeling perspective, it is convenient to view the operation of the OU-CSMA/CA protocol as follows. A node begins transmitting a local packet as soon as it detects a void. The transmission is interrupted when a transit packet arrives from an upstream node (i.e., the void is not large enough), and the packet returns to the queue waiting for the next void. At detection of the next void, the node starts again transmitting the local packet whose transmission was interrupted. This process is repeated until the packet is successfully transmitted (i.e., a large enough void is found). Thus, for performance modeling purposes, a transmission of a packet in the OU-CSMA/CA protocol may be viewed as a number of interrupted (preempted) transmission attempts followed by one successful transmission.

The behavior of the real network is very similar to that of a priority queuing system, in which a single server (i.e. the shared wavelength) services N queues (i.e. the N bus nodes) with Preemptive-Repeat-Identical (PRI) priority service discipline. Each node in this system defines a separate priority level according to its position on the bus. Nevertheless, the queuing system with PRI priority discipline does not exactly match the operation of the real network under study. Indeed, in the queuing system with PRI discipline, a low level (a downstream node) can start transmitting if and only if there is no client packet at higher levels (upstream nodes). This means that the server (bandwidth) viewed by a low level client remains occupied until all higher level clients have been successfully serviced. But in the real network, the bandwidth viewed by a downstream node is occupied only during the *successful transmission periods* of client packets at any upstream node, and the bandwidth remains available during *interrupted transmission periods* of client packets at any upstream node (i.e. client packets at downstream nodes may be serviced even if upstream nodes are attempting to transmit their client packets). More precisely, when an upstream node detects a void and cannot use it for transmission (i.e. transmission attempt is interrupted) because the void length is not long enough, this void (or, in other words, the time elapsed from the moment where the node attempts to transmit the packet till the transmission is interrupted) may be used by a downstream node for transmitting smaller packets whose lengths fits this void. Thus, from queuing model perspective, the real network behavior corresponds to a queuing system with a special priority service discipline, which appears more complicated than the PRI discipline. Note, however, that if the network supports only fixed length packets, then the real network behavior perfectly matches that of the queuing system with PRI discipline, because in this case a void shorter than the packet length is unusable for any node.

In this paper, we approximately represent the OPS bus-based network supporting variable length packets as a queuing system with the PRI discipline. The case with more complex priority discipline is left for future work. Starting with most upstream (and highest priority) node, we number the nodes *1* through N so that an upstream node i has priority over a downstream node j, $1 \leq i < j \leq N$. We assume that each node has an infinite buffer, and client packets stored in the buffer are serviced in First-Come-First-Serve (FCFS) order.

Since the network under consideration is supposed to be a future metro network receiving a high aggregation of traffic coming from high-speed access networks such as Fiber To The Home (FTTH), we can reasonably assume that local packets at node i arrive according to a Poisson process with rate λ_i, and that their service times are mutually independent and have a general distribution with known finite mean (m_i) and variance (Var_i). Thus our network model is an M/G/1 system with N priority levels and PRI FCFS service discipline.

To solve this model, we propose a new approach, different from classical approaches for solving the M/G/1 priority queue, which allows us to approximately compute not only the mean queue length but also the steady-state queue length distribution at each node. In our method, we analyze bus nodes one by one, and we use a specific state description to represent the interference from upstream nodes (if any). In the following sections, we present a model of the general service distribution at each node, highlighting the effect of PRI discipline on the service distribution at downstream nodes.

4.2.2.1 Modeling of Original Service Time Distribution

To model the service time distribution at each node, we use a Coxian distribution [19] with the minimum number k of exponential stages needed to match the first two moments of the service time distribution as shown in Fig. 4.2 (k-stages Coxian system modeling service time distributions at each node). The resulting form of the Coxian depends on the coefficient of variation Cv of the distribution being represented. In the case where $Cv \geq 1$, we use a two-stage Coxian distribution ($k = 2$ in Fig. 4.2) with three parameters μ_1, μ_2 and p_2. For $1/\sqrt{k} \leq C_v < 1/\sqrt{(k-1)}$, we use a k-stage hypoexponential distribution [20] ($p_2 = 1$ in Fig. 4.2) with two parameters μ_1 and μ_2. As the number of stages k increases, the Cv of this distribution tends to zero, which corresponds to the case of fixed-length packets. The parameters of these Coxian distributions are readily derived from the first and second moments of the original service-time distributions as follows.

Consider the case with $Cv \geq 1$. We use a two-stage Coxian distribution (Fig. 4.2 with $k = 2$). Let $t_1 = 1/\mu_1$ and $t_2 = 1/\mu_2$ be the mean service times at each stage, we have the following equations:

$$\begin{aligned} m_i &= t_1 - p_2 t_2, \\ Var_i &= t_1^2 + p_2(2 - p_2)t_2^2 = m_i^2 C_v^2. \end{aligned} \tag{4.1}$$

Set $t_1 = \gamma m_i$, where $\gamma (0 < \gamma < 1)$ is a constant. We are able to obtain the values for the Coxian distribution parameters

$$p_2 = \frac{2(1-\gamma)^2}{C_v^2 + (1-\gamma)^2 - \gamma^2}, \text{ and } t_2 = \frac{m_i(1-\gamma)}{p_2}, \tag{4.2}$$

In practice, we may choose γ such that $t_1 < t_2$ (i.e. $0 < \gamma \leq 0.5$). Similarly, we can derive the parameters for the other cases with $1/\sqrt{k} \leq C_v < 1/\sqrt{(k-1)}$.

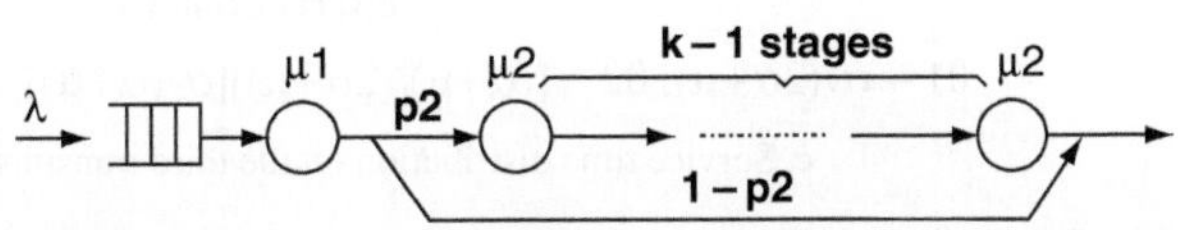

Fig. 4.2 k-stages Coxian system modeling service time distributions at each node

4.2.2.2 Modeling of Interrupted Service Time Distribution

As stated earlier, the service at nodes $i > 1$ may be interrupted due to arrivals of client packets at upstream nodes. Thus, a node may have to attempt the transmission of a packet several times, each time the transmission being interrupted by transit traffic from higher priority upstream nodes until a long enough void comes along (no interruption). Clearly, after every interrupted transmission, the node reattempts the transmission of the same packet at the next void. Somewhat paradoxically, to describe the fact that it is the same packet whose transmission is reattempted, we need to represent a potentially different packet length (and hence service time) distribution (for packets whose transmission got interrupted) at each consecutive attempt.

To understand what is going on, perhaps the simplest example is to examine the case where the original service times are exponentially distributed with parameter μ and interruptions arrive from a Poisson source with rate α.

On the first transmission attempt of any client packet, the service time distribution is the original exponential distribution with parameter μ (Fig. 4.3.a: System

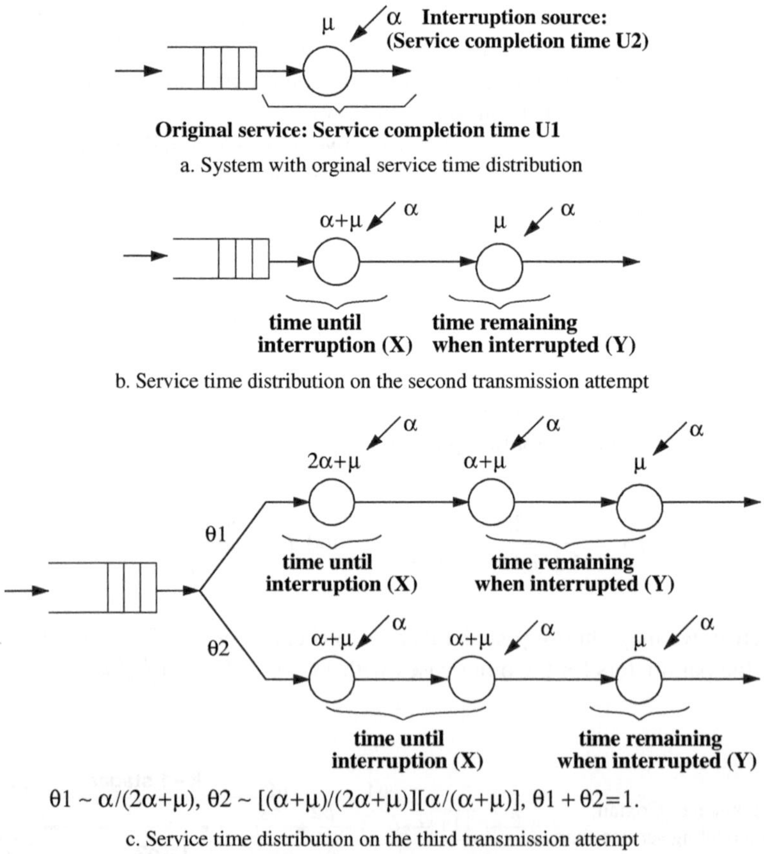

Fig. 4.3 Evolution of service time distribution following interruptions

with original service time distribution). This service may be interrupted by the Poisson source with rate α. Note that on the first attempt, we are dealing with the whole population of client packets.

On the second attempt, we are dealing with a subset of client packets whose transmissions got interrupted for the first time (i.e., we exclude all client packets that have been successfully transmitted at the first attempt). To derive the service time distribution of this subset of client packets, we consider the time until the interruption (X) and the time remaining when the interruption occurred (Y). Consider a small interval of time $(t, t+\Delta t]$. The probability that a first service interruption will happen during this interval can be expressed as $\Delta t \alpha e^{-\alpha t} e^{-\mu t} + o(\Delta t)$ where $o(\Delta t)/\Delta t \underset{\Delta t \to 0}{\to} 0$. The overall probability that a first service will be interrupted is simply the probability that an exponentially distributed process with parameter α (the Poisson interruption arrivals) finishes before the exponentially distributed service process with parameter μ, which is readily seen to be $\frac{\alpha}{\alpha+\mu}$. Hence, the conditional density of the time to interruption given that the service is interrupted is $(\alpha + \mu)e^{-(\alpha+\mu)t}$, i.e., the time until interruption X is exponentially distributed with parameter $\alpha + \mu$. Because of the memoryless property of the original service time distribution, the remaining service time Y at the point of interruption is exponentially distributed with the original parameter μ. Therefore the resulting service time distribution of the subset of client packets after the first interruption (or, in other words, on the second transmission attempt) is the hypoexponential distribution shown in Figure 4.3(b) (Service time distribution on the second transmission attempt).

On the third attempt, we are dealing with a subset of client packets whose transmissions got interrupted for the second time (i.e., we exclude all client packets that have been successfully transmitted on the first and the second attempt). In other words, we are dealing with the two-stage hypoexponential distribution in Fig. 4.3(b) interrupted by a Poisson arrival process with rate α. This interruption could have taken place while the service was in the first or the second stage of the two-stage hypoexponential. Thus, as shown in Fig. 4.3(c) (Service time distribution on the third transmission attempt), with probability θ_1, the interruption could have taken place while the service was in the first stage of the two-stage hypoexponential. This results in an exponentially distributed time to interruption with parameter $2\alpha + \mu$, followed by an exponentially distributed residual of the first stage (parameter $\alpha+\mu$) and the full second stage. With probability θ_2 the interruption could have taken place while the service was in the second stage. Then the time before interruption consists of the full first stage, followed by the part of the second stage preceding the interruption (exponential with parameter $\alpha + \mu$), and the time after the interruption is the exponentially distributed residual with parameter μ. An analogous process continues at subsequent interruptions.

In a similar way we can derive and represent the distribution of service times at each interruption when $Cv > 1$ and $Cv < 1$. We observe that, with the obvious exception of a constant packet length (hence, service time), for all distributions of packet lengths, the mean increases while the coefficient of variations decreases on each subsequent transmission attempt. In our exponential service example, after the first interruption the mean nearly doubles when α is small. We also observe that both

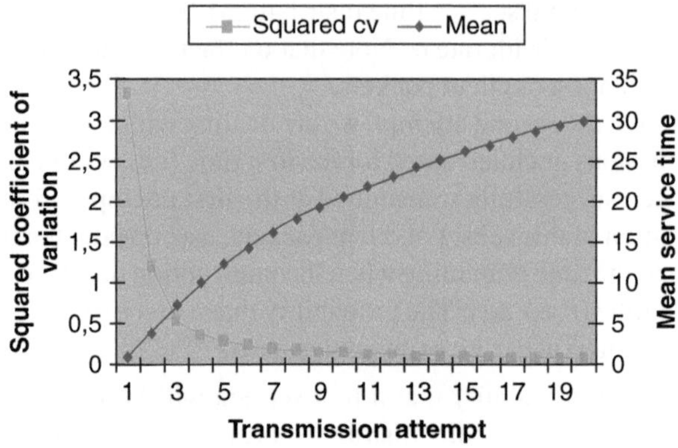

Fig. 4.4 Mean and square coefficient of variation of the service time as a function of the number of transmission attempts

the increase in the mean and the decrease in the coefficient of variation slow down at each subsequent attempt. This makes perfect physical sense: as transmissions are attempted, shorter packets are more likely to be successfully transmitted and longer packets need more attempts. The elimination of shorter packets accounts for both the increase in the mean and the decrease in variability of the packet length of the subsets of client packets. In the limit, we expect the packet length to tend to the maximum packet length (MTU) at subsequent attempts at the given node with variance tending to zero.

Figure 4.4 (Mean and square coefficient of variation of the service time as a function of the number of transmission attempts) illustrates the evolution of the mean and the squared coefficient of variation of the service time at each transmission attempt for an initial distribution with an initial squared coefficient of variation of 3.3. For the example considered, at the second attempt the mean service time is close to 4 times the initial average while the squared coefficient of variation is less than half the initial value. At the third attempt, the mean is over 7 times the original value while the squared coefficient of variation drops to less than 20% of the original.

4.3 Outline of Model Solution

In order to obtain a tractable approximate solution to our model in the steady state, we analyze the bus nodes one by one, starting with node *1* for which there is no upstream transit traffic. We focus on node i ($i = 1 \ldots N$), and, to simplify our notation, we omit the node subscript i whenever this does not create confusion.

4.3.1 Solution for Node 1

4.3.1.1 Balance Equations Derivation

Since node *1* always "sees" the bus bandwidth free, we simply describe the equilibrium behavior of this node by the joint steady-state probability $p(n, l)$, where

n ($n \geq 1$) is the number of packets at this node and l refers to the stage of service of the Coxian service time distribution shown in Fig. 4.2. We denote by $p(n)$ the marginal distribution for the number of packets at the node, and by $p(l|n)$ the corresponding conditional probability for l given n. Using the fact that $p(n, l) = p(n)p(l|n)$, we are able to readily obtain the equations for $p(l|n)$ and $p(n)$.

4.3.1.2 Fixed-Point Iteration Method for Solution of the Balance Equations

We detail in the Appendix 1A a simple recurrent solution using fixed point iteration method for these equations. This solution is based on the general notion of state equivalence [21], and its specific application to M/G/1-like queues [22]. This solution allows us to compute the conditional rate at which packets are served (i.e., effectively transmitted) given n, which we denote by $u(n)$, and, hence, $p(n)$.

The computation scheme can be described by the following pseudo code:

```
factor = sum = 1.0;
mean_occupancy = 0.0;
for (n= 1; n< n_max; n++) {
        Solve equations for conditional probabilities p(l|n);
        Compute u(n);
        factor *=λ / u(n);
        sum+= factor;
        mean_occupancy+= n*factor;
        if(| u ( n) - u ( n- 1) <ε |) break;
}
Complete computation of ''infinite part'' of sum and mean_occupancy;
sum= 1.0/sum;
mean_occupancy= sum;                    /* normalize*/
prob_node_idle= sum;
Compute server reappearance rate for node 2, i.e. β2;
```

In our computation, we used $\varepsilon = 10^{-10}$ for the test of convergence of $u(n)$ to its limiting value. The computation of the infinite part of the normalizing constant, as well as the mean node occupancy, is straightforward given the geometric-series form of the tail of the node occupancy distribution [22].

4.3.2 Solution for Node $i > 1$

4.3.2.1 Balance Equations Derivation

As stated earlier, in our PRI model, a node $i > 1$ may find the server (bandwidth) available or occupied. In our model, the server is available if and only if there are no client packets at all nodes $1 \ldots i-1$, and is occupied otherwise. When the server is available, it serves client packets with a constant rate R which is the wavelength bit rate. Thus, viewed from node $i > 1$, an available server becomes occupied whenever a client packet arrives at an upstream node $j < i$, hence interrupting the service at node i; and an occupied server becomes available whenever the last client packet at the upstream nodes $i - 1$ has been successfully transmitted (recall that, in our model, client packets at node $i - 1$ are serviced if and only if all upstream node queues are empty).

Let α_i and β_i respectively be the disappearance and reappearance rates of the server viewed by node $i > 1$. Since the arrivals to each node come from a Poisson source, the disappearance rate α_i of the server is exactly given by $\sum_{j<i} \lambda_j$. The reappearance rate β_i can be expressed in terms of the conditional transmission rate $u(n)$. As an example, for node 2 we have $\beta_2 \approx u_1(1)p_1(1)/[1 - p_1(0)]$ (see Appendix 4 for more details).

Since the service distribution changes at consecutive transmission attempts in our model, to represent the Preemptive-Repeat-Identical discipline at node $i > 1$, we choose a state description that explicitly accounts for possible service interruptions and retrials at the node. The parameters of the service time distribution change with each transmission attempt as discussed above. We denote by k_j the number of exponential stages required to represent the service time distribution at the j-th transmission attempt ($j = 1, 2 \ldots$). We describe the state of node i by the triple (n, j, l) where n is the current number of packets at the node, j is the transmission attempt, and l is equal to the current number of the service stage at this attempt $(1, \ldots, k_j)$ or 0 if the server (bandwidth) is unavailable. Recall that, in our model, the server becomes unavailable with rate α_i, and available again with rate β_i.

4.3.2.2 Fixed Point Iteration Method for Solution of the Balance Equations

Using the fact that $p(n, j, l) = p(n)p(j, l|n)$, we are able to transform the balance equations for $p(n, j, l)$ into equations for the conditional probability $p(j, l|n)$. We then derive a recurrent solution using a fixed-point iteration method for increasing values of $n \geq 1$. To limit the size of the state space for each n, we explicitly compute the parameters of the service time distributions at transmission attempts up to a certain value j^*, and we use "catch all" average values for the parameters of the service time distribution at transmission attempts above j^*. As stated above, the mean value of the service time distribution increases and its coefficient of variation decreases as the number of transmission attempts increases. Theoretically, for a Coxian distribution this mean value might increase to infinity. But, in our real network, this mean value is naturally limited by the service time of the maximum transmission unit (MTU) of the transmission protocol used. Thus, if j^* is chosen large enough, all service time distributions at transmission attempts above j^* may be replaced by a constant service which is the service time of the maximum length packet MTU. We also limit the number of stages in the Coxian distribution to k^* (so that the minimum value of Cv in our model is $1/\sqrt{k^*}$).

From the conditional probability $p(j, l|n)$ computed using the above recurrent solution, we readily obtain the conditional rate of transmission completions $u(n)$, and, hence, the marginal probability of the number of packets at the node $p(n)$, as well as an approximate value for β_{i+1}, the rate at which the server becomes idle, i.e. available for downstream nodes. We give in the Appendices 2 and 3 additional details of our solution, and in Appendix 4 an outline of the estimation of the value of β_{i+1}.

The computation scheme can be described by the following pseudo code:

Compute *rate of server disappearance for this node, i.e.* α_i ;
Compute $p\,(\,l{=}0 \mid 0\,) = \alpha_i/[\alpha_i + \beta_i + \lambda]$;
$factor = sum = 1.0$;
$mean\_occupancy = 0.0$;
$for\,(n = 1;\, n < n\_max;\, n++)$ {
  Solve *equations for conditional probabilities* $p(j, l|n)$;
  {
  $let\ \ x = p(j = 0, l = 1|n)\ \ and\ \ y = u(n)$;
  $For\ \ all\ \ j = 1, \ldots, j^* - 1,$ **Express** $p(j, l|n)$
    $as\ \ p(j, l|n) a_{j,l} x + b_{j,l} y$;
  Use
  $$\sum_{j\geq 1} \sum_{l=0}^{k_j} [a_{j,l}x + b_{j,l}y] = 1$$
  $$y = \sum_{j\geq 1} \sum_{l=0}^{k_j} [a_{j,l}x + b_{j,l}y] \mu_l^j (1 - \bar{q}_l^j)$$
  to **determine** $p(j = 0, l = 1|n)\ \ and\ \ u(n)$;
  }
  $factor^* = \lambda/u(n)$;
  $sum += factor$;
  $mean\_occupancy += n^* factor$;
  $if\ \ (|u(n) - u(n-1) < \varepsilon|)$ **break**;
}
Complete *computation of "infinite part" of sum and mean_occupancy*;
$sum = 1.0/sum$;
$mean\_occupancy = sum$;   /*normalize*/
$prob\_node\_idle = sum$;
Compute *server reappearance rate for next node (if applicable)*;

In the next section, we present numerical results obtained from our model, and compare them with results obtained from a detailed network simulation, as well as from another analytical model [13].

4.4 Results

4.4.1 Model Accuracy

In our approach to the solution of a model of an optical bus, outlined in the preceding section, we are able to approximately analyze each node one by one. At node $i > 1$, the presence of upstream nodes is represented as the server (bandwidth) becoming busy with rate α_i and then available with rate β_i. Since we are only able to compute the rate β_i approximately, this is one potential source of inaccuracy of the model.

Another potentially important point is the fact that we match only the first two moments of the distribution in our representation of the service times. For the M/G/1 Preemptive Resume, as well as for Head-of-Line Non-Preemptive priority queue, it is well known (e.g. [23]) that the expected number of customers of each class in the system depends on the service time distribution only through its first two moments. To assess the effect of higher moments in our Preemptive-Repeat-Identical model, we focus on the simple case of the first two nodes.

We consider a full two-class PRI queuing model, as well as a model of node 2 with the approximate server reappearance rate β_2 computed from the recurrent solution of node 1. In this way, in addition to the potential influence of higher moments, we are able to study the effect of the approximate computation of the rate with which the server becomes available (by comparing the results for node 2 in both models). We use discrete-event simulations of both models for two different random two-stage Coxian distributions with the same mean and variance. The parameters of these distributions are given as follows. The parameters for Distribution I are $\mu_1 = 1.9606$, $\mu_2 = 0.4915$, $p_2 = 0.328$. For Distribution II the corresponding parameters are $\mu_1 = 9.8573$, $\mu_2 = 0.6316$, $p_2 = 0.5802$. Both distributions have a mean of 1.0201 and a variance of 2.0755, but Distribution I has a third moment of 17.4364 (and hence a skewness of 3.3521), while Distribution II has a third moment of 14.787 for a skewness of 2.4594. Table 4.1 (Influence of high moments of service time distribution in PRI model) illustrates the results for the expected number of packets at node 2 obtained as the rate of packet arrivals to both nodes increases. All simulation results in Table 4.1 include confidence interval estimated at 95% confidence level using 7 independent replications of 800,000 successful packet transmissions each.

We observe that the inaccuracy caused by the approximate server reappearance rate β_2 seems to be quite limited (on the order of a few percent, and, in several cases, the confidence intervals for both models overlap). As the packet arrival rate increases, the shape of the distribution of service times beyond the first two moments appears to have a much greater influence: over 25% relative difference in the expected number of packets at node 2 for the example considered.

Table 4.1 Influence of high moments of service time distribution in PRI model

Packet rate ($\lambda_1 = \lambda_2$)	Model	Mean packets number at node 2 (Distribution I)	Mean packets number at node 2 (Distribution II)
0.06733	Full two-node model	0.1073 ± 0.0002	0.1058 ± 0.0002
	Model of node 2 alone	0.1043 ± 0.0003	0.1023 ± 0.0003
0.13466	Full two-node model	0.4179 ± 0.0091	0.4292 ± 0.0123
	Model of node 2 alone	0.4009 ± 0.0026	0.4038 ± 0.0038
0.20199	Full two-node model	2.5726 ± 0.3549	1.6053 ± 0.0750
	Model of node 2 alone	2.6417 ± 0.4280	1.4507 ± 0.1157

Table 4.2 Two random different packet length distributions with the same mean and variance

Distribution	Mean (μs)	C_v^2	Packet length distribution
III	2.56	0.4375	63.64% packets of 400 bytes 36.36% packets of 1500 bytes
IV	2.56	0.4375	36.36% packets of 100 bytes 63.64% packets of 1200 bytes

Table 4.3 Influence of higher moments of service time distribution in real network

Mean packet arrival rate (packets/μs)	Mean response time (μs) at node 4	
	Distribution III	Distribution IV
0.058	20.04 ± 0.1316	19.50 ± 0.0849
0.068	51.39 ± 0.6901	47.78 ± 0.3457
0.078	1633.00 ± 204.2	965.80 ± 90.43

Interestingly, we have also observed the effect of higher moments of service time (packet length) distribution in the real network environment. For example, we simulate the network with 4 nodes transmitting on one wavelength at 2.5 Gbps, and all nodes are subjected to the same packet arrival process. We consider two different random packet length distributions with the same mean and variance as shown in Table 4.2 (Two different packet length distributions with the same mean and variance). Table 4.3 (Influence of higher moments of service time distribution in real network) shows the mean response time observed at the last node (node 4) as the packet arrival rates at all nodes increase. Clearly, in the real network, the shape of the packet length distribution (or service time distribution) beyond the first two moments also has a significant impact on the mean response time at bus nodes when the packet arrival rate increases (hence, the network load increases). In this example, the relative difference in mean response time at node 4 is about 40%.

There are also some potential sources of inaccuracy of the model related to numerical computation. The use of a "catch all" average service time distribution for transmission attempts $j > j^*$ is one potential source. This one is more likely to have an effect when a larger number of interruptions can be expected (e.g. low priority node, heavier bandwidth utilization, service time distributions with higher variability ...). Our limitation on the maximum number of stages k^* in the Coxian representation of the service time at a given transmission attempt may also introduce some inaccuracies, notably when C_v becomes very small (e.g. $C_v < 1/\sqrt{k^*}$).

4.4.2 Performance Evaluation

We now attempt to analyze the accuracy of our PRI model in evaluating the performance of the OPS unslotted bus-based network discussed earlier. We use discrete-event network simulator tool (NS 2.1b8 [24]) to simulate the network with 8 nodes transmitting on one wavelength at 2.5 Gbps. The simulation results are then

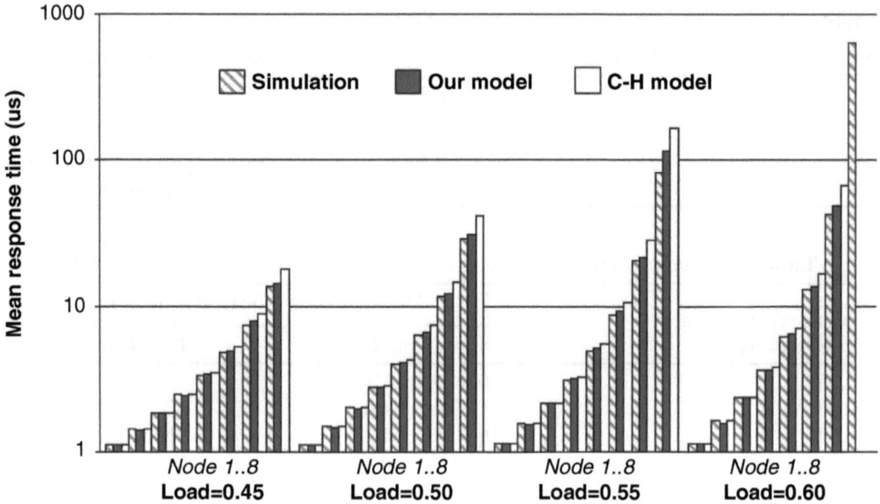

Fig. 4.5 Mean response time at each node as a function of offered network load for service time distribution 4

compared to the results obtained from our PRI model, and from the model proposed in [13] using Jaiswal's results on priority queues (we refer to it as the C-H model). All mean values in our simulation results are estimated with confidence intervals of no more than a few percent wide around the midpoint at 95% confidence level using Batch Means method [25] (i.e., each mean response time value is computed by collecting at least 7 batches of 800,000 successful packet transmissions each). In the first set of results, illustrated in Figs. 4.5 and 4.6, we assume that all nodes share

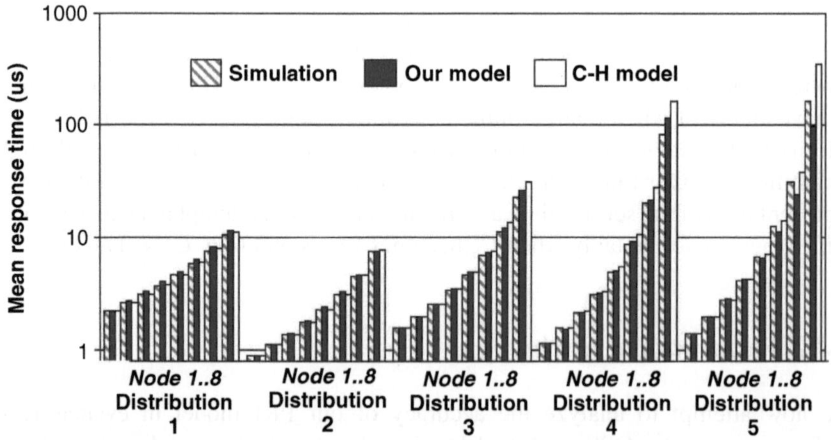

Fig. 4.6 Mean response time at each node as a function of service time distribution for offered network load of 0.55

Table 4.4 Original service time distributions used for performance study

Distribution	Mean (μs)	C_v^2	Packet length distribution
1	2.13	0.125	67% packets of 500 bytes 33% packets of 1000 bytes
2	0.8418	0.7295	53% packets of 50 bytes 47% packets of 500 bytes
3	1.434	1.182	45% packets of 50 bytes 40% packets of 500 bytes 15% packets of 1500 bytes
4	1.02	1.994	64% packets of 50 bytes 26% packets of 500 bytes 10% packets of 1500 bytes
5	1.23	2.524	77% packets of 50 bytes 23% packets of 1500 bytes

the same rate of packet arrivals and service time distribution (uniform traffic profile at all bus nodes).

Table 4.4 (Original service time distributions used for performance study) describes service time distributions (packet size mixes) used in this study. The square coefficients of variation of these distributions vary from close to zero to higher than 1, implying the use of both two-stage Coxian and hypoexponential distributions to model the service time distribution in our PRI model.

Notice that distribution 4 is closest to the "real life" service time distribution of Internet traffic (e.g. [26]). Figure 4.5 (Mean response time at each node as a function of offered network load for service time distribution 4) shows the mean response time of the system at each bus node for packet length distribution 4 as the offered network load increases. The offered network load is defined as the ratio of the sum of traffic volume offered to all nodes to the network transmission capacity. The mean response time at a node is defined as the expected time elapsed from the moment when a client packet arrives at the queue of the node until it is successfully transmitted on the bus. For this experiment, we first observe that both simulation and our analytic model capture the expected behavior of mean response time in OPS bus-based networks: the mean response time is likely to increase rapidly as the node's priority decreases. Moreover, the mean response time tends to "explode" at the lowest priority nodes as the offered network load increases. For instance, for packet size mix distribution 4, simulation results show that the mean response time at node 8 is about 13.7 μs with offered network load of 0.45, and some 640 μs with offered network load of 0.60. An explanation for these results is that the transmission at low priority nodes may be delayed (in our model, interrupted) by the arrivals of packets from higher priority nodes, thus a successful transmission at low priority nodes takes on average a longer time than at higher priority nodes. The number of service interruptions becomes more and more important as the offered network load increases, leading to excessive response time at the lowest priority nodes.

Additionally, we observe in this experiment that when the offered network load is low, the results obtained with our analytical model are very close to those obtained with simulation. For instance, the difference between analytical and simulation results is on the order of only a few percent for all offered network loads below 0.55. But this difference becomes more significant at the last bus nodes as the network load increases (e.g. some 28% relative difference in the mean response time at node 8 when the offered network load is 0.55). We also notice that in comparison with simulation results, our model provides remarkably better results than those obtained with C-H mode at most downstream nodes (node 5–8). On the contrary, the C-H model provides results lightly closer to simulation results than our model at first upstream nodes (node 1–4).

Note that, with the network load at 0.6, the precise shape of the service time distribution (in terms of moments higher than the first two) starts playing an important role at the most downstream node where the bandwidth (server) is close to saturation. While the network simulation indicates a mean response time of some 640 μs at node 8, both analytical models peg the node as unstable. Interestingly, a direct simulation of the PRI system with the same service time distribution as in the analytical model (i.e., matching only the first two moments of the service time distribution in the network model) also shows that the node 8 becomes overloaded.

It is worth noting that the saturation at the most downstream node is not due to the lack of physical bandwidth (server) capacity, because the bus is actually loaded at merely 60% of its transmission capacity. As stated at the beginning of the paper, this saturation is mostly due to the fact that the physical bandwidth has been fragmented into small segments of bandwidth (voids) between asynchronous transmissions of packets at upstream nodes. Those voids are unusable for most downstream nodes to insert big client packets (i.e., in our model, this is equivalent to a large number of interruptions during the service of big client packets), leading to the "head-of-the-line" (HOL) blocking phenomenon. Clearly, the effect of this phenomenon on most downstream nodes depends not only on the network load, but also on the shape of the packet length distribution. We now specifically study the impact of packet length distribution (or, equivalently, service time distribution) on the network performance.

To assess the behavior of our PRI model with respect to the service time distribution, we focus on the analysis of the mean response time obtained with the same offered network load but with different service time distributions. Figure 4.6 (Mean response time at each node as a function of service time distribution for offered network load of 0.55) illustrates the mean response time at each node with the offered network load of 0.55 as a function of service time distribution. In this study, we set the offered network load at 0.55 because, as shown by the results of our preceding experiment, beyond this load level the stability condition might not be satisfied for some nodes. As before, the workload is uniform, i.e., statistically identical for all nodes.

We observe for the uniform workload considered that the first few nodes on the bus experience little queuing time and few interruptions (for nodes other than the first one). This means that the server viewed by the first few upstream nodes is highly available, servicing clients at those nodes rapidly. Therefore, the mean response time (which is the sum of mean queuing time and mean service time) at these nodes depends mostly on the mean of the service time (and not higher moments). Indeed, in Fig. 4.6, we notice that for the first upstream nodes (up to node 4), distribution 1 with highest mean of service time (but lowest variance) provides highest mean response time, followed by distribution 3, 4 and 2. For lower priority nodes, the mean response time becomes clearly more sensitive to higher moments of the service time distribution. In particular, when the variance of the service time distribution is high, we observe very high response time at the lowest priority nodes compared to the response time of higher priority nodes. For example, in Fig. 4.6, for distribution 5 with highest C_v^2 of 2.524 (but not highest mean), the simulation shows that the mean response time at node 8 is the highest compared to other distributions, and is some five times higher than the mean response time at node 7, and nearly 120 times higher than the response time at node 1.

The above effect of service time distribution with high variability on the mean response time at most downstream nodes is readily explained by the bandwidth fragmentation phenomenon. The high variability of service time distributions in our experiments means that there is an important percentage of small/medium packets (e.g. 50/500 bytes) and a smaller percentage of big packets (e.g. 1500 bytes) in the offered traffic (see Table 4.4). From physical perspective, this translates into the fact that the available and usable bandwidth for low priority nodes is strongly reduced because it becomes considerably fragmented into small voids due to the asynchronous insertion of small/medium packets at higher priority nodes. In reality, when an upstream node detects a void, it may insert a packet at the beginning, at the middle or at the end of this void depending on whether a packet is available in the queue at that moment. The insertion of a small/medium packet into a big void will break the void into two small voids, which may be unusable for the transmission of bigger packets at lower priority nodes. Thus the high variability of service time distribution leads to high probability of the HOL blocking phenomenon at the most downstream node, resulting in excessive mean response time at that node. Notice that HOL blocking may not occur if downstream nodes have small packets only.

As far as the accuracy of our model is concerned, we observe in Fig. 4.6 that the difference between our model and simulation results is limited to a few percent when the service time is not highly variable (e.g. distribution 1 and 2), but it becomes larger as the service time becomes more variable (e.g. 11% and 28% relative difference for distribution 3 and 4 respectively). However, our model still provides significantly better result than the C-H model for highly variable service times. For instance, the relative difference between the C-H model and simulation results is about 26% and 50% for distribution 3 and 4 respectively. As mentioned earlier, part of reason for the behavior of these analytical models (which only match the first two moments of the service time distribution) may be due to the influence of moments higher than the first two.

In the numerical results shown in Fig. 4.7 (Mean response time at each node as a function of traffic patterns at offered network load of 0.55), we study the effect of varying patterns of the offered traffic on the network performance. We consider the "real life" service time distribution 4, and we set the offered network load to 0.55. In addition to the uniform traffic considered before, we include the case where the traffic increases uniformly as we move downstream on the bus, the case

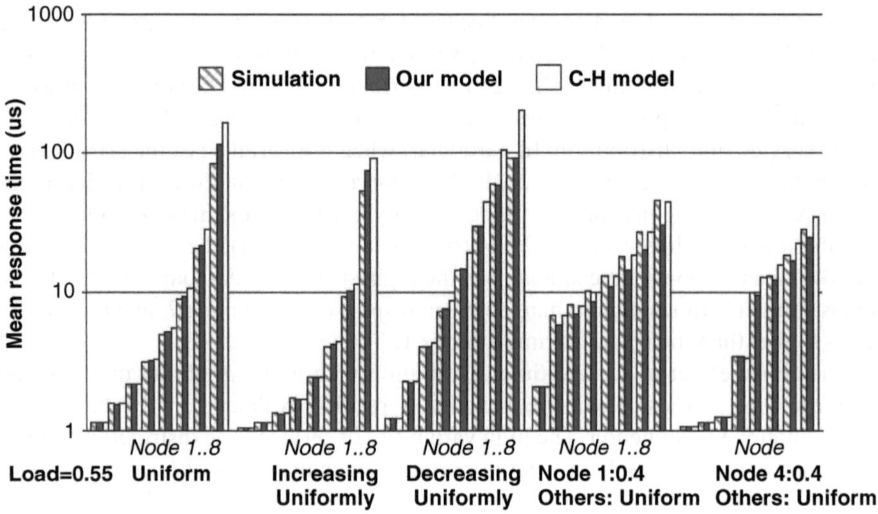

Fig. 4.7 Mean response time at each node as a function of traffic patterns at offered network load of 0.55

where traffic decreases uniformly as the node priority decreases, the case where the highest priority node carries 70% of the traffic the remaining nodes sharing uniformly the remaining 30% of the load, and, finally, the case where node 4 carries 70% of the traffic while other network nodes share the remaining 30%. We show the mean response time estimated from the network simulation, obtained from our approximate PRI model, as well as from the C-H model.

We observe that uniformly decreasing traffic (lower priority nodes carry less traffic) and the uniform traffic pattern appear most penalizing in terms of mean response time at the lowest priority nodes. Interestingly, uniformly increasing traffic and the case where the middle node (node 4) dominates the network seem to fare best. We also observe that, with the possible exception of the last node, the results of our model tend to closely track simulation results, and are in most cases closer to simulation than those of the C-H model (in this experiment, the C-H model provides results closer to simulation results than our model only in the case where the first node dominates the network).

One of the advantages of our approach is that it produces the approximate marginal distribution for the number of packets at each node in the simple product form akin to that of an M/M/1-like queue with a state-dependent service time. Such a distribution can then be used to dimension buffers at each node, as well as to assess packet loss ratios. In our approach, we analyze nodes one by one, representing upstream nodes through rates at which the server vanishes and reappears. Clearly, one might be concerned that for lower priority nodes, accumulated approximations might excessively distort the queue-length distribution. In Fig. 4.8a–e (Queue length distribution at the three lowest priority nodes for different traffic patterns at offered network load of 0.55), we compare the distribution of the number of packets at the

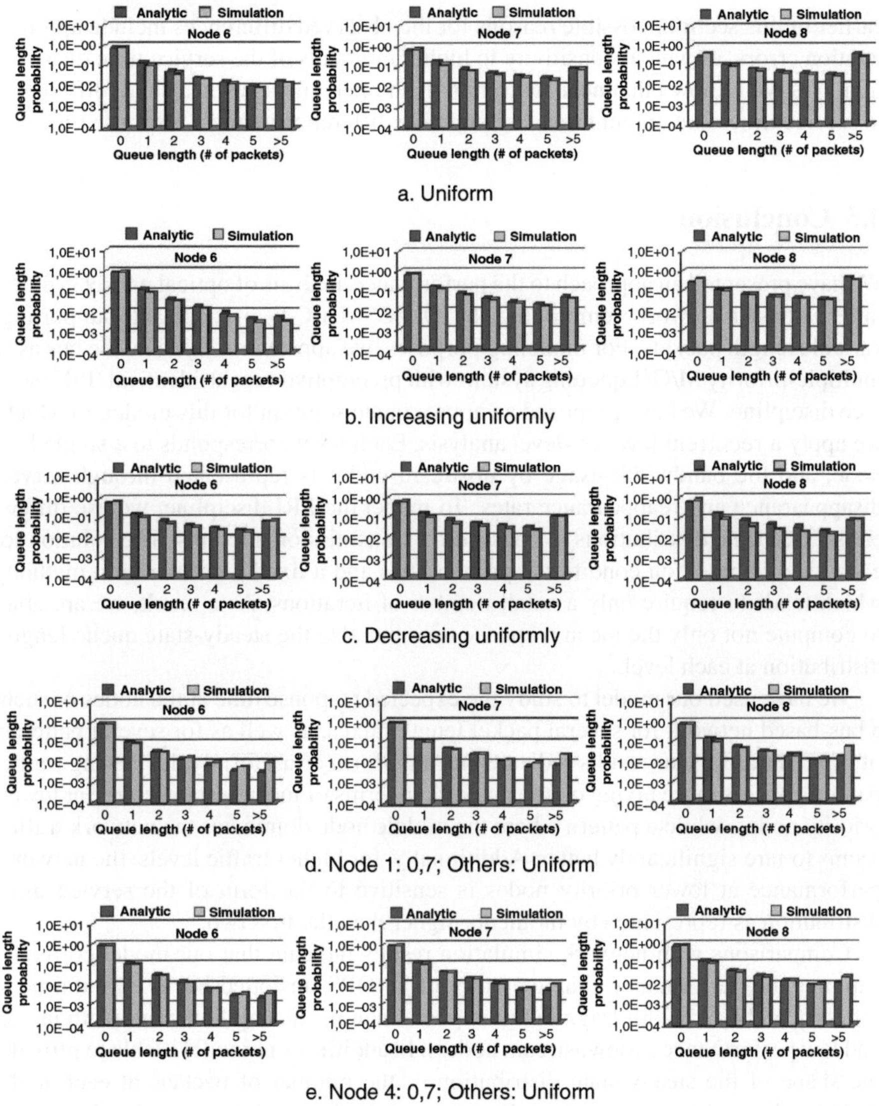

Fig. 4.8 Queue length distribution at the three lowest priority nodes for different traffic patterns at offered network load of 0.55

three lowest priority nodes for the different traffic patterns considered in Fig. 4.7. We observe that, even for the lowest priority node, our model produces results remarkably close to those obtained from network simulation.

Overall, we think that our model correctly captures the performance characteristics of an OPS bus-based network, including the shape of the stationary distribution of the number of packets at each node. The results of our model may on occasion deviate from simulation results (typically, close to node saturation). As discussed earlier in this section, possible reasons for the observed differences include approximation errors, as well as sensitivity to higher moments of the service time distribution. It is well known that near saturation in an open queue, even a small difference in service times can amount to a large relative difference in mean response times.

4.5 Conclusion

We have presented an approach to the performance analysis of optical packet switching bus-based networks employing the OU-CSMA/CA protocol and supporting variable length packets. For modeling purposes, we approximately view the bus as a multiple-priority M/G/1 queuing system with preemptive-repeat-identical (PRI) service discipline. We have proposed an approximate solution for this model, in which we apply a recurrent level-by-level analysis. Each level corresponds to a single bus node, and the bandwidth usage by upstream nodes is represented through server disappearance and reappearance rates. To model the PRI discipline, we use different service time distributions at consecutive transmission attempts. The solution to each level is based on conditional probabilities and a fixed point iteration method, which tends to require only a small number of iterations. As a result, we are able to compute not only the mean response time but also the steady-state queue length distribution at each level.

We have used our model to study the expected response time at the nodes of such a bus-based network for several packet length mixes, as well as for several patterns of offered traffic. Our results indicate that a uniform or uniformly decreasing traffic pattern appears more taxing on the network in terms of mean response time at lower priority nodes, while a pattern where the middle node dominates the network traffic seems to fare significantly better. Additionally, for higher traffic levels, the network performance at lower priority nodes is sensitive to the form of the service time distribution as represented by moments higher than the first two.

Comparisons with network simulation results indicate that our model correctly captures the performance characteristics of an OPS unslotted bus (i.e., unfairness property and bandwidth fragmentation phenomenon causing low bandwidth usage and low performance at downstream nodes). In addition, our model is able to provide the shape of the steady-state distribution of the number of packets at each node that closely tracks simulation results, even for the lowest priority node. Compared with other models proposed in the literature such as the C-H model [13], our model in most cases provides better results (i.e., closer to simulation results) than those obtained with the C-H model.

Occasionally, the results of our model may deviate from simulation results. This appears most likely close to node saturation when the service time distribution is highly variable. We have identified approximation errors, as well as sensitivity to higher moments of the service time distribution as possible causes for the observed differences.

In our model, we assume that packets arrivals come from a Poisson source, each node has unlimited buffer space, and we match only the first two moments of the service time distribution. Future work includes an improved matching of the distribution of packet lengths, as well as a possible extension of our approach to different packet arrival patterns and finite buffer sizes.

Appendix 1: Recurrent Solution for Node 1

Balance Equations Derivation

With the service time distribution represented by the Coxian distribution of Fig. 4.2, the conditional rate of packet transmissions given the number of packets at node 1 can be expressed as

$$u(n) = p(1|n)\mu_1(1 - p_2) - p(k|n)\mu_k, \tag{4.3}$$

where k is the total number of exponential stages in the Coxian distribution, $p(l|n)$ is the conditional probability for the current service stage given n, for $l = 1, \ldots, k$ and $n = 1, \ldots$

It is not difficult to show that the steady-state probability distribution for n can be expressed as (where G is a normalizing constant):

$$p(n) = \frac{1}{G}\prod_{i=1}^{n} \lambda/u(i). \tag{4.4}$$

We show the proof of (4.4) in Appendix 3. Using (4.4) and the fact that $p(n, l) = p(n)p(l|n)$ in the balance equations for $p(n, l)$, we obtain the following equations for $p(l|n)$ when $n = 1$.

$$p(2|n)[\lambda - \mu_2] = p(1|n)\mu_1 p_2,$$

$$p(l|n)[\lambda - \mu_1] = p(l - 1|n)\mu_{l-1},\ l = 3, \ldots, k.$$

Hence, for $n = 1$ we have

$$p(2|n) = p(1|n)\mu_1 p_2/[\lambda - \mu_2],$$

$$p(l|n) = p(l - 1|n)\mu_{l-1}/[\lambda - \mu_l],\ \text{for } l = 3, \ldots, k, \tag{4.5}$$

where $p(1|n)$ is readily determined from the normalizing condition $\sum_{l=1}^{k} p(l|n) = 1$ that must hold for all values of $n = 1, \ldots$ For values of $n > 1$, we have

$$\begin{aligned} p(2|n)[\lambda - \mu_2] &= p(1|n)\mu_1 p_2 - p(2|n - 1)u(n), \\ p(1|n)[\lambda - \mu_1] &= p(l - 1|n)\mu_1 - p(l|n - 1)u(n), \end{aligned} \tag{4.6}$$

for $l = 3, \ldots, k$.

Fixed-Point Iteration Method for Solution of the Balance Equations

Starting with the known solution for $n = 1$, together with the normalizing condition $\sum_{l=1}^{k} p(l|n) = 1$, we can solve (4.6) as recurrence for increasing values of $n = 2, \ldots$. In theory, since there is no upper limit to the values of n, there would be an infinite number of equations to solve. In practice, for the service time distributions considered, the conditional probabilities $p(l|n)$ and the conditional rate of packet transmissions $u(n)$ quickly reach limiting values as n increases. In the examples explored, convergence was typically achieved for values of n on the order of a few tens. Clearly, knowing $u(n)$ we can use (4.4) to compute $p(n)$ and any performance measures derived from it.

Appendix 2: Recurrent Solution for Nodes *i > 1*

Balance Equations Derivation

We describe the state of a downstream node by the triple (n, j, l) where n is the current number of client packets at the node, j is the transmission attempt, and l is equal to the current number of the service stage at this attempt $(1, \ldots, k_j)$, or equal to 0 if the server (bandwidth) is unavailable. Let α be the rate with which the server becomes unavailable, and β be the rate with which the server returns from unavailability. The balance equations for $p(n, j, l)$ are readily derived.

$$p(n, j = 1, l = 0)[\lambda + \beta] = p(n - 1, j = 1, l = 0)\lambda;$$

$$p(n, j, l = 0)[\lambda + \beta] = \sum_{l=1}^{k_j - 1}(n, j - 1, l)\alpha + p(n - 1, j, 1 = 0)\lambda,$$
$$j = 2, \ldots, j^* - 1;$$

$$p(n, j, l = 1)[\lambda + \mu_1^j + \alpha] = p(n, j - 1, l = 0)\beta + p(n - 1, j, l = 1)\lambda;$$

$$p(n, j, = 1, l)[\lambda + \mu_l^j + \alpha] = p(n, j = 1, l - 1)\mu_{l-1}^j \bar{q}_{l-1}^j + p(n - 1, j = 1, l)\lambda, 1 > 1;$$

$$p(n, j^*, l = 0)[\lambda + \beta] \sum_{l=1}^{k_{j^*-1}} p(n, j^* - 1, l)\alpha + \sum_{l=1}^{k_{j^*}} p(n, j^*, l)\alpha + p(n - 1, j^*, l = 0)\lambda;$$

$$p(n, j^*, l = 1)[\lambda + \mu_1^{j^*} + \alpha] = p(n, j^* - 1, l = 0)\beta + p(n, j^*, l = 0)\beta + p(n - 1, j^*, l = 1)\lambda;$$

$$p(n, j^*, l)[\lambda + \mu_l^{j^*} + \alpha] = p(n, j^*, l - 1)\mu_{l-1}^{j^*} \bar{q}_{l-1}^{j^*} + p(n - 1, j^*, l)\lambda, \ l > 1.$$

In the above equations, we denote by μ_l^j the parameter of stage l of the Coxian representation of the service time at transmission attempt j, and by $\bar{q}_l^j$ the corresponding probability that stage l will be followed by another service stage, where $l = 1, \ldots, k_j$. For the "catch all" value j^*, we use "average" service parameters values set up so as to maintain the proper average number of transmission attempts.

For $n = 1$, the first balance equation becomes

$$p(n = 1, j = 1, l = 0)[\lambda - \beta] = p(n = 0, l = 0)\lambda,$$

where $p(n = 0, l = 0)$ is the probability that there are no packets at the node and the server is unavailable. In the remaining equations for $n = 1$, the term involving $n - 1$ is simply absent.

The conditional rate of packet transmissions given the number of packets at the node can be expressed as:

$$u(n) = \sum_{j \geq 1} \sum_{l=1}^{k_j} p(j, l|n)\mu_l^j(1 - \bar{q}_l^j). \tag{4.7}$$

As for node 1, the steady-state probability distribution for n can be expressed as (G is a normalizing constant):

$$p(n) = \frac{1}{G} \prod_{i=1}^{n} \lambda/u(i). \tag{4.8}$$

Using (4.8) together with the fact that $p(n, j, l) = p(n)p(j, l|n)$, we transform the balance equations for $p(n, j, l)$ into equations for the conditional probability $p(j, l|n)$:

$$p(j = 1, l = 0|n)[\lambda + \beta] = p(j = 1, l = 0|n - 1)u(n);$$

$$p(j, l = 0|n)[\lambda + \beta] = \sum_{l=1}^{k_j - 1} p(j - 1, l|n)\alpha + p(j, l = 0|n - 1)u(n),$$

$$j = 2, \ldots j^* - 1;$$

$$p(j, l = 1|n)[\lambda + \mu_1^j + \alpha] = p(j - 1, l = 0|n)\beta + p(j, l = 1|n - 1)u(n);$$

$$p(n, j, = 1, l)[\lambda + \mu_l^j + \alpha] = p(n, j = 1, l - 1)\mu_{l-1}^j \bar{q}_{l-1}^j + p(n - 1, j = 1, l)\lambda, \ l > 1;$$

$$p(j^*, l = 0|n)[\lambda + \beta] = \sum_{l=1}^{k_{j^*-1}} p(j^* - 1, l|n)\alpha + \sum_{l=1}^{k_{j^*}} p(j^*, l|n)\alpha + p(j^*, l = 0|n - 1)u(n);$$

$$p(j^*, l = 1|n)[\lambda + \mu_1^{j^*} + \alpha] = p(j^* - 1, l = 0|n)\beta + p(j^*, l = 0|n)\beta + p(j^*, l = 1|n - 1)u(n);$$

$$p(j^*, l|n)[\lambda + \mu_l^{j^*} + \alpha] = p(j^*, l - 1|n)\mu_{l-1}^{j^*} \bar{q}_{l-1}^{j^*} + p(j^*, l|n - 1)u(n), \ l > 1.$$

For $n = 1$, the first equation becomes:

$$p(j = 1, l = 0|n = 1)[\lambda - \beta] = p(1 = 0|0)u(1),$$

where $p(l = 0|0)$ is the conditional probability that the server is unavailable given that there are no packets to be transmitted at the node. Note that we must have

$$\sum_{j\geq 1}\sum_{l=0}^{k_j} p(j, l|n) = 1, \tag{4.9}$$

for all values of $n \geq 1$.

For $n = 0$, the only possible states correspond to the availability of the server, and we easily get for the probability that the server is unavailable

$$p(l = 0|0) = \alpha/[\alpha - \beta - \lambda] \tag{4.10}$$

Fixed-Point Iteration Method for Solution of the Balance Equations

Armed with the known value of $p(l = 0|0)$, we consider the above equations for $p(j, l|n)$ for increasing values of $n = 1, 2, \ldots$ For each n, we express all $p(j, l|n)$ in terms of $p(j = 0, l = 1|n)$ and $u(n)$. Then, these two unknowns are determined from the normalizing condition (4.9) and from the definition of $u(n)$ in (4.7). As was the case for node 1, although in theory there is an infinite number of values of n (and hence an infinite number of equation sets to solve), in practice, the conditional probabilities $p(j, l|n)$ and the conditional rate of packet transmissions $u(n)$ quickly reach limiting values as n increases.

Knowing $u(n)$ we readily obtain $p(n)$ from (4.8). The steady-state probability distribution for the number of transmission attempts at node i can then be expressed as

$$r(j) = \frac{1}{H}\sum_{n} p(n)\sum_{l} p(j, l|n)\mu_1^j(1 - \bar{q}_l^j),$$

where H is a normalizing constant.

The expected number of packets at the node is given by $\sum_{n=1}^{\infty} np(n)$, and the expected number of interruptions per transmission (due to void too small, in actual network) at node other than the first one is approximately $\sum_{n>0} p(n)\sum_{j}\sum_{l>0} p(j, l)\alpha/\lambda$.

Appendix 3: Proof of Product-Form for Coxian Distribution

Consider the general Coxian system in Fig. 4.9 (General Coxian distribution). We describe the equilibrium behavior of this system by the joint steady state probability $p(n, l)$, where n ($n \geq 1$) is the number of packets in the system and $l = 1 \ldots k$

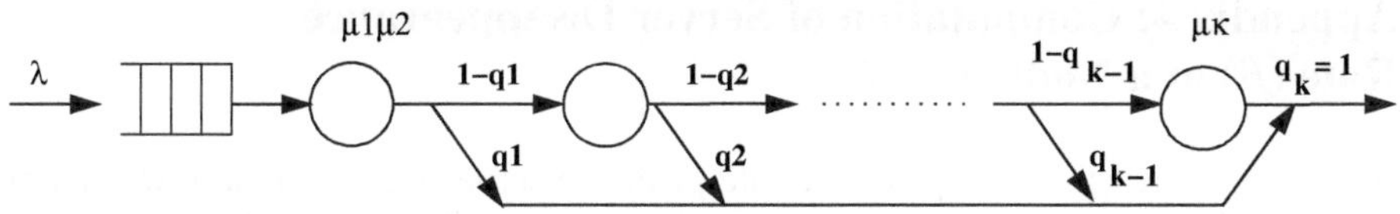

Fig. 4.9 General Coxian distribution

refers to the stage of service of the Coxian distribution. The balance equations for $p(n, l)$ are readily derived as follows:

$$p(n, 1)[\lambda + \mu_1] = p(n-1, 1)\lambda + \sum_{l=1}^{k} p(n+1, l)\mu_1 q_1, \tag{4.11}$$

$$p(n, l)[\lambda - \mu_l] = p(n-1, l)\lambda - p(n, l-1)\mu_{l-1}(1 - q_{l-1}), \text{ for all } l = 2,3, \ldots, k. \tag{4.12}$$

Using the definition of conditional probability $p(n, l) = p(n)p(l|n)$ in (4.11) and (4.12), and then summing equations (4.11) and (4.12) for all $l = 1, \ldots, k$ while taking into account the normalizing condition $\sum_{l=1}^{k} p(l|n) = 1$ for all $n = 1, 2, \ldots$, we obtain:

$$p(n)[\lambda + \sum_{l=1}^{k} p(l|n)\mu_l] = p(n-1)\lambda + p(n+1)\sum_{l=1}^{k} p(l|n+1)\mu_l q_l + p(n)\sum_{l=1}^{k-1} p(l|n)\mu_l(1 - q_l).$$

Simplifying the above equation, we have:

$$p(n)[\lambda + \sum_{l=1}^{k} p(l|n)\mu_l q_l] = p(n-1)\lambda + p(n+1)\sum_{l=1}^{k} p(l|n+1)\mu_l q_l. \tag{4.13}$$

Set $u(n) = \sum_{l=1}^{k} p(l|n)\mu_l q_l$ for $n = 1, 2, \ldots$, (4.13) becomes:

$$p(n)[\lambda - u(n)] = p(n-1)\lambda - p(n-1)u(n-1). \tag{4.14}$$

We observe that equation (4.14) is identical to the steady state balance equation of an M/M/1 queuing system with arrival rate λ and service rate *u(n)*. Thus, we are able to readily obtain the product-form for the general Coxian system: $p(n) = \frac{1}{G}\prod_{i=1}^{n} \lambda/u(i)$, where G is a normalizing constant: $G = 1 + \sum_{n\geq 1}\prod_{i=1}^{n} \lambda/u(i)$.

Note that this result is obviously applicable for the k-stage Coxian distribution shown in Fig. 4.2 since it is a special case of the general Coxian distribution.

Appendix 4: Computation of Server Disappearance Rate (β) at a Nodes $i > 1$

As stated earlier, the disappearance rate β_i of the server viewed by a downstream node i is computed approximately in our solution. For the node 2, this rate can be expressed in terms of conditional probabilities as follows:

$$\beta_2 = p(n_1 = 1|n_2, U_2) \sum_{l=1}^{k} \mu_l q_l p(l|n_1 = 1, n_2, U_2), \tag{4.15}$$

where n_i is the current number of packets at node i, U_i (A_i respectively) indicates the server at node i is unavailable (available respectively), and k, l, μ_l, q_l are the parameters of the service time distribution as shown in Fig. 4.9. Using the fact that $p(U_2) = p(n_1 \geq 1) = 1 - p(n_1 = 0)$ and $u(n) = \sum_{l=1}^{k} p(l|n)\mu_l q_l$, equation (4.15) can be approximately computed based on known parameters (i.e. parameters computed from the solution of the preceding node) as follows:

$$\beta_2 \approx p(n_1 = 1|U_2) \sum_{l=1}^{k} \mu_l q_l p(l|n_1 = 1) \approx \frac{p(n_1 = 1)u(n_1 = 1)}{1 - p(n_1 = 0).} \tag{4.16}$$

Regarding nodes $i > 2$, the exact expression of β_{i+1} is:

$$\begin{aligned}\beta_{i+1} = {} & p(n_i = 1|n_{i+1}, U_{i+1}) \sum_{l=1}^{k} \mu_l p_l p(l|n_i = 1, n_{i+1}, U_{i+1}) \\ & + p(n_i = 0|n_{i+1}, U_{i+1}) p(U_i|n_i = 0, n_{i+1}, U_{i+1}) \beta_i(n_i).\end{aligned} \tag{4.17}$$

We notice in the right hand side of this expression that in addition to the first term similar to the one in (4.15), we introduce the second term that represents the case where the server at the preceding node i has been occupied and its queue was empty. This case does not exist for node 2 because the node 1 (which is the preceding node of node 2) always finds server available. Similar to the case for node 2, we can compute the value of β_{i+1} approximately:

$$\begin{aligned}\beta_{i+1} & \approx p(n_i = 1|U_{i+1}) \sum_{l=1}^{k} \mu_l p_l p(l|n_i = 1) + p(n_i = 0|U_{i+1}) p(U_i|n_i = 0)\beta_i \\ & \approx \frac{p(n_i = 1)u(n_i = 1) + p(n_i = 0)p(U_i|n_i = 0)\beta_i}{p(U_{i+1})}.\end{aligned} \tag{4.18}$$

Here, the probability that the server is unavailable at node i + 1 (i > 1) is computed as follows:

$$p(U_{i+1}) = 1 - p(n_i = 0, A_i) = 1 - p(A_i | n_i = 0)p(n_i = 0).$$

Using the fact that $p(A_i|n_i = 0) = 1 - p(U_i|n_i = 0)$ with $p(U_i|n_i = 0)$ computed from (4.10), we are able to determine the value of β_{i+1} from (4.18):

$$\beta_{i+1} \approx \frac{p(n_i = 1)u(n_i = 1) - p(n_i = 0)p(U_i|n_i = 0)\beta_i}{1 - \{1 - p(U_i|n_i = 0)\}\, p(n_i = 0)}. \tag{4.19}$$

References

1. Nguyen, V.H., Ben Mamoun, M., Atmaca, T., et al. Performance evaluation of Circuit Emulation Service in a metropolitan optical ring architecture. *In Proceedings of the Telecommunications and Networking – ICT 2004*. LNCS 3124 vol., pp. 1173–1182, Fortaleza – Brazil, August 2004.
2. Le Sauze, N., Dotaro, E., Dupas, A., et al. DBORN: A Shared WDM Ethernet Bus Architecture for Optical Packet Metropolitan Network. *In Proceedings of Photonic in Switching Conference*. July 2002.
3. White I.M., A New Architecture and Technologies for High-Capacity Next Generation Metropolitan Networks, *Ph.D. Dissertation*, Department of Electrical Engineering of Stanford University, CA, August 2002.
4. LaMaire, R.O. An M/G/1 Vacation Model of an FDDI Station. *IEEE Journal on Selected Areas in Communications*. Vol. 9, Issue 2, Feb. 1991, pp.257–264.
5. Rubin, I., and WU, J.C.H. Analysis of an M/G/1/N Queue with Vacations and its Application to FDDI Asynchronous Timed-Token Service Systems. *In Global Telecommunications Conference (GLOBECOM '92)*. Communication for Global Users, vol. 3, pp. 1630–1634.
6. Ghafir, H.M., and Silio, C.B. Performance Analysis of a Multiple-Access Ring Network. *IEEE Transactions on Communications*. Vol. 41, Issue 10, Oct. 1993, pp. 1494–1506.
7. Mukherjee, B., and Banerjee, S. Alternative Strategies for Improving the Fairness in and an Analytical Model of the DQDB Network. *IEEE Transactions on Computers*. Vol. 42, Issue 2, Feb. 1993, pp. 151–167.
8. Starvrakakis, I., and Landry, R. Delay Analysis of the DQDB MAN Based on a Simple Model. *IEEE International Conference on Communications*. Vol. 1, June 1992, pp. 154–158.
9. Mukherjee, B., and Meditch, J. The p_i-Persistent Protocol for Unidirectional Broadcast Bus Networks. *IEEE Transactions on Communications*. Vol. 36, Issue 12, Dec. 1988, pp. 1277–1286.
10. Miller, G.J., and Paterakis, M. A Dynamic Bandwidth-Allocation-Based Priority Mechanism for the p_i-Persistent Protocol for MAN's. *IEEE Journal on Selected Areas in Communications*. Vol. 11, Issue 8, October 1993.
11. Takine, T. Y., Takahashi, and Hasegawa, T. An Approximate Analysis of a Buffered CSMA/CD. *IEEE Transactions on Communications*. Vol. 36, Issue 8, Aug. 1988, pp. 932–941.
12. Matsumoto, Y., Takahashi, Y., and Hasegawa, T. The Effects of Packet Size Distributions on Output and Delay Processes of CSMA/CD. *IEEE Transactions on Communications*. Vol. 38, Issue 2, Feb. 1990, pp. 199–214.
13. Castel, H., and Hébuterne, G. Performance Analysis of an Optical MAN Ring for Variable Length Packet Traffic. *In Proceedings of Photonic in Switching Conference*. 2003.
14. Castel, H., Chaitou, M., and Hébuterne, G. Preemptive Priority Queues for the Performance Evaluation of an Optical MAN Ring. *In Proceedings of Performance Modeling and Evaluation of Heterogeneous Networks (Het-Net'05)*. 2005.

15. Hu, G., Gauger, C.M., and Junghans, S. Performance Analysis of the CSMA/CA MAC Protocol in the DBORN Optical MAN Network Architecture. *In the Proceedings of the 19th International Teletraffic Congress (ITC 19)*. 2005.
16. Bouabdallah, N., Beylot, A.L., Dotaro, E., and Pujolle, G. Resolving the Fairness Issue in Bus-Based Optical Access Networks. *IEEE Journal on Selected Areas in Communications*. Vol. 23, Issue 8. August 2005.
17. Jaiswal. *Priority Queues*. New York: Academic 1966.
18. Takagi, H. *Queuing Analysis*. Vol. 1, pp. 365–373. North-Holland, 1991.
19. Cox, D. R., and Smith, W. L. *Queues*. John Wiley, New York, 1961.
20. Trivedi, K.S. *Probability and Statistics with Reliability, Queuing, and Computer Science Applications*. Prentice-Hall, Inc., Englewood Cliffs, NJ 07632.
21. Brandwajn, A. Equivalence and Decomposition in Queueing Systems – A Unified Approach. *Performance Evaluation*. Vol. 5, pp. 175–186, 1985.
22. Brandwajn, A., and Wang, H. A Conditional Probability Approach to M/G/1-like Queues. Submitted for publication, available as a technical report, 2006.
23. Allen, A.O. *Probability, Statistics, and Queuing Theory*. Academic Press, 2nd edition, 1990.
24. Network Simulator. Available at HTTP: http://www.isi.edu/nsnam/ns/.
25. McDougall, M.H. *Simulating Computer Systems: Techniques and Tools*. The MIT Press, 1987.
26. CAIDA, "IP packet length distribution", [Online document] 2000, Available at HTTP: http://www.caida.org/analysis/AIX/plen_hist.

Chapter 5
A Novel Early DBA Mechanism with Prediction-Based Fair Excessive Bandwidth Allocation Scheme in EPON

I-Shyan Hwang, Zen-Der Shyu, Liang-Yu Ke and Chun-Che Chang

Abstract In this paper, we propose a novel *Early dynamic bandwidth allocation* (E-DBA) mechanism incorporated with a prediction-based fair excessive bandwidth allocation (PFEBA) scheme in Ethernet Passive Optical Networks (EPONs). The E-DBA mechanism can reduce the *idle period* in the traditional DBA mechanism. On the other hand, the PFEBA scheme can provide more accurate prediction to ensure the fairness of each ONU and improve the overall system performance. The proposed model makes prediction for different traffic classes according to the variation in traffic for each ONU in the EPON. The PFEBA scheme includes the *unstable degree list*, predictions made using *linear estimation credit* and the fair excessive bandwidth allocation scheme. The simulation results show that the proposed E-DBA mechanism with PFEBA scheme can improve the system performance of well-known DBA algorithms in terms of wasted bandwidth, wasted bandwidth improved percentage, downlink data available bandwidth, throughput, average end-to-end delay and average queue length, especially under high traffic load.

Keywords E-DBA · PFEBA · EPON · Fairness · System performance

5.1 Introduction

The high-capacity feature of multi-access optical fiber networks compared with other access network technologies is the main motivation behind advances in optical technology. The passive optical network (PON) is regarded as a promising solution for the next-generation broadband access network because it is simple, cost-effective and scalable. The PON architecture, shown in Fig. 5.1, comprises a centralized optical line terminal (OLT), splitters, and connects a group of associated optical network units (ONUs) over point-to-multipoint topologies to deliver broadband packet and reduce cost relative to maintenance and power.

I-S. Hwang (✉)
Department of Computer Engineering and Science, Yuan-Ze University, Chung-Li, Taiwan, 32026

M. Ma. (ed.), *Current Research Progress of Optical Networks*,
DOI 10.1007/978-1-4020-9889-5_5,

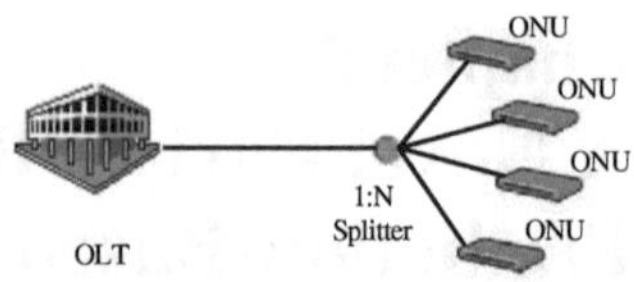

Fig. 5.1 Tree-based PON topology

Two standard organizations, *ITU-T* (International Telecommunications Union Standardization Sector) and *IEEE* (Institute of Electrical and Electronics Engineers), have led the discussion of PON specifications. In *ITU-T*, a series of ATM-based Broadband PON (i.e., ATM-PON, BPON and GPON), have been recommended [1]. On the other hand, Ethernet PON (EPON) has been discussed in *IEEE 802.3ah* as one of the extensions of Gigabit-Ethernet [2]. The main difference between EPON and ATM-based Broadband PON is that EPON carries all data encapsulated according to the IEEE 802.3 Ethernet frame format between the OLT and ONUs. Low maintenance cost, compatibility compared with existing networks, and minimal protocol overhead make EPON a promising solution for the next-generation broadband access networks. Moreover, the EPON is the primary type of PON technology that reduces fiber deployment dramatically while preserving the merits of Ethernet networks.

The EPON provides bi-directional transmissions, one is downstream transmission from OLT to ONUs; the other is upstream transmission from ONUs to OLT in sequence. In the downstream transmission of the EPON, all the control messages and the data packets are carried and broadcasted from the OLT to each ONU through the entire bandwidth of one wavelength as a downstream channel. Each ONU discards or accepts the incoming Ethernet frames depending on the packet header addressing. In the upstream direction, all ONUs share the common transmission channel towards the OLT, only a single ONU may transmit data in its time slots to avoid data collision. Hence, a robust mechanism is needed for allocating time slots and upstream bandwidth for each ONU to transmit data. In EPONs, the mechanism is called *multi-point control protocol* (MPCP) involving both GATE messages and REPORT messages. The OLT allocates upstream bandwidth to each ONU by sending GATE messages with the form of a 64-byte MAC control frames. GATE messages contain a timestamp and granted time slots which represent the periods that ONU can transmit data. ONUs may send REPORT messages about the queue state of each ONU to the OLT, so that the OLT can allocate the upstream bandwidth and time slots to each ONU accordingly.

In other words, the EPON can be regarded as a multipoint-to-point network in the upstream direction where multiple ONUs share the same transmission channel and transmit data to the OLT. Hence, an important issue for emerging research is how to access the shared bandwidth allocation by medium access control (MAC) protocols to prevent collision and share the channel capacity fairly among ONUs to provide better system performance. The bandwidth allocation schemes can be divided into two categories: *fixed bandwidth allocation* (FBA) and *dynamic bandwidth allocation* (DBA). In the FBA scheme, each ONU is pre-assigned a fixed time slot (TDMA scheme) to send its backlogged packets at the full capacity of the link. This will lead to inefficient bandwidth utilization when the traffic of ONUs is light. Contrast to the FBA, the DBA assigns the bandwidth according to the bandwidth requested by each ONU. Therefore, the DBA scheme can provide more efficient bandwidth allocation for each ONU to share the network resources and offer better Quality-of-Service (QoS) for end-users than the FBA scheme.

First, this chapter proposes an *Early DBA* (E-DBA) *mechanism* for reducing the idle period in the traditional DBA mechanism. Second, the E-DBA sorts the sequence of each ONU according to the variance in historical traffic required and arranges some REPORT messages of ONUs with the violent variance in traffic required to precise DBA time. Therefore, the OLT can get the fresh queue information to make more accurate prediction for the next cycle. Furthermore, the efficient and robust prediction-based fair excessive bandwidth allocation (PFEBA) scheme is incorporated to consider the fairness of excessive bandwidth allocation among ONUs in the EPON to improve system performance. For the concept of fairness, not only the heavily-loaded ONUs, but also the lightly-loaded ONUs are considered in the proposed scheme.

The proposed model makes prediction for different traffic classes according to the variation in traffic for each ONU in the EPON. In this paper, we discuss an EPON architecture that supports differentiated services and classify services into three priorities as defined in IETF RFC 2475 [3], namely the best effort (BE), the assured forwarding (AF), and expedited forwarding (EF). While EF services require bounded end-to-end delay and jitter specifications, AF is intended for services that are not delay-sensitive but require bandwidth guarantees. Finally, BE applications are not delay-sensitive and do not require any jitter specifications. Simulation results show that the proposed E-DBA mechanism with PFEBA scheme outperforms other existing well-known DBA algorithms in high traffic load.

The rest of this chapter is organized as follows. Section 5.2 describes the related work. Section 5.3 proposes an E-DBA mechanism which incorporates the PFEBA scheme for dealing with fairness involving prediction. Section 5.4 shows the simulation results in terms of average packet delay, average queue length, wasted bandwidth, downlink available data bandwidth and throughput. Finally, Section 5.5 draws conclusions and offers suggestions.

5.2 Related Work

Dynamic bandwidth allocation without prediction mechanism, such as limited bandwidth allocation (LBA), has been studied by Glen [4, 5]. In the LBA, the time-slot length of each ONU is upper bounded by the maximum time-slot length, B_{max}, which could be specified by service level agreement (SLA). When the reported queue size is less than B_{max}, the OLT grants the requested bandwidth; otherwise, B_{max} is granted. The drawback of LBA is that no more bandwidth is granted to ONUs already assigned a guaranteed bandwidth B_{max}, regardless whether other

ONUs have excessive bandwidth. The LBA has poor utilization for the upstream bandwidth and restricts aggressive competition for the upstream bandwidth, especially under non-uniform traffic [6]. Because the amount of time slots requested is different for each ONU, the authors [7] classify the requests of different ONUs into lightly-loaded and heavily- loaded according to the amount of traffic requests. In each transmission cycle, some ONUs may have less traffic requests to transmit and thus need smaller bandwidth than the minimum guaranteed bandwidth (thus called lightly-loaded ONUs), while other ONUs may have more traffic requests to transmit and need larger bandwidth (thus called heavily-loaded ONUs).

It is observed that there might be some lightly-loaded ONUs with bandwidth requirement less than the limit in LBA. When the guaranteed bandwidth is more than the demand for lightly-loaded ONUs, there exists excessive bandwidth while other heavily-loaded ONUs are allocated insufficiently. The sum of the underexploited bandwidth of lightly-loaded ONUs is called *excessive bandwidth* B_{excess}. As an extension of the LBA, the excessive bandwidth reallocation (EBR) [7, 8] redistributes the available bandwidth to heavily-loaded ONUs in proportion to each request and results in better performance in terms of packet delay. Then, the heavily-loaded ONU_i obtains an additional bandwidth $B_{add,i}$ from B_{excess} as follows

$$B_{add,i} = \frac{B_{excess} \times R_i}{\sum_{h \in H} R_h}, \tag{5.1}$$

where H is the set of heavily-loaded ONUs, h is a heavily-loaded ONU in H and R_i is the bandwidth requested by ONU_i.

Unfortunately, the drawbacks of EBR are unfairness and excessive bandwidth allocated to ONUs than that requested [9], which is redefined as *redundant bandwidth problem* in our research. Xiaofeng et al. [10] proposed another DBA scheme that maintains well the fairness mechanism of excessive bandwidth allocation operation for heavily-loaded ONUs. This operation reallocates all excessive bandwidth from lightly-loaded ONUs to heavily-loaded ones, but ignores the fairness of lightly-loaded ONUs. The reason is that the request R_i by the lightly-loaded ONU_i in the EBR scheme does not consider the possible packets arriving during the waiting time before data transmission, which is shown in Fig. 5.2. Those packets cannot be transmitted in the next cycle because excessive bandwidth has been reallocated to the heavily-loaded ONUs. This will result in longer packet delay and is unfair for the lightly-loaded ONUs. Therefore, one feasible method involving prediction is to grant more bandwidth than that requested by lightly-loaded ONUs to improve overall fairness.

The prediction-based schemes are studied in order to decrease packet delay and allocate more granted time slots efficiently. In the predictive schemes, the measured and predicted aggregated traffic are employed to update the allocated bandwidth to meet the QoS requirements. Accurate traffic predictor is required to avoid over- or under-estimation, which will result in longer packet delay to degrade the network performance [4, 11–15].

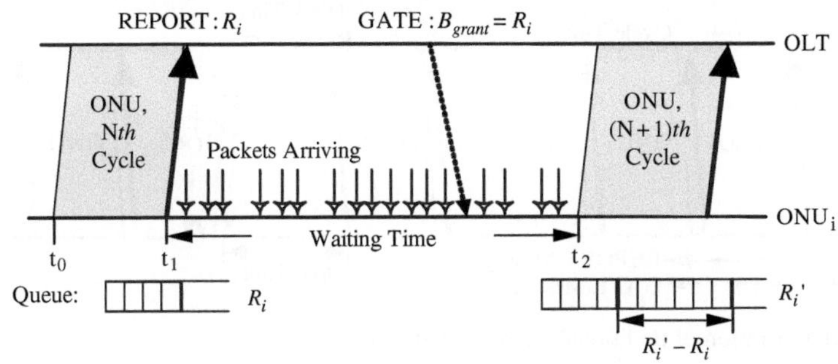

Fig. 5.2 Queue state between waiting time

The credit-based bandwidth allocation (CBA) takes some precedent transmitted frames into consideration [4], and it adds a credit into the requirement of each ONU when the OLT allocates the upstream bandwidth. The bandwidth granted to each ONU is illustrated as $B_{grant} = B_{queue} + C$, where B_{grant} is the bandwidth granted to an ONU, B_{queue} denote the queue of frames in the buffer, and C is the credit which could be a constant or linear credit. The CBA grants the requested window plus a credit that is proportional to the requested window. Some packets do not have to wait for the next grant to arrive; they can be transmitted with the current grant and the average packet delay can be reduced.

The DBA with multiple services (DBAM) [11] is a prediction-based LBA that executes prediction according to the *linear estimation credit*. The linear estimation credit of each ONU$_i$ is obtained according to the ratio of the ONU$_i$ waiting time (i.e. $t_2 - t_1$) over the time length of current interval (i.e. $t_2 - t_0$), which is shown in Fig. 5.2. The OLT allocates the time slots for multiple services among ONUs according to each bandwidth required and the SLA limits. In fact, packet delay will be improved by the DBAM in uniform traffic flows. However, the performance is deteriorated in non-uniform traffic flows because the prediction model suffer serious inaccuracy in the DBAM for some ONUs with high variations in traffic.

5.3 PFEBA Scheme in Early DBA Mechanism

In this section, we first address the operation of *Early DBA* (E-DBA) *mechanism*, followed by the prediction-based fair excessive bandwidth allocation (PFEBA) scheme embedded in the E-DBA. The proposed E-DBA mechanism, shown in Fig. 5.3(b), can improve the packet delay time by early execution of the DBA mechanism to reduce the idle period. In the E-DBA mechanism, the bandwidth is allocated to each ONU according to the decreasing order of unstable degree list. The reason is that the prediction will be more accurate if more information is obtained during waiting time for unstable traffic ONUs. Furthermore, the PFEBA scheme is incorporated in the E-DBA mechanism to improve the fairness for all ONUs and system performance. There are three steps in the PFEBA scheme. First, the

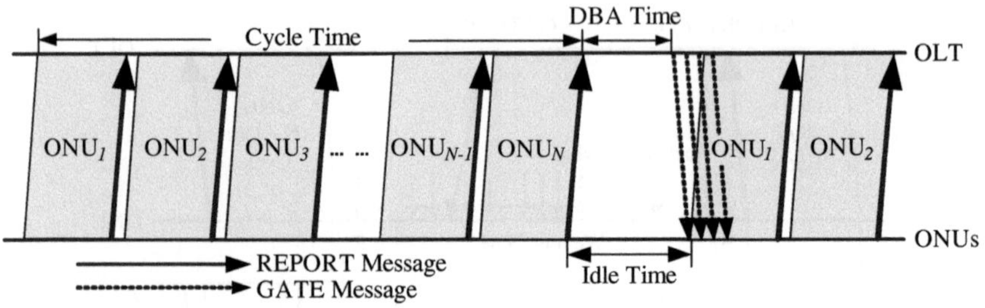

Fig. 5.3 (a) Operation of traditional DBA mechanism

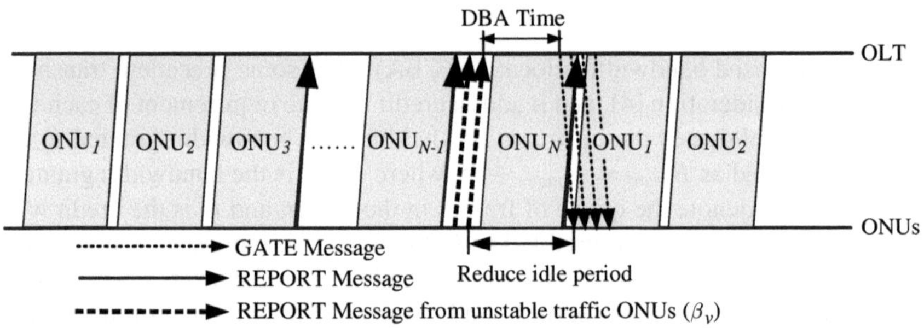

Fig. 5.3 (b) Operation with the proposed *Early DBA mechanism* (E-DBA)

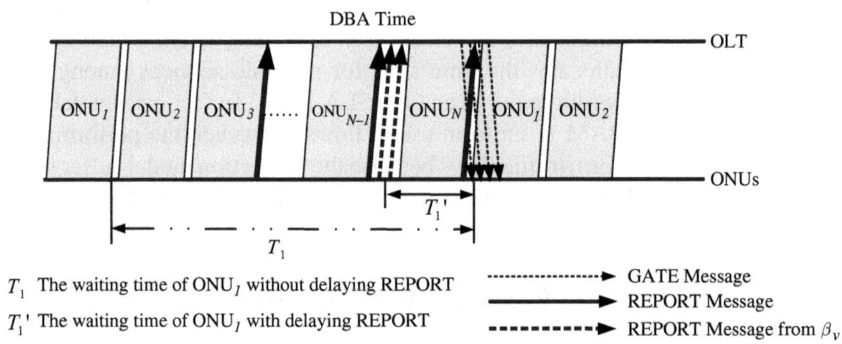

Fig. 5.3 (c) Improve prediction accuracy by shortening the waiting time of unstable ONUs

unstable degree list is calculated using variance in historical traffic required for each ONU. Prediction is then made according to the inference results obtained in the first step to improve prediction accuracy. Finally, the fair excessive bandwidth allocation scheme is implemented to improve bandwidth utilization and reduce packet delay time. Table 5.1 summarizes the definition of parameters. The terms time slots and bandwidth are used interchangeably.

Table 5.1 The Definition of Parameters

Parameters	Definition
N_H	Number of historical REPORT messages recorded
V_i	Variance of ONU_i
$\overline{V}$	Mean variance of all ONUs
β_V	Set of ONUs with higher variance in unstable degree list and $V_i > \overline{V}$
N_V	Number of ONUs in β_V
T_{cycle}	Maximum cycle time
N	Number of ONUs
$C_{capacity}$	Link capacity of OLT (bits/sec)
$B_{i,n}^c$	Requested BW of ONU_i in nth cycle, where $c \in \{EF, AF, BE\}$
$R_{i,n}^c$	Requested BW of ONU_i after prediction in nth cycle, where $c \in \{EF, AF, BE\}$
S_i^c	Guaranteed BW from SLA in ONU_i, where $c \in \{EF, AF, BE\}$
$G_{i,n+1}^c$	Granted upload BW of ONU_i in $(n+1)th$ cycle, where $c \in \{EF, AF, BE\}$

5.3.1 The Operation of Early DBA Mechanism

The traditional DBA scheme, shown in Fig. 5.3(a) piggybacks REPORT messages in data time slots and starts the bandwidth allocation sequence after collecting all REPORT messages. The idle period is the sum of computation time of DBA and round-trip time between OLT and each ONU [2]. Reducing the idle period can improve bandwidth utilization and system performance.

The *Early DBA* (E-DBA) *mechanism* comprises two operations. First, the OLT executes the DBA mechanism after the REPORT messages from β_V are received at the end of ONU_{N-1}, shown in Fig. 5.3(b), instead of ONU_N in the traditional DBA mechanism shown in Fig. 5.3(a). At the same time, the ONU_N can transmit data simultaneously. The operation reduces the idle period in the traditional DBA mechanism and gathers the fresh queue information for unstable traffic ONUs to make more accurate prediction in the next cycle. Second, the bandwidth for each ONU in the next cycle is allocated according to the traffic variation of all ONUs in decreasing order, and β_V is updated by assigning some unstable traffic ONUs with higher variations. This operation will alleviate variance by shortening waiting time before transmitting data for unstable traffic ONUs, shown in Fig. 5.3(c), to maintain better prediction accuracy.

The unstable degree list, prediction made using *linear estimation credit* and the fair excessive bandwidth allocation scheme involved in the proposed prediction-based fair excessive bandwidth allocation (PFEBA) scheme are described in Section 5.3.2.

5.3.2 PFEBA Scheme

5.3.2.1 Unstable Degree List

The PFEBA calculates the variance of each ONU using the historical traffic required, and the variance of each ONU is sorted in decreasing order according to the *unstable degree list*. The variance of ONU$_i$, V_i, can be expressed as

$$V_i = \frac{1}{N_H} \sum_{n \in historical\ cycle} (B_{i,n}^{Total} - \overline{B}_i)^2, \tag{5.2}$$

$$B_{i,n}^{Total} = B_{i,n}^{EF} + B_{i,n}^{AF} + B_{i,n}^{BE} \text{ and } \overline{B}_i = \frac{1}{N_H} \sum_{n=1}^{N_H} B_{i,n}^{Total}$$

where $B_{i,n}^{Total}$ is the sum of differential traffics of ONU$_i$ in the n*th* cycle, $\overline{B}_i$ is the mean of $B_{i,n}^{Total}$, and N_H is the number of historical REPORT messages piggybacked.

β_V denotes a set of ONUs with higher variance traffic required, say, in one-eighth of the unstable degree list and each variance is greater than the mean variance $\overline{V}$, where

$$\overline{V} = \frac{1}{N} \sum_{i=1}^{N} V_i.$$

After obtaining the unstable degree list, the bandwidth prediction of each ONU is described as follows. Unlike the algorithm that piggybacks all REPORT messages in the data time slot, the E-DBA mechanism shifts the REPORT messages of β_V between (N-1)*th* and N*th* ONU, shown in Fig. 5.3(c). The PFEBA needs the fresh queue information of unstable traffic ONUs to avoid prediction inaccuracy, which deteriorates system performance.

It is observed that guard time needed between the (N-1)*th* and the N*th* ONU increases with increase in number of REPORT messages of β_V, thus lengthening packet delay and deteriorating system performance. Our future work will be to determine the optimal number of REPORT messages needed to obtain the best system performance.

5.3.2.2 Prediction According to Unstable Degree List

After the sequence of all ONUs from the unstable degree list is uploaded, the PFEBA predicts the traffic bandwidth needed according to the unstable degree list. The predicted request, $R_{i,n+1}^c$, for different traffic classes of all ONUs is defined as follows:

$$R_{i,n+1}^c = (1+\alpha) B_{i,n}^c,\ c \in \{EF, AF, BE\}, \tag{5.3}$$

where $B_{i,n}^c$ is the requested bandwidth of ONU$_i$ in the n*th* cycle, for different traffic classes $c \in \{EF, AF, BE\}$, and α is the *linear estimation credit* modified from the DBAM [11], i.e.,

$$\alpha = \begin{cases} 0 & \textit{if } \text{ONU}_i \in \beta_V \\ 0.5 \times \frac{T_{i,n}^W}{T_{i,n}} & V_i > \overline{V} \textit{ and } \text{ONU}_i \notin \beta_V \\ \frac{T_{i,n}^W}{T_{i,n}} & \textit{otherwise} \end{cases}$$

If $ONU_i \in \beta_V$, when ONU_i reports the latest information, then α of each $ONU_i \in \beta_V$ is 0 to reduce prediction inaccuracy. If $V_i > \overline{V}$ and $ONU_i \notin \beta_V$, then α is $0.5 \times T_{i,n}^W / T_{i,n}$, where $T_{i,n}^W$ is waiting time of the ONU_i (i.e., $t_2 - t_1$) and $T_{i,n}$ is the time length of current interval (i.e. $t_2 - t_0$); otherwise, α is $T_{i,n}^W / T_{i,n}$.

The waiting time of ONU_i ($T_{i,n}^W$) in DBAM is a certain value, which is the sum of the transmission time slots of ONUs in the interval of ONU_i (i.e. $t_2 - t_1$), shown in Fig. 5.2. However, in the PFEBA, $T_{i,n}^W$ is undecided. The reason is that the transmission time slots of ONUs in the interval of ONU_i are not granted in this step. Therefore, $T_{i,n}^W$ will be redefined as the bandwidth requested by ONUs after prediction in the interval of ONU_i. In order to mitigate the drawback of predicting too much bandwidth, the time slots of each ONU in the interval of ONU_i are limited to the minimum guaranteed time slots as follows:

$$T_{i,n}^W = \sum_{\substack{k \in \text{ONUs} \\ \text{in the interval of ONU}i}} \min\left(R_k^{Total}, S_k\right),$$

where R_k^{Total} is the sum of differential traffics after being predicted, R_k^c, $c \in \{EF, AF, BE\}$ of ONU_k in the interval of ONU_i, S_k is the sum of S_k^c, the minimum guaranteed time slots for the *EF*, *AF* and *BE* traffic determined by service level agreement (SLA). R_k^{Total} and S_k can be expressed as

$$R_k^{Total} = \sum_c R_k^c \text{ and } S_k = \sum_c S_k^c,\ c \in \{EF, AF, BE\}.$$

5.3.2.3 Excessive Bandwidth Allocation

After having predicted the bandwidth needed for each ONU, the PFEBA then executes the EBR to assign uplink bandwidth to each ONU as illustrated in Fig. 5.4. The proposed PFEBA scheme can provide fairness for excessive bandwidth allocation according to the guaranteed bandwidth instead of requested bandwidth [7, 8], with no partiality and increase in bandwidth utilization. The operation of fair EBR in the proposed PFEBA is described as follows.

First, $R_{i,n}^{Total}$ of all ONUs is calculated and the available bandwidth, $B_{available}$, initialized is expressed as

$$B_{available} = C_{capacity} \times \left(T_{cycle} - Ng - N_V g\right) - N \times 512, \tag{5.4}$$

where $C_{capacity}$ is the OLT link capacity (bits/sec), T_{cycle} is the maximum cycle time, g is the guard time, N is the number of ONUs and N_V is the number of ONUs in β_V with control message length of 512 bits (64 bytes).

Then, the proposed PFEBA will select the ONU_i with the maximal residue bandwidth, i.e., $\max\left(S_i - R_{i,n}^{Total}\right)$, from unassigned ONUs. The bandwidth granted for ONU_i, $G_{i,n+1}^{Total}$, in the next cycle is given as follows:

$$G_{i,n+1}^{Total} = \min\left(B_{available} \times \frac{S_i}{\sum_{k \in unassigned} S_k},\ R_{i,n}^{Total}\right), \tag{5.5}$$

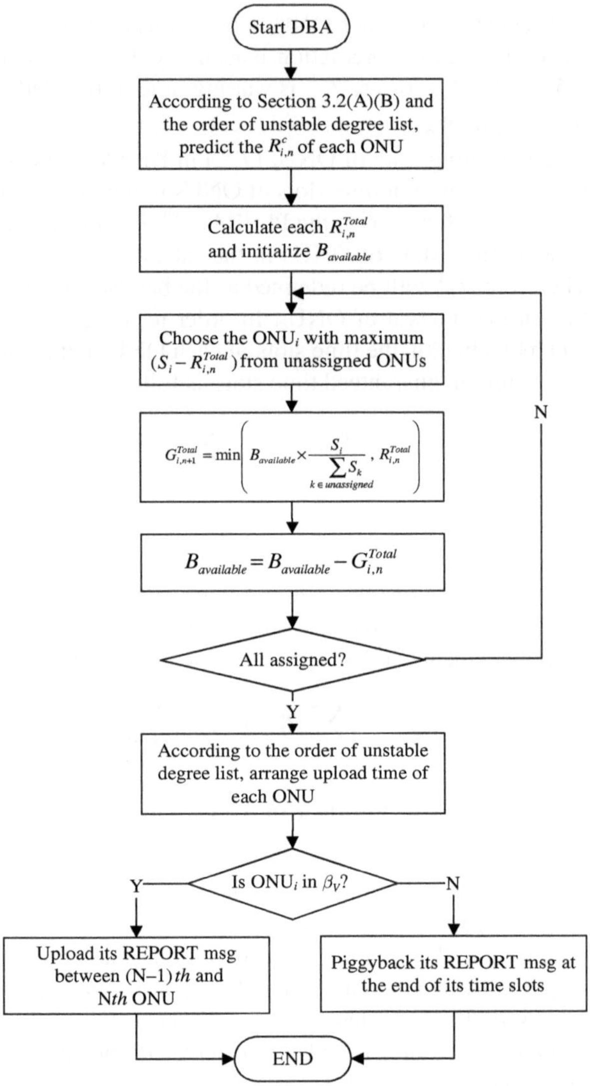

Fig. 5.4 Flowchart of PFEBA

where $R^{Total}_{i,n}$ is the sum of differential traffics after being predicted, $R^c_{i,n}$, $c \in \{EF, AF, BE\}$ of ONU$_i$ in the nth cycle. Furthermore, the granted bandwidth for EF, AF and BE classes are as follows:

$$\begin{cases} G^{EF}_{i,n+1} = \min\left(G^{Total}_{i,n+1}, R^{EF}_{i,n}\right) \\ G^{AF}_{i,n+1} = \min\left(G^{Total}_{i,n+1} - G^{EF}_{i,n+1}, R^{AF}_{i,n}\right) \\ G^{BE}_{i,n+1} = G^{Total}_{i,n+1} - G^{EF}_{i,n+1} - G^{AF}_{i,n+1} \end{cases}$$

In the final, the available bandwidth becomes $B_{available} = B_{available} - G_{i,n+1}^{Total}$. The process continues until all ONUs have been assigned, and the PFEBA will arrange the upload time sequence in order and report the time of each ONU in the unstable degree list.

5.4 Performance Analysis

In this section, the system performance of the proposed E-DBA mechanism with DBAM and EBR schemes are compared in terms of wasted bandwidth, wasted bandwidth improved percentage, downlink data available bandwidth, throughput, average end-to-end delay and average queue length. For ease of exposition, the DBAM [11], EBR1 [7] and EBR2 [8] are employed to represent the existing schemes.

The system model is set up in the OPNET simulator with one OLT and 32 ONUs. The downstream and upstream channels are both 1 Gb/s. The distance from an ONU to the OLT is assumed to range from 10 to 20 km and each ONU has infinite buffer. The service policy is in first-in first-out (FIFO) discipline. For the traffic model considered here, an extensive study shows that most network traffic can be characterized by self-similarity and long-range dependence (LRD) [16]. This model is utilized to generate highly bursty BE and AF traffic classes with the Hurst parameter of 0.7, and packet sizes are uniformly distributed between 64 and 1518 bytes. On the other hand, high-priority traffic (e.g., voice applications) is modeled using a Poisson distribution and packet size is fixed to 70 bytes [4]. The traffic profile is as follows: 20% of the total generated traffic is considered for high-priority traffic, and the remaining 80% is equally distributed between low- and medium-priority traffic [8, 17]. The simulation scenario is summarized in Table 5.2.

Table 5.2 Simulation scenario

Number of ONUs	32
Upstream/downstream link capacity	1 Gbps
OLT - ONU distance (uniform)	10–20 km
Maximum transmission cycle time	2 ms
Guard time	5 μs
Computation time of DBA	10 μs
The Number of ONUs in β_V	4
Control message length	0.512 μs(= 64bytes)

5.4.1 Wasted Bandwidth and Improved Percentage

The wasted bandwidth is due to the prediction model suffer serious inaccuracy. Figure 5.5 compares the wasted bandwidth and wasted bandwidth improved percentage vs. traffic load between the proposed scheme and DBAM. The wasted bandwidth problem is not considered in both EBR1 and EBR2 because no prediction mechanism is used. Figure 5.5 (a) shows that both E-DBA and DBAM schemes waste too

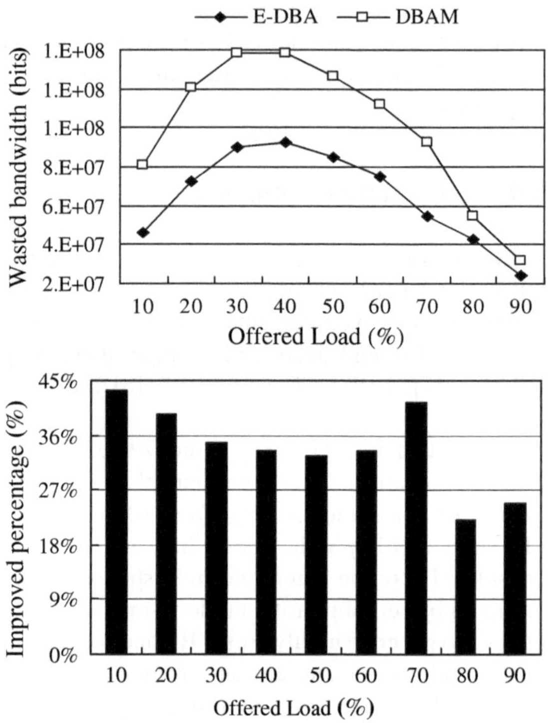

Fig. 5.5 (**a**) Wasted bandwidth (**b**) Wasted bandwidth improved percentage

much bandwidth when traffic loads range between 30% and 60%, especially for the DBAM scheme. For traffic load exceeding 70%, the wasted bandwidth is decreased because no more available bandwidth can be granted to ONUs when the system traffic load is high.

The improved percentage of waste bandwidth is defined as

$$\frac{\text{Bandwidth}_{\text{DBAM}}^{\text{waste}} - \text{Bandwidth}_{\text{PEFEBR}}^{\text{waste}}}{\text{Bandwidth}_{\text{DBAM}}^{\text{waste}}}.$$

The improved percentage of waste bandwidth exceeds 30% when the traffic load is below 70%, which is shown in Fig. 5.5 (b). When the offered load increases up to 70%, the E-DBA can still have 20–30% improved percentage more than the DBAM. The reason is that the E-DBA has better prediction accuracy than the DBAM, especially when some ONUs have big traffic variation.

5.4.2 Throughput

Figure 5.6 compares the throughput vs. traffic load among the proposed scheme, EBR1, EBR2 and DBAM. The proposed E-DBA has almost the same throughput as that of EBR1, EBR2 and DBAM until the traffic load exceeds 70%. The DBAM has the worst throughput because the inaccurate prediction problem and limit bandwidth

Fig. 5.6 Throughput

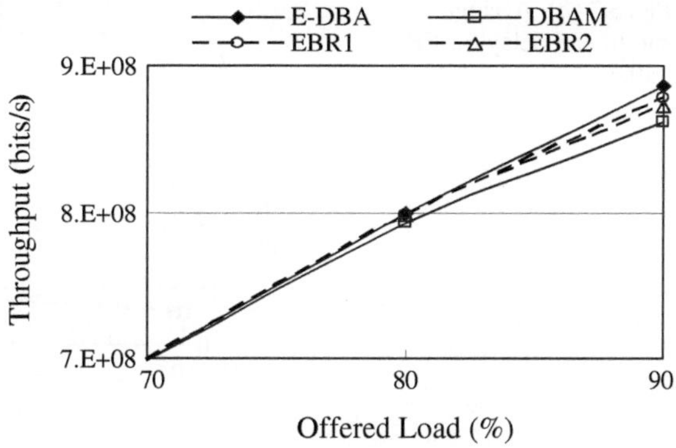

allocation (LBA) are proven to have lower throughput under non-uniform traffic[6]. Both EBR1 and EBR2 allocate more bandwidth to ONUs than that requested[9] (redundant bandwidth problem), thus lowering system throughput.

5.4.3 End-to-End Delay

Figure 5.7 compares the average end-to-end delay vs. traffic load among the proposed scheme, EBR1, EBR2 and DBAM. The results of AF and BE are very similar because the traffic characteristics are the same. Figure 5.7 (a) shows that the proposed E-DBA outperforms the other three schemes when the traffic load is high. The DBAM has the worst performance because of serious prediction inaccuracy when the traffic has high variation. However, the DBAM is more suitable for stable traffic like EF traffic, shown in Fig. 5.7 (b). In Fig. 5.7 (c), the E-DBA can handle varying traffic, such as AF and BE. On the other hand, as shown in Fig. 5.7 (b), the allocated bandwidth of EF traffic is limited by the prediction scheme of E-DBA. Therefore, the EF result of E-DBA is not as good as that of DBAM when the traffic load is high. Both EBR1 and EBR2 can adjust the excessive bandwidth, but cannot avoid redundant bandwidth problem that results in longer end-to-end delay. The packet delay time has three components: polling delay, grant delay and queuing delay [18].

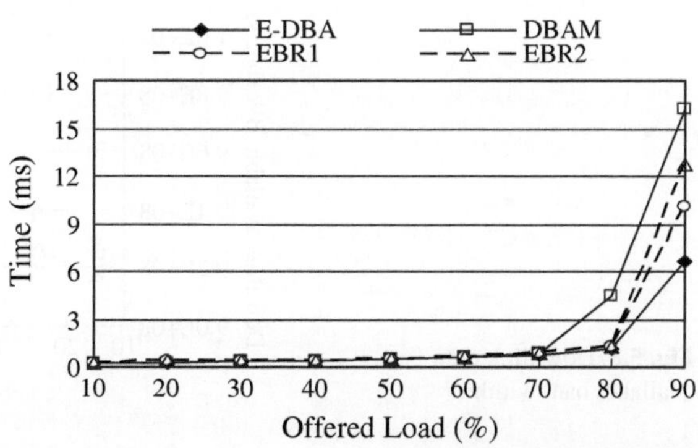

Fig. 5.7 (a) Average end-to-end delay for total traffic

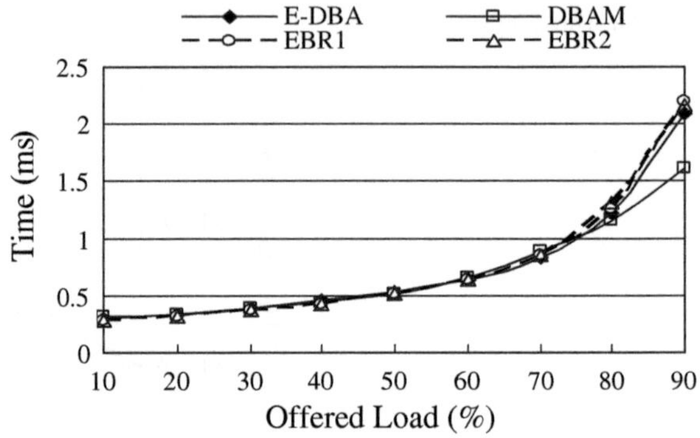

Fig. 5.7 (b) Average end-to-end delay for EF traffic

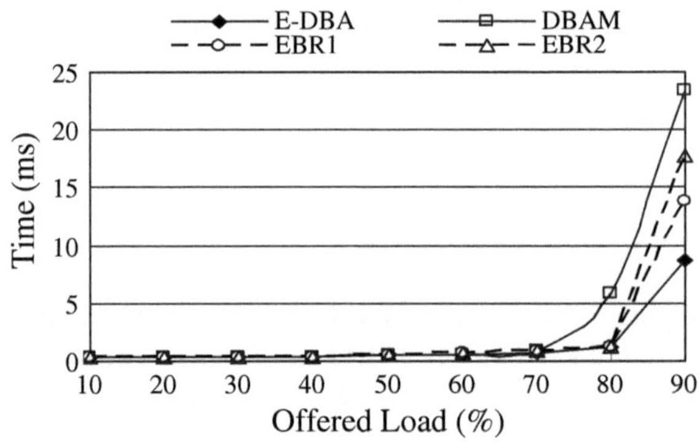

Fig. 5.7 (c) Average end-to-end delay for BE traffic

Both EBR1 and EBR2 reduce the polling delay by shorter polling cycle, but increase the flow of control messages which results in diminishing downlink data available bandwidth as shown in Fig. 5.8. Prediction-based schemes, E-DBA and DBAM, will decrease more queuing delay than polling delay. Therefore, E-DBA and DBAM can reduce more packet delay and the traffic of control messages.

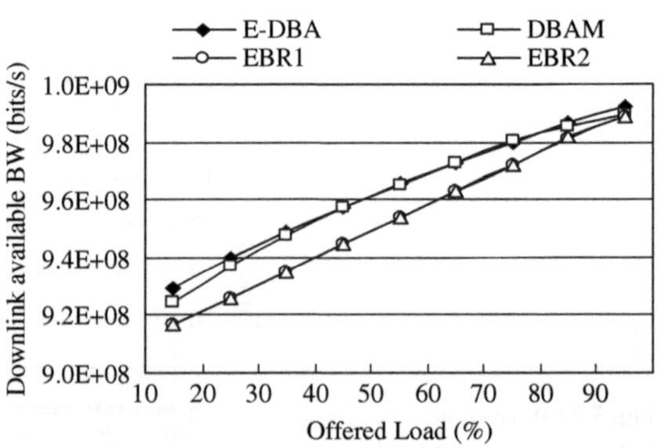

Fig. 5.8 Downlink data available bandwidth

5.4.4 Downlink Data Available Bandwidth

Figure 5.8 compares the downlink data available bandwidth vs. traffic load among the proposed scheme, EBR1, EBR2 and DBAM. All downlink data available bandwidth increases when the traffic load increases. The proposed E-DBA has more downlink data available bandwidth than EBR1 and EBR2, and is close to that of the DBAM scheme. It is because that the E-DBA and DBAM have variable longer cycle time for data transmission than the fixed cycle time scheme of EBR1 and EBR2 because less GATE messages of the PFEBA and DBAM are needed.

5.4.5 Average Queue Length

Figure 5.9 compares the average queue length vs. traffic load among the proposed scheme, EBR1, EBR2 and DBAM. The result is similar to that shown in Fig. 5.7 (a). The E-DBA has the best performance than the other three schemes when the traffic load exceeds 80%. The DBAM without excessive bandwidth allocation scheme yields the longest average queue length when the traffic load is above 70%. Owing to the redundant bandwidth problem, neither EBR1 nor EBR2 can reallocate excessive bandwidth sufficiently. The average queue length is accumulated dramatically when the offered load is more than 80%.

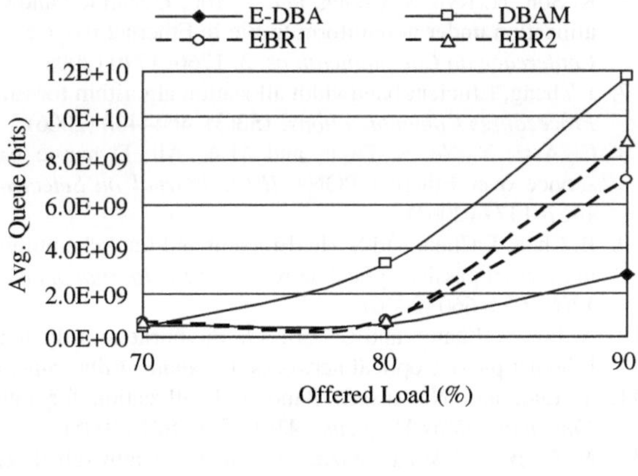

Fig. 5.9 Average queue length

5.5 Conclusions

The proposed *Early DBA mechanism* with the PFEBA scheme integrates an efficient DBA scheme to improve the prediction accuracy of unstable traffic ONUs of β_V by shortening the waiting time and to reduce the idle period. The proposed algorithm outperforms other existing well-known DBA algorithms, DBAM and EBR when the network is under high traffic load. As compared with the DBAM, the proposed E-DBA mechanism with PFEBA scheme can improve the wasted bandwidth improved percentage from 24% up to 44%. In the downlink data available bandwidth, the proposed algorithm performs better than the EBR. It is because that the E-DBA has variable longer cycle time for transmitting more data than that of fixed cycle time in the EBR. In the throughput, the proposed algorithm performs better than the EBR and is close to the DBAM. As compared with the EBR, the E-DBA can reduce end-to-end packet delay and average queue length of about 30% to 50% when the traffic load is high. The E-DBA uses the guaranteed bandwidth ratio not only to alleviate the redundant bandwidth problem but also provides fairness in the excessive bandwidth allocation scheme. The optimal number of ONUs in β_V and the value of the linear estimation credit, α, for obtaining the best system performance is our future research.

References

1. ITU-T Recommendations. Available: http://www.itu.int/ITUT/publications/recs.html.
2. IEEE 802.3ah task force home page. Available: http://www.ieee802.org/3/efm.
3. S. Blake, D. Black, M. Carlson, E. Davies, Z. Wang, and W. Weiss, "An Architecture for Differentiated Services," IETF RFC 2475, 1998.
4. G. Kramer, B. Mukherjee, and G. Pesavento, IPACT: A dynamic protocol for an Ethernet PON (EPON), *IEEE Communications Magazine*, **40**(2), 74–80 (2002).
5. G. Kramer, B. Mukherjee, S. Dixit, Y. Ye, and R. Hirth, Supporting differentiated classes of service in Ethernet passive optical networks, *Journal of Optical Networks*, **1**(8), 280–298 (2002).
6. K. Son, H. Ryu, S. Chong, and T. Yoo, Dynamic bandwidth allocation schemes to improve utilization under nonuniform traffic in Ethernet passive optical networks, *IEEE International Conference on Communications*, **3**, 1766–1770 (2004).
7. J. Zheng, Efficient bandwidth allocation algorithm for Ethernet passive optical networks, *IEE Proceedings Communications*, **153**(3), 464–468 (2006).
8. C. Assi, Y. Ye, S. Dixit, and M.A. Ali, Dynamic bandwidth allocation for quality-of-service over Ethernet PONs, *IEEE Journal on Selected Areas in Communications*, **21**(9), 1467–1477 (2003).
9. B. Chen, J. Chen, and S. He, Efficient and fine scheduling algorithm for bandwidth allocation in Ethernet passive optical networks, *IEEE Journal of Selected Topics in Quantum Electronics*, **12**(4), 653–660 (2006).
10. X. Bai, A. Shami, and C. Assi, On the fairness of dynamic bandwidth allocation schemes in Ethernet passive optical networks, *Computer Communications*, **29**(11), 2123–2135 (2006).
11. Y. Luo, and N. Ansari, Bandwidth allocation for multiservice access on EPON, *IEEE Communications Magazine*, **43**(2), S16–S21 (2005).
12. N. Sadek, and A. Khotanzad, A Dynamic bandwidth allocation using a two-stage fuzzy neural network based traffic predictor, *IEEE International Conference on Neural Networks*, **3**, 2407–2412 (2004).
13. M. Wu, R.A. Joyce, H.S. Wong, L. Guan, and S.Y. Kung, Dynamic resource allocation via video content and short-term traffic statistics, *IEEE Transactions on Multimedia*, **3**(2), 186–199 (2001).
14. N. Sadek, A. Khotanzad, and T. Chen, ATM dynamic bandwidth allocation using F-ARIMA prediction model, Proceedings of International Conference on Computer Communications and Networks, 359–363 (2003).
15. Y. Luo, and N. Ansari, Limited sharing with traffic prediction for dynamic bandwidth allocation and QoS provisioning over Ethernet passive optical networks, *Journal of Optical Networking*, **4**(9), 561–572 (2005).

16. W. Willinger, M.S. Taqqu, and A. Erramilli, A bibliographical guide to self-similar traffic and performance modeling for modern high-speed networks, Stochastic Networks: Theory and Applications. In Royal Statistical Society Lecture Notes Series, Oxford University Press, **4**, pp.339–366 (1996).
17. X. Bai, and A. Shami, Modeling Self-Similar Traffic for Network Simulation, Technical report No. NetRep-2005-01, 2005.
18. H. Naser, and H.T. Mouftah, A joint-ONU interval-based dynamic scheduling algorithm for Ethernet passive optical networks, *IEEE/ACM Transactions on Networking*, **14**(4), 889–899 (2006).

Chapter 6
Overview of MAC Protocols for EPONs

Yongqing Zhu and Maode Ma

Abstract Ethernet Passive Optical Network (EPON) has been regarded as the best candidature for next-generation access network, because it represents the convergence of the inexpensive Ethernet equipment and high-bandwidth fiber infrastructure. In EPON networks, multiple Optical Network Units (ONUs) share the upstream bandwidth to transmit data packets to the Optical Line Terminal (OLT). An efficient Medium Access Control (MAC) protocol is required in EPONs to arbitrate upstream transmissions. Many efforts have been put on the design of MAC protocols, especially Dynamic Bandwidth Allocation (DBA) schemes, in EPONs. This chapter aims to present a comprehensive survey on the up-to-date DBA schemes for EPON networks. We have provided a categorization method to classify numerous DBA schemes into corresponding groups. Besides the description and comments for each individual scheme, this chapter has also provided common features as well as the merits and shortcomings for each category of DBA schemes.

Keywords Ethernet Passive Optical Network (EPON) · Medium Access Control (MAC) · Dynamic Bandwidth Allocation (DBA) · Quality of Service (QoS)

6.1 Introduction

6.1.1 Evolution of Access Networks

While bandwidth in the backbone network is increasing dramatically through the use of Wavelength Division Multiplexing (WDM) and other new technologies, the access network has experienced little changes in recent years. At the same time, Local Area Networks (LANs) have grown up from 10 Mb/s to 100 Mb/s then to 1 Gb/s; even 10 Gb/s speed is available for the residential subscribers now. The result is a growing gulf between high-capacity LANs and backbone networks with

Y. Zhu (✉)
Data Storage Institute, A*STAR, Singapore

M. Ma. (ed.), *Current Research Progress of Optical Networks*,
DOI 10.1007/978-1-4020-9889-5_6, © Springer Science+Business Media B.V. 2009

the bottleneck of access networks. Once called the "last mile", the access network between the backbone and residential subscribers has been renamed as the "first mile" to express its importance. Because the subscribers have an increasing demand for the Internet traffic with a variety of types, a powerful technology is needed for the "first mile" to provide broadband access to the backbone. It is also expected to be inexpensive, simple, scalable, and capable of delivering integrated voice, data, and video services to end users over a single network.

Two most widely deployed broadband access solutions today are Digital Subscriber Line (DSL) and Hybrid Fiber Coax (HFC) networks [1], which depend on the reuse of the existing infrastructure. Using the twisted pairs as the transmission medium, DSL is deployed mainly by the traditional Plain Old Telephone Service (POTS) providers. The point-to-point configuration requires a DSL modem at the customer premises and a DSL Access Multiplexer (DSLAM) in the Central Office (CO). With the typical data rate of 128 kb/s $\sim$ 1.5 Mb/s, DSL is hardly to be considered broadband to support integrated voice, data and video applications. In addition, the distance that a CO can cover with DSL is limited to less than 5.5 km, which only covers approximately 60% [1] of the potential subscribers. HFC networks are preferred by cable television providers to deliver the data services together with the television signals over the existing CATV infrastructure. HFC combines the optical fiber and coaxial cable in the transmission path with the point-to-multipoint configuration. The drawback of HFC is that each optical node has less than 36 Mbps effective data throughput in the network, which is shared by up to 2000 users [1]. The resulting slow speed is unable to provide enough bandwidth for the emerging services like Video on Demand (VoD), interactive gaming and two-way video conferencing.

Optical fiber is an ideal transmission medium that can deliver bandwidth-intensive, integrated voice, data and video services at distances around 20 km in the subscriber access network. In order to alleviate the bandwidth bottleneck at the access network, it is required to deploy the optical fiber and optical nodes with deeper penetration. Especially, the optical fiber should be deployed throughout the "first mile" to meet the bandwidth requirements.

A simple way to deploy the optical fiber in the access network is to use a point-to-point topology with a dedicated fiber from the CO to each end-user subscriber [2]. Although a simple architecture, the point-to-point topology is cost prohibitive because it requires significant fiber deployment and connector termination space in the local exchange. Considering M subscribers at an average distance L km from the CO in Fig. 6.1(a), a point-to-point design requires $2M$ transceivers and M^*L fiber length totally (assuming that a single fiber is used for bi-directional transmissions).

A remote curb switch can be deployed close to the neighborhood in order to reduce the fiber deployment (Fig. 6.1(b)). Such architecture will reduce the fiber consumption to only L km (assuming negligible distance between the switch and customers). However, because there is one more link added to the network, the number of transceivers will increase to $(2M + 2)$. In addition, the curb-switched

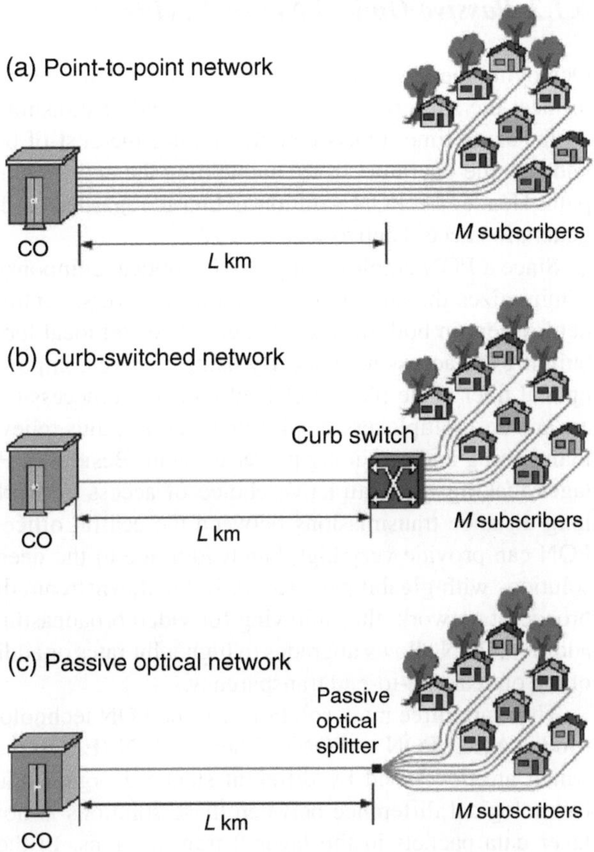

Fig. 6.1 Deployment of optical fiber in the access network

network architecture requires electrical power as well as back-up power at the curb switch, which will increase the cost of Local Exchange Carriers (LECs).

It is reasonable and logical to replace the active curbside switch with an inexpensive passive optical splitter, which results in the solution of Passive Optical Network (PON). A PON is a point-to-multipoint optical network with no active elements in the signal path from the source to destination. The only interior elements used in a PON are passive optical components, such as optical fiber, splices and splitters. PONs can minimize the number of optical transceivers, CO terminations and fiber deployment. An access network based on a single-fiber PON only requires $(M + 1)$ transceivers and L km fiber length as shown in Fig. 6.1 (c).

PON has been viewed by many as an attractive solution for the "first mile" access networks, because it can support gigabit speed at low cost comparable to the DSL and HFC solutions. With PON, the subscriber access network can be effectively implemented into the Fiber-To-The-Home (FTTH), Fiber-To-The-Building (FTTB) and Fiber-To-The-Curb (FTTC), which is one of the objectives of next-generation access networks.

6.1.2 Passive Optical Networks (PONs)

Deploying the high-capacity optical fiber, a PON can provide very large bandwidth to meet the subscribers increasing demand of transmitting various types of traffic. At the same time, PONs can also reduce the cost of building the access networks and free the operators from maintaining the active components in the transmission path. Hence, the PON solution is highly regarded as the best choice for the next-generation access networks.

Since a PON employs only passive optical components in the transmission path, it minimizes the amount of optical transceivers, central office terminals, and fiber deployment in both local exchange office and local loop, thus reducing the cost of building the access network. By using passive components deployed as part of the optical fiber cable plant, PONs eliminate the necessity to install active multiplexers and de-multiplexers at splitting locations, thus relieving network operators from maintaining and providing power to them. Besides, a PON has many other advantages making it an attractive choice of access networks. It allows around 20 km long distance transmissions between the central office and customer premises. A PON can provide very high bandwidth due to the deep fiber penetration, offering solutions with gigabit per second. In the downstream direction, PON operates as a broadcast network, thus allowing for video broadcasting and other applications. In addition, PON allows upgrades to higher bit rates or additional wavelengths because of the optical end-to-end transparency.

There are three main solutions for the PON technology: Asynchronous Transfer Mode (ATM) PON (APON), Ethernet PON (EPON), and Gigabit PON (GPON), which are developed by different standard organizations respectively. The main technological difference between these solutions is how to encapsulate the upper layer data packets in the layer 2 transmissions. In the APON solutions, packets are carried in the ATM cells for transmission. While in EPON, Ethernet frames are the carrier in layer 2 to take the data packets. Finally, GPON Encapsulation Method (GEM) is used to encapsulate packets in GPON. APON and GPON permit fragmentation of data packets, which can achieve strict service requirements for the applications. While, it is also required to reconstruct the fixed frames at the destination, which leads to the extra complexity. No segmentation is allowed in the EPON solutions, the packet reconstruction is neither needed. Hence, the individual frame length in EPON is variable.

APON was defined by the Full Service Access Network (FSAN) in the mid-1990s and has been accepted by International Telecommunication Union-Telecommunication (ITU-T) as standard series G.983 [3]. However, ATM has many shortcomings that prevent it from an ideal technology for PONs to transmit the predominant Internet Protocol (IP) data traffic in the future networks. Since the ATM cell has the fixed size of 53 bytes, the IP packet with variable length should be fragmentized to segments in order to be carried by ATM cells. That will impose a high overhead over IP packet transmissions. In addition, a dropped or corrupted ATM cell will invalidate the entire IP packet; however other ATM cells carrying the same IP packet will propagate further to the destination, thus consuming network resources unnecessarily [2]. Moreover, the cost of ATM equipment has not declined as expected, which inhibits the deployment of the APON solutions.

Due to the technological and economic considerations, APON has lost the dominant position for next-generation access networks. The FSAN then developed another Broadband PON (BPON) technology [4], GPON, which has been standardized as ITU-T G.984 series [5]. GPON supports transport of various native protocols including ATM and Ethernet. Furthermore, it expands APON's transmission capacity from Megabit to Gigabit. Although achieving the comparative bandwidth with the EPON solutions, GPON needs more complex operations and frame structure to provide backward compatibility with the legacy technologies including APON. Moreover, the necessities to segment and reconstruct packets [6] also result in the additional complexity of GPON.

Supported by the IEEE community, EPON has just been proposed and developed in several years. The IEEE 802.3ah Ethernet in the First Mile (EFM) Task Force [7] has made a serial of standardizations for EPON including Multi-Point Control Protocol (MPCP) arbitration mechanism. Being a low-cost technology, Ethernet has been universally accepted and can be interoperable with a great deal of legacy equipments. Since Ethernet can achieve 10-Gigabit and higher capacity, an EPON network can easily provide high transmission rate with Gigabit per second for both the upstream and downstream transmissions [8]. Besides, newly adopted QoS techniques have made Ethernet networks capable of supporting the integrated voice, data, and video services efficiently. These techniques include the full-duplex transmission mode, prioritization (P802.1p), and Virtual LAN (VLAN) tagging (P802.1Q) [2]. Ethernet has become a perfect choice of PONs for delivering IP packets.

Among the three PON solutions, EPON is highly regarded as the best candidature for the next-generation access networks as it represents the convergence of the inexpensive Ethernet equipments and the high-speed fiber infrastructure. In an EPON network, multiple Optical Network Units (ONUs) access to the shared fiber channel to reach the Optical Line Terminal (OLT) through a passive optical splitter. To arbitrate the multiple access from ONUs, an effective Medium Access Control (MAC) protocol is required to allocate bandwidth among ONUs. Although the IEEE 802.3ah EFM Task Force has been standardizing for EPONs to ensure the interoperability of products from different vendors, it has not specified the particular bandwidth allocation scheme. Instead, designing and developing multiple access schemes for upstream transmissions is an open and hot topic in the EPON area.

In the recent years, a variety of bandwidth allocation schemes have been proposed for EPON networks. This chapter aims to present a comprehensive survey on the up-to-date bandwidth allocation schemes for EPONs. A detailed description on EPONs will be given in Section 6.2, including the topologies, transmission principles, and related standards. Section 6.3 will provide a review of MAC protocols for the upstream transmissions in EPON networks. Then a thorough survey will be provided in Section 6.4 for the centralized bandwidth allocation schemes with time-sharing for EPONs. The whole chapter will be summarized in Section 6.5.

6.2 Ethernet Passive Optical Networks (EPONs)

An EPON network is a PON-based network where data packets are encapsulated into Ethernet frames for transmission. The EPON solution has been regarded as the best candidature for the next-generation access networks because it merges the virtues of Ethernet and PON. An EPON network deploys a point-to-multipoint architecture where an OLT is connected with multiple ONUs through a passive optical component.

There are several topologies suitable for the point-to-multipoint access networks, including tree, tree-and-branch, ring, and bus. Using 1:2 optical tap couplers and 1:M optical splitters, EPONs can be flexibly configured in any of these topologies (Fig. 6.2). Additionally, an EPON can be deployed into the redundant configuration, i.e. double rings or double trees. The redundancy may be added only to a part of the EPON, such as the trunk of the tree [2].

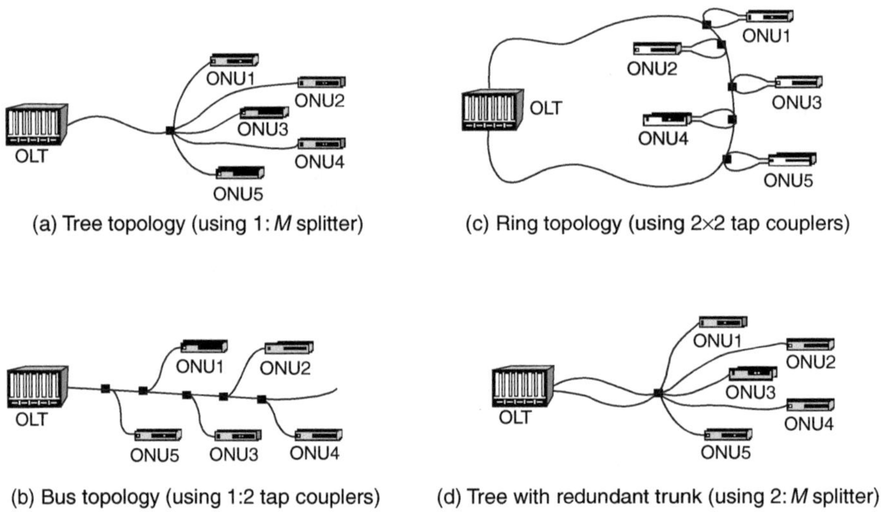

Fig. 6.2 Topologies of EPON networks

6.2.1 Transmission Principle

In an EPON, all transmissions are performed between the OLT and ONUs, which are located at the local exchange (central office) and end user locations, respectively. The OLT is responsible for connecting the optical access network to the backbone. The ONU takes charge of conveying the broadband voice, data, and video services between the end users and the OLT.

In the downstream direction (from the OLT to ONUs), an EPON is a point-to-multipoint network that can easily operate by broadcasting. Ethernet frames transmitted by the OLT pass through a *1:M* passive splitter and reach each ONU. Ethernet fits perfectly with the EPON architecture due to its broadcast property

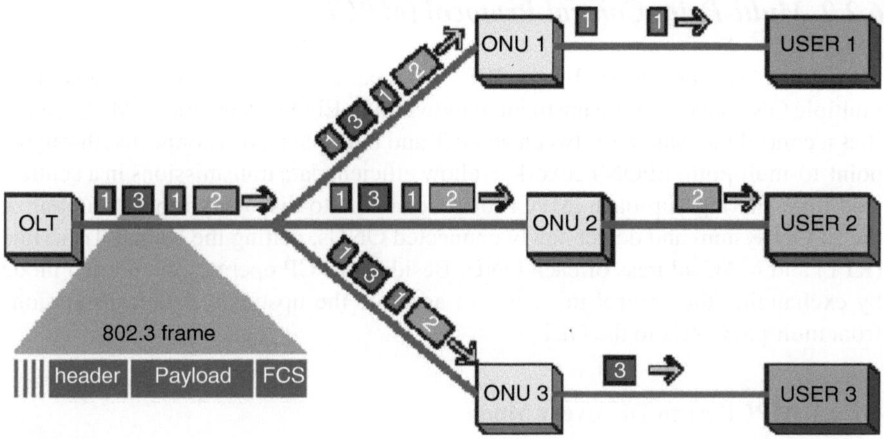

Fig. 6.3 Downstream transmissions in EPON networks

(Fig. 6.3). Frames are broadcast by the OLT, carrying the destined ONU's MAC address. If the destination address matches with the ONU's MAC address, the ONU will extract the frame and deliver to the end users; otherwise the ONU will ignore and discard the frame.

In the upstream direction (from ONUs to the OLT), an EPON is a multipoint-to-point network where different ONUs share the upstream channel to transmit data. Data frames from any ONU will only reach the OLT due to directional properties of the passive combiner (optical splitter). Since data frames from different ONUs being transmitted simultaneously may collide over the channel, an efficient MAC protocol is required for the EPON to perform the bandwidth allocation among multiple ONUs. Figure 6.4 illustrates the upstream data transmissions from different ONUs to the OLT.

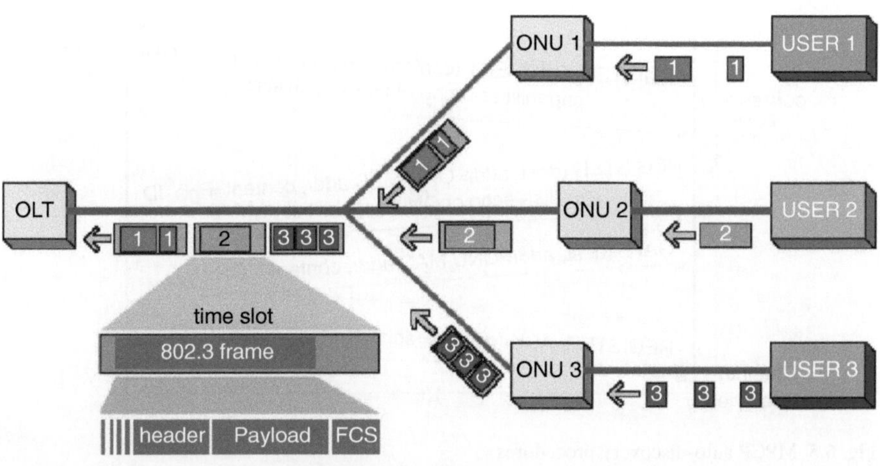

Fig. 6.4 Upstream transmissions in EPON networks

6.2.2 Multi-Point Control Protocol (MPCP)

MPCP has been developed by the IEEE 802.3ah EFM Task Force [7] to support multiple ONUs sharing the upstream bandwidth in EPONs. Generally, MPCP specifies a control mechanism between an OLT and multiple ONUs connected through a point-to-multipoint EPON network to allow efficient data transmissions in a centralized time-sharing approach. MPCP operates the auto-discovery mode to initialize the EPON system and detect newly connected ONUs, getting the Round-Trip Time (RTT) and MAC address of each ONU. Besides, MPCP operates the normal mode by exchanging the control messages to arbitrate the upstream data transmissions from multiple ONUs to the OLT.

6.2.2.1 MPCP Auto-Discovery Mode

The OLT periodically reserves the discovery window for auto-discovery. It broadcasts the discovery GATE message to detect whether there is any new ONU connected to the EPON network. Only the un-initialized ONUs respond to the discovery GATE message and set the local time according to the timestamp contained in the arriving message. An un-initialized ONU will transmit a REGISTER_REQUEST message to request registration, which includes the ONU's address and local time. The OLT will calculate the ONU's RTT upon receiving the message. After the relative message procedures (Fig. 6.5), the ONU will be initialized and the channel between the ONU and OLT is established.

Since there will be multiple ONUs requesting initialization at one time, auto-discovery is a contention-based procedure. The ONU whose REGISTER_REQUEST message collides with others is considered failure during the auto-discovery procedure. It can attempt to request initialization again in every following discovery

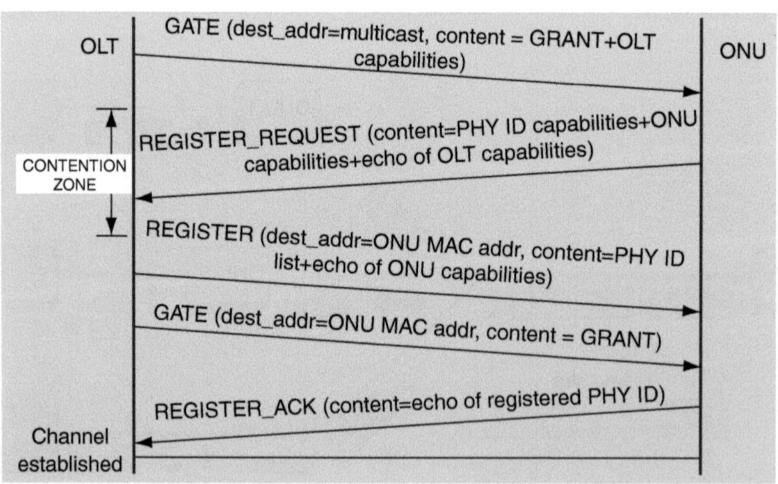

Fig. 6.5 MPCP auto-discovery procedures

window or skip random number of discovery windows (i.e. using exponential backoff) before next request.

6.2.2.2 MPCP Normal Mode

In the normal mode, MPCP controls the upstream data transmissions. Two control messages, GATE and REPORT, are defined in MPCP and transmitted between the OLT and ONUs. The OLT performs the bandwidth allocation algorithm to get the transmission grants for ONUs. A GATE message is generated by the OLT and sent to an ONU granting it to transmit data over the upstream channel at an appropriate time. The 64-byte GATE messages can carry up to six grants to a particular ONU, each corresponding to a queue within the ONU. Illustrated in Fig. 6.6, the GATE message contains the information of timestamp when GATE is sent out, the start time that the ONU is granted for transmission, and the stop time of the transmission. On receiving the GATE message from the OLT, an ONU will update its local clock according to the timestamp, and wait until the start time for transmission. The transmission may include multiple Ethernet frames, depending on the size of the granted transmission window and the number of waiting packets at the ONU. The ONU should ensure that no fragmentation is allowed during the data transmissions. The frame that is too large to be accommodated in the assigned timeslot will be deferred to the next timeslot.

ONUs can send REPORT messages to the OLT automatically or on-demand to request bandwidth allocation. REPORT is transmitted together with the upstream data frames in the assigned timeslot. It can be either transmitted at the beginning or at the end of the timeslot, depending on the bandwidth request approach implemented by the ONU. The 64-byte REPORT message must contain the timestamp

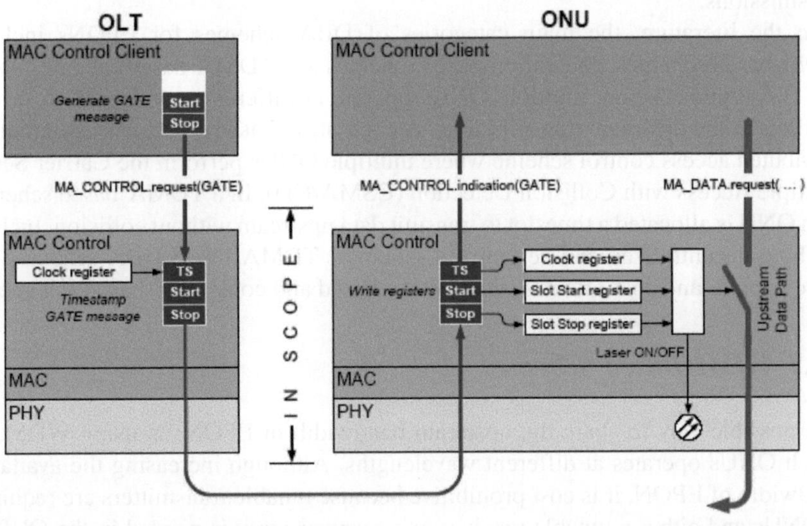

Fig. 6.6 MPCP GATE operations

used by the OLT to adjust the RTT for the ONU. It may contain the desired size of the next timeslot based on the ONUs buffer occupancy. For the ONU with multiple traffic queues, it can report the entire buffer occupancy in the REPORT message, or report up to eight queues status to request individual grant. Depending on the bandwidth allocation scheme deployed, the OLT can choose to issue one grant for an ONU, or issue multiple grants in the same GATE message for different queues.

It is important to notice that MPCP does not specify any particular bandwidth allocation scheme for EPON upstream transmissions. Rather, it supports the implementation of various schemes as long as they conform to the MPCP framework. Recently a great number of bandwidth allocation schemes have been proposed for EPONs compatible with MPCP, which will be investigated in the following parts.

6.3 Review of MAC Protocols in EPONs

In the upstream direction of EPON networks, effective MAC protocols are required to allocate bandwidth to multiple ONUs for accessing the shared trunk fiber channel. The main functionality of a MAC protocol should be avoiding collisions of packets from different users. In addition, a MAC protocol should also be capable of introducing low overhead, making efficient use of the resources and guaranteeing the requested QoS for different types of traffic.

Because of the traffic bursty nature in the access networks, the bandwidth requirements change greatly from time to time. Hence, it is inefficient to allocate bandwidth to ONUs or individual traffic queues in a static manner. A Dynamic Bandwidth Allocation (DBA) scheme is more efficient for EPONs where the instantaneous requirements are considered in the bandwidth allocation. Hereafter we will provide a review on the DBA schemes proposed for EPON networks for the upstream transmissions.

In the literature, the main categories of DBA schemes for EPONs include WDM-based schemes, contention-based schemes and TDMA-based schemes. With a WDM-based scheme, multiple ONUs operate at different wavelengths to avoid conflicts in the upstream transmissions. A contention-based scheme is essentially a distributed access control scheme where multiple ONUs perform the Carrier Sense Multiple Access with Collision Detection (CSMA/CD). In a TDMA-based scheme, each ONU is allocated a timeslot to transmit data upstream without collision. Including both decentralized and centralized schemes, TDMA-based DBA schemes are more popular and cost-effective than WDM-based and contention-based schemes.

6.3.1 WDM-Based Schemes

One possible way to share the upstream bandwidth in EPONs is using WDM, in which ONUs operates at different wavelengths. Although increasing the available bandwidth of EPON, it is cost prohibitive because tunable transmitters are required in ONUs and either a tunable receiver or a receiver array is needed in the OLT. A more serious problem is that multiple types of ONUs based on their laser wavelengths would be required. Some schemes are proposed to combine WDM with other methods such as TDMA and Code Division Multiple Access (CDMA). By the combination, EPONs can achieve the high capacity without employing one wavelength for each ONU. However, the cost of a WDM-based EPON is still too high to be accepted widely.

In [9], the TDMA mechanism is combined into a WDM-based EPON system where multiple wavelength channels are established in both upstream and downstream directions. The system employs a WDM Interleaved Polling with Adaptive Cycle Time with a Single polling Table (WDM IPACT-ST) scheme based on the interleaved polling. The OLT can predict when transmissions will finish on all upstream channels, thus to schedule the next ONU to transmit packets over the first available channel. A Weighted Fair Queuing (WFQ) scheme is adopted to support the Differentiated Services (DiffServ) by reserving different weighted proportions for various types of traffic from cycle to cycle.

Reference [10] evaluates the efficiency of the hybrid time/wavelength/code division approach for EPONs. The EPON system employs *N* wavelength channels for transmission, where the optical CDMA is combined in order to increase the network capacity. All ONUs are divided into *N* groups, each with one wavelength. Within the group, each ONU randomly picks an Optical Orthogonal Code (OOC) from the multiple OOCs to encode the data packets for transmission. The ONU should make an announcement on the control channel about the selected wavelength, OOC sequence, and the amount of packets to be sent.

Reference [11] recommends an evolutionary upgrade at the architecture and extension of MPCP protocol for WDM EPONs. It then compares two main paradigms, online and offline scheduling, for dynamically allocating grants for upstream transmissions on different wavelengths. Authors have found that online scheduling, which make bandwidth allocation based on individual ONU request, tends to result in lower packet delays at medium and high traffic loads. Offline scheduling, which makes bandwidth allocations based on requests from all ONUs, may introduce extra delay to packets.

Reference [12] presents various dynamic wavelength and bandwidth allocation algorithms for WDM WPONs. The first scheme, Static Wavelength Dynamic Time (SWDT), allocates wavelength statically among ONUs and assigns bandwidth in time domain dynamically to each ONU. The second scheme, Dynamic Wavelength Dynamic Time (DWDT), allows dynamic allocation for different ONUs in both wavelength and time domains. With DWDT, the OLT allocate a wavelength with the least waiting time for transmission to an ONU. Authors then propose three variants of DWDM to determine the length of transmission window allocated to this ONU, where the OLT assigns the window in the manner of online, offline, and a hybrid of online and offline.

6.3.2 Contention-Based Schemes

According to [2], the contention-based medium access (similar to CSMA/CD) is difficult to implement in EPONs, because ONUs cannot detect a collision at the OLT due to directional properties of the optical splitter. A collision can only be detected by the OLT and be informed to ONUs by the contention signal. However, propagation delays in EPONs greatly reduce the efficiency of such a scheme. In order to achieve the prompt collision detection in a contention-based scheme, several proposals consider including additional components in the EPON architecture. Accordingly, the cost of the access network will increase and the extra maintenance should be taken for the components.

References [13] and [14] propose an implementation of optical CSMA/CD scheme for EPONs employing a $3 \times N$ Star Coupler (SC). A redirection mechanism is introduced where a portion of the optical power transmitted upstream is redirected back and distributed to all ONUs for carrier sensing and collision detecting. If an ONU has sensed that the upstream wavelength is unoccupied, it can transmit the packets. Otherwise, the packets are backed off. If a collision is detected, transmission is promptly aborted and the collided packets are backed off. The OLT is removed from the implementation of the scheme. [15] extends the scheme to EPON networks with the $N \times N$ SC architecture.

A Hybrid-Slot Decentralized Control (HSDC) scheme is presented in [16] to provide QoS for various types of traffic and simplify the network control mechanism. HSDC divides the transmission window into fixed time slot part and contention part. At the beginning of each cycle, ONUs send the high priority traffic in the respective fixed time slots whose sizes are statically set according to the subscription rates. In the remaining contention part of the transmission window, multiple ONUs contend to transmit the low priority traffic using the CSMA/CD scheme. HSDC is not compatible with MPCP as there is no control message exchange between the OLT and ONUs for data packet transmissions.

6.3.3 TDMA-Based Schemes

Time-sharing is a more popular and attractive method for optical channel sharing in an EPON network. It allows for a single upstream wavelength and a single transceiver in the OLT, resulting in a cost-effective solution. In the TDMA-based schemes, each ONU is allocated a timeslot, and each timeslot is capable of carrying multiple Ethernet frames. An ONU should buffer frames received from the users until its timeslot. When its timeslot arrives, the ONU would "burst" out frames at full channel capacity.

Recently, numerous schemes based on TDMA have been proposed for EPONs including both decentralized and centralized schemes. In a decentralized scheme, the OLT is excluded from the implementation of bandwidth allocation. ONUs perform the allocation mechanism to share the upstream transmission link. While in a centralized scheme, the OLT is the central controller to assign the bandwidth for all ONUs. Following are some decentralized TDMA-based schemes proposed for EPONs in the literature.

The Full Utilization Local Loop Request Contention Multiple Access (FULL-RCMA) scheme is proposed for EPONs in [17] and [18] as a decentralized time-sharing scheme. The EPON network requires that the splitter reflect the upstream data back to all ONUs for collision detection, which needs two fibers per ONU. FULL-RCMA contains a contention-based request period and a contention-free data period for each cycle. ONUs contend to submit requests during the request period in the random time slots. Each ONU knows the results of the request period by monitoring the echoes from the splitter, and transmits data packets accordingly in the data period without collisions.

References [19] and [20] introduce another collision-free scheme for the distributed EPON networks. It also requires part of the upstream optical power be redirected back to all ONUs. Each ONU runs the same cycle-based algorithm with the identical bandwidth allocation results. The cycle is divided into three periods: the static update period for each ONU transmitting the control message, the fixed waiting period to process the control messages and get the ONUs transmission assignments, and the dynamic transmission period for ONUs sending data packets over the upstream channel following the assignments without collisions. The order of ONUs transmissions is dynamic for different cycles based on the traffic demand.

The decentralized approaches require the connectivity/communicability between ONUs that may have some security issues. This also imposes some constraints on EPON topologies that only a ring or a broadcasting star can be deployed [2]. Since a preferred scheme should support any point-to-multipoint EPON topology, the decentralized approach is not a good choice for TDMA-based DBA schemes.

Comparatively, a centralized scheme is an ideal choice because it ensures only the connectivity between the OLT and each ONU, which can be employed in all EPON topologies. The OLT knows the state of the entire network and can flexibly switch from one allocation scheme to another according to the information. The entire EPON network is more robust and scalable, and the ONU can be very cheap and simple without any intelligence. The centralized TDMA-based DBA schemes have attracted more and more interests from researches, and there are much more centralized schemes than decentralized schemes proposed for EPONs to date. Next part we will provide a comprehensive survey on the centralized DBA schemes based on TDMA.

6.4 Literature Survey on TDMA-Based Centralized DBA Schemes for EPONs

In the literature, numerous centralized DBA schemes with time-sharing have been proposed for EPON networks. Most of these schemes are compatible with the MPCP protocol. In this section they will be surveyed in much detail. Considering the large number of related DBA schemes, a proper classification is necessary for the investigation.

Since most of the DBA schemes are based on polling, we can simply classify them by the ways of polling. One classification is to divide the DBA schemes into group of individual polling and group of joint polling. These two kinds of polling differentiate from each other by which ONUs requests are exploited for the bandwidth allocation. In the individual polling, an ONU's individual request

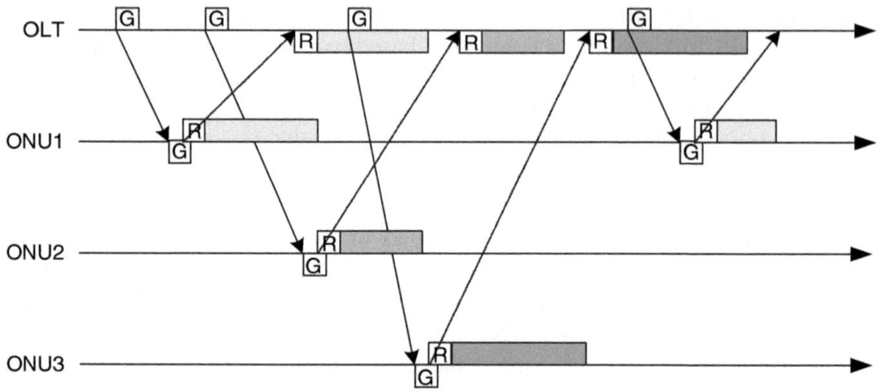

Fig. 6.7 Individual polling

alone can decide the transmission grant to itself (Fig. 6.7). The OLT instantaneously determines the grant for an ONU upon receiving its REPORT message. In contrast, in the joint polling shown in Fig. 6.8, the OLT collects REPORT messages from all ONUs in each cycle. Then it decides the transmission grants jointly by all ONUs' requests instead of any ONU's single request.

A better way of classification has been presented in [21], where the DBA schemes are categorized into schemes with statistical multiplexing and schemes with QoS assurances. Then the schemes with QoS assurances are grouped into absolute assurances and relative assurances. We decide to follow this classification because it can be extended to the comprehensive category, where each bandwidth allocation scheme will be grouped appropriately. Based on this method, we further divide the schemes with statistical multiplexing into interleaved polling schemes and non-interleaved schemes. Besides, the schemes with relative QoS assurances are grouped into OLT-ONU decision and OLT decision (Fig. 6.9).

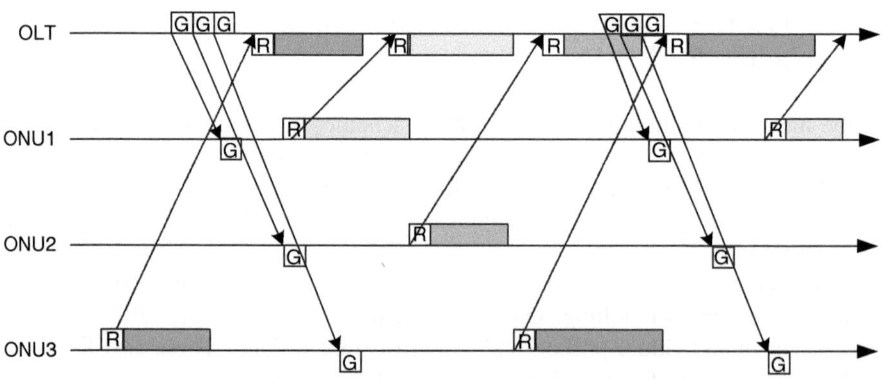

Fig. 6.8 Joint polling

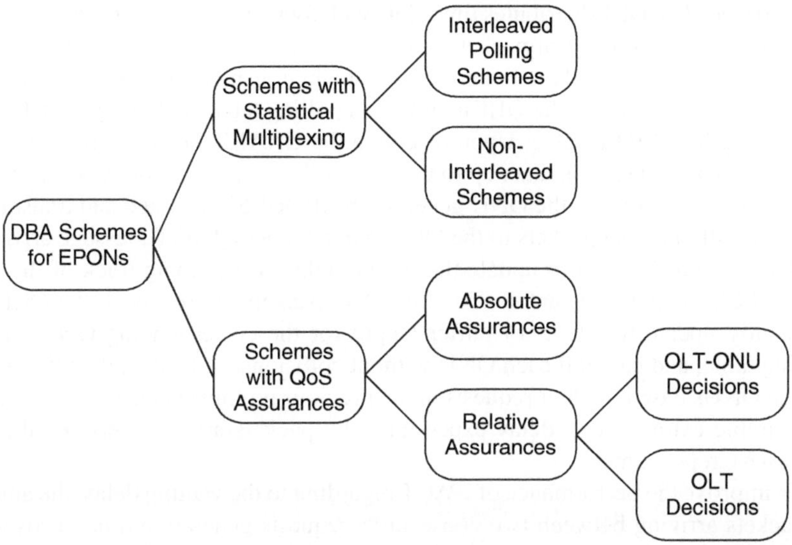

Fig. 6.9 Categories of centralized DBA schemes for EPONs

In EPON networks, multiple ONUs can statistically share the upstream transmission link without any privilege, or the individual ONU can request for the specific bandwidth according to its transmission requirements. Accordingly, the DBA schemes for EPONs are classified into schemes with statistical multiplexing and schemes with QoS assurances.

6.4.1 DBA Schemes with Statistical Multiplexing

According to the transmission mechanism of the downstream GATE messages, the DBA schemes with statistical multiplexing can be further divided into interleaved polling schemes and non-interleaved schemes. In the interleaved schemes, the GATE message transmission is overlapped with the upstream data transmissions. It will not take extra time to transmit the downstream polling messages in such a scheme. Contrarily, there is no interleaved transmission for GATE messages in the non-interleaved scheme. The OLT sends out the GATE message simply upon receiving the REPORT message from the corresponding ONU. So the ONUs will have some idle time to wait for the polling messages transmitted from the OLT.

6.4.1.1 Interleaved Polling Schemes

A typical interleaved scheme called Interleaved Polling with Adaptive Cycle Time (IPACT) is proposed in [22] and [23] for the dynamic bandwidth allocation in EPONs. This OLT-based polling scheme is similar to the hub polling, where next ONU is polled by a GRANT message before data packets from the previous ONU have arrived. The OLT distributes time slots with dynamic window size according to the instantaneous amount of packets buffered in ONUs reported by the REQUEST messages. All ONUs in IPACT share the upstream transmission link statistically.

In the IPACT scheme, the OLT maintains a polling table containing each ONU's buffer length and RTT value. Upon receiving the GRANT message from the OLT, the ONU starts sending its data up to the size of the granted window. At the end of its transmission window, the ONU generates a REQUEST message and transmits it together with the data packets to the OLT. When the REQUEST message arrives at the OLT, it will be used to update the polling table. By keeping track of the time when GRANT messages are sent out and data packets are received, the OLT can constantly update the items of buffer length for the corresponding ONUs in the polling table, and can poll each ONU without collisions. In IPACT, the OLT grants for the ONUs based on their requests in the previous polling cycle. This leads to the undesirable extra waiting delay experienced by packets arriving between the two successive report times.

To improve the performance of IPACT regarding to the waiting delay, the amount of packets arriving between two consecutive requests generation time needs to be estimated by the ONU and reported to the OLT. Then the OLT can decide the granted transmission size for the ONU in the next cycle based on this estimation. With estimation, the grant size can be more approximate to the ONU's buffer occupancy when the ONU is polled for transmission on receiving the GRANT message. And packets waiting delay and buffer occupancy can be shortened accordingly.

One such scheme, named Estimation-Based DBA (EB-DBA), has been proposed in [24]. In EB-DBA, the OLT grants the transmission window to the ONU based on the buffer occupancy and grant size at the previous cycle. Although packets can experience shorter waiting delays than in IPACT, the estimation in EB-DBA is only based on the history values of the grant size and queue length, which is not effective enough to approximate the real buffer occupancy when the ONU is polled. [25] and [26] propose another enhancement to IPACT, IPACT with Grant Estimation (IPACT-GE). The estimation method in IPACT-GE is based on the network traffic's characteristics of self-similarity. It estimates the amount of packets arriving between two pollings according to the observed arrival rate, which reflects the real-time traffic arrival rate. By this method, the grant size can be very close to the instant buffer occupancy when the ONU is polled for transmission. IPACT-GE can improve the performances of IPACT on the average waiting delay and buffer occupancy effectively.

To achieve the higher utilization and support the max-min fairness under the non-uniform traffic, [27] proposes a scheme to allocate the timeslot in consideration of other ONUs queue occupancy. The authors develop two algorithms accordingly. The first algorithm makes use of the recently granted timeslot of other ONUs. It allows the OLT to grant the unused timeslot of other ONUs to the current ONU to meet its request. This algorithm relaxes the maximum timeslot restriction to improve the utilization. While, this algorithm may lead to the fairness problem because the new coming ONU will get much less bandwidth than existing ONUs. The second algorithm aims to achieve the max-min fairness for ONUs. It uses the latest requested queue length of other ONUs and allows the OLT to grant before all requests in the current cycle are collected. This algorithm assumes that there is little difference of the queue information between two successive cycles. This assumption is quite doubtable in EPONs with burst traffic or highly variable traffic.

Reference [28] introduces a simple work for the multiple access control in EPONs. Similar to IPACT, the GRANT message transmission is interleaved with the data packet transmission. The fixed Maximum Transmission Window (MTW) allocation is provided as a simple example. Besides, it builds the detailed models for all components in EPONs for the experimental simulation.

Instead of polling ONUs in a round-robin manner, the authors in [29] consider scheduling the transmission order of multiple ONUs based on the instantaneous traffic condition at each ONU. Two algorithms are proposed in the paper. The Longest-Queue-First (LQF) algorithm schedules the transmission order of different ONUs based on the queue length of each ONU, and polls ONUs in a descending order of the queue lengths in the polling table. The Earliest-Packet-First (EPF) algorithm schedules the transmission order of different ONUs based on the first packet's arrival time of each ONU, and polls ONUs in an ascending order of the arrival time in the polling table.

6.4.1.2 Non-Interleaved Schemes

A polling mechanism with threshold is proposed in [30] for the FTTH EPON networks. Since the propagation time is relatively low due to the short physical distance ($<= 1\ km$) between the OLT and ONUs, the authors suggest that it is unnecessary to deploy the interleaved scheme in the FTTH EPONs. In the proposed mechanism, the OLT sends a GATE frame to the ONU after receiving the data and End of Transmission (EOT) frame from the previous ONU. Each ONU uses the threshold to limit the transmission amount of packet in each cycle. This mechanism is very simple to be implemented in FTTH EPONs. While considering the distance between the OLT and ONUs in EPONs can reach 20 km according to the IEEE 802.3ah EFM Task Force, such a mechanism is not suitable for EPON networks spread to long distances.

In the star topology in [31], a Transmission Upon Reception (TUR) scheme is presented to ensure both downstream and upstream transmissions in EPONs. It allows the ONU to transmit data packets with the amount proportional to the amount it receives from the OLT. In TUR, each ONU keeps a start point indicator to ensure the collision-free access to the upstream transmission link. Thus the control of the OLT is loose in TUR compared to other centralized DBA schemes. Although TUR takes some actions for the OLT to periodically detect the buffer status of ONUs, it still creates the unavoidable fairness issue because the ONUs may have more data to transmit than to receive from the OLT. Moreover, TUR is not compatible with the MPCP protocol because it does not support synchronization between the OLT and ONUs.

In conclusion, the interleaved schemes can effectively improve the channel utilization by reducing the ONUs' waiting time for the GATE messages. They require the OLT to carefully set the time for sending out the GATE messages, so that ONUs can be activated for transmission by the messages without idle time. For a non-interleaved scheme, the operation of the OLT is relatively simple because no interleaving is required for the GATE message transmission. However, such schemes cannot achieve the high utilization because it will take the additional time to transmit downstream GATE messages to ONUs.

Consisting of both the interleaved and non-interleaved schemes, the DBA schemes with statistical multiplexing provide the common service to all ONUs without difference. This group of DBA schemes is easy to be designed and implemented for they don't require meeting the different transmission requirements of end users. However, such schemes are not suitable for the future emerging access networks, where different ONUs request for differentiated service requirements.

6.4.2 DBA Schemes with QoS Assurances

Although next-generation access networks will mainly carry the IP data traffic, traditional services, such as T1/E1s, ISDN, POTS, analog video, etc., will still exist in the foreseeable future. Hence, it is crucial for EPONs to provide both IP-based services and jitter-sensitive and time-critical legacy services that have not been the focus of Ethernet traditionally. That means, EPONs should be designed to transmit the best-effort data together with the time-critical voice and video traffic and fulfill their transmission requirements. Therefore, DBA schemes for EPONs should be able to support the Differentiated Services (DiffServ) and provide Quality of Service (QoS) guarantees to traffic flows with specific service requirements. Recently, a variety of DBA schemes with QoS assurances have been proposed in the literature, some of which can guarantee the absolute service for ONUs or users. While most of the schemes only consider providing the relative QoS assurances.

The schemes with absolute QoS assurances usually consider the Service Level Agreement (SLA) contracts between the service provider and end users. They can support the accurate service just as the contracts specify; in other words, end users will be satisfied with the specified transmission requirements, such as the minimum bandwidth allocation and the maximum tolerable waiting delay. In the contrary, the schemes with relative QoS assurances can only provide end users with the probable service without the quantitative guarantee, because they seldom take SLA into account for the bandwidth allocation.

6.4.2.1 Schemes with Absolute QoS Assurances

A Bandwidth Guaranteed Polling (BGP) scheme has been proposed for EPONs [32–34], which takes the SLA contracts into consideration in the bandwidth allocation scheme design. In BGP, the ONUs are divided into two disjoint groups as bandwidth guaranteed (BG) ONUs and bandwidth non-guaranteed (non-BG) ONUs according to the SLA between the service provider and users. The OLT maintains a scheduling Entry Table and a List to determine the polling sequence of entries and the polling order of non-BG ONUs respectively. Table entries are similar to time slots in a TDM system, which are either allocated to BG ONUs or dynamically assigned to non-BG ONUs. The bandwidth guarantee is achieved for a BG ONU by allocating one or more bandwidth units/entries to it based on the SLA contracts. While, the non-BG ONUs can only be provided with the best-effort service without bandwidth guarantee. Entries in the Table that are not occupied by BG ONUs can be assigned to non-BG ONUs dynamically. In addition, a superfluous transmission window (unused portion of the bandwidth unit/entry) that is occupied yet unconsumed by an ONU may also be assigned to a non-BG ONU dynamically.

BGP deploys the individual polling in the bandwidth allocation, where the OLT grants the transmission window to an ONU without waiting for the REPLY messages from all ONUs. The BGP scheme ensures the BG ONUs receiving the bandwidth guarantee specified by the SLA contracts. It can also achieve the statistical multiplexing for traffic of non-BG ONUs in the unallocated bandwidth units/entries and the superfluous transmission windows. By combining the priority-based mechanisms employed by the ONUs, [35] and [36] extend the BGP scheme to support the DiffServ for EPONs.

References [37] and [38] develop the duel DEB-GPS scheduler for EPONs where the OLT and ONUs employ the Deterministic Effective Bandwidth (DEB) admission control and resource allocation together with the Generalized Processor Sharing (GPS) scheduling. There are multiple queues including QoS and best-effort queues in each ONU. The incoming QoS traffic is constrained by leaky bucket parameters. By using the GPS scheme, each QoS traffic achieves a minimum service rate DEB with delay-constraint and without loss. The surplus bandwidth can be assigned to the best-effort traffic based on the weight. DEB-GPS uses the joint polling, where the master scheduler in the OLT collects requests from all ONUs to calculate the fixed bandwidth assigned to all QoS traffic and the dynamic bandwidth to the best-effort traffic in each ONU. The slave scheduler in the ONU further decides the bandwidth allocated to each queue according to the weight.

In DEB-GPS, the individual QoS traffic can get the deterministic guaranteed service with bounded-delay and lossless transmission. The surplus bandwidth can also be utilized by the best-effort traffic efficiently. While there is the increased complexity to conduct the admission control and update the weight of the traffic's DEB value, leading to the extra overhead to EPON networks.

In [39], the OLT allocates bandwidth to each ONU according to the contracted rate. If an ONU requests less bandwidth than its contracted rate, its remainder bandwidth will be allocated to other ONUs proportional to their contracted rates. In this mechanism, the ONU's REQUEST transmission is separate from its data transmission, which is not compatible with MPCP. Furthermore, the separate transmission of REQUEST and data packets leads to the low upstream utilization due to the double propagation time per ONU.

By incorporating the SLA contracts into bandwidth allocation decision, the DBA schemes with absolute QoS assurances provide the deterministic services to ONUs that can satisfy their specific transmission requirements. In the future access networks, both the service provider and end users will prefer such DBA schemes to ensure the contracted services. However, these schemes may lead to the implementation complexity with regard to the accurate scheduling design. In the following section, the DBA schemes with relative QoS assurances will be addressed whose implementation is generally simpler compared to the schemes with absolute QoS assurances, but they will only provide the rough QoS guarantee to end users.

6.4.2.2 Schemes with Relative QoS Assurances

In EPONs with multiple types of traffic, the ONU receives traffic from the connected users and transmits different types of traffic when it is polled by the OLT. The OLT can make the bandwidth allocation among the multiple types of traffic for the ONU; either, the ONU can schedule the transmission for its own traffic. Hence, there are two kinds of DBA schemes with relative QoS assurances: the schemes with OLT-ONU decision and the schemes with OLT decision. In the schemes with OLT-ONU decision, both the OLT and ONUs jointly decide the bandwidth allocated to each type of traffic. The OLT only needs to issue one grant per ONU in each polling cycle and the ONU further assigns the bandwidth to different types of traffic. In the schemes with OLT decision, ONUs are not the allocation determinant and the OLT alone decides the bandwidth allocated to each ONU as well as to each individual traffic. The OLT will issue multiple grants to each ONU corresponding to the multiple types of traffic. Actually, some DBA schemes allow applying both the OLT-ONU and OLT decision for the bandwidth allocation. Such schemes are difficult to be grouped. Here we only classify the relative DBA schemes into the two groups roughly.

6.4.2.2.1 Schemes with OLT-ONU Decision

Reference [40] extends the IPACT scheme to support differentiated classes of service in EPONs. It focuses on how the message transmission mechanism (MPCP) and the bandwidth allocation scheme (IPACT) can be combined with the priority scheduling. The MAC mechanism is divided into inter-ONU scheduling and intra-ONU scheduling, where IPACT is adopted as the inter-ONU scheduling scheme. The Strict Priority Queuing (SPQ) scheme is deployed for the intra-ONU scheduling in each ONU. With SPQ, packets in each ONU are transmitted strictly according to their priorities. The lower-priority traffic will only be served after the higher-priority traffic being transmitted. Besides, a newly arriving higher-priority packet can preempt lower-priority packets when the finite buffer is not enough to accommodate it. SPQ will result in the performance polarization between different classes of traffic in the same ONU, where the higher-priority traffic gets better-than-required services while the lower-priority traffic starves at the high load. Furthermore, the unexpected network performance of light-load penalty occurs, where the delays for some classes of traffic increase with the decrease of the network load.

To overcome the light-load penalty, the authors suggest two mechanisms named the Two-Stage Queuing (TSQ) scheme and the Constant-Bit-Rate Credit (CBRC) scheme for the intra-ONU scheduling. In TSQ, packets are sequenced in the first stage of priority queues when arriving at the ONU. When it is the turn for transmission, the ONU advances packets from the first stage to the second stage of First-In-First-Out (FIFO) queue before transmitting them over the upstream link. The new arriving packet is not allowed to replace the existing packets in the FIFO queue. The CBRC scheme predicts the number of higher-priority packets arriving between the consecutive REPORT and GATE message, so that the granted window will be large enough to carry all reported packets in the buffer. Both schemes can avoid the light-load penalty effectively.

To enhance the performance of the extended IPACT scheme described in [40], [41], proposes the Dynamic Polling Order Arrangement (DPOA) scheme for the inter-ONU scheduling and Priority with Insertion Scheduling (PIS) scheme for the intra-ONU scheduling, respectively. In DPOA, the polling order is variable for different cycles, which is decided by the current queue length at each ONU. The OLT arranges the polling order at the end of each cycle. The variable polling order can achieve efficient bandwidth utilization under the unbalanced traffic from different ONUs. In PIS, the ONU does not schedule transmission among the traffic flows merely based on the strict priority. Being polled by the OLT, the ONU first sends all the highest priority packets. Then it considers transmitting some packets of the non-real-time traffic before packets of the real-time traffic, as long as the real-time traffic can be satisfied with the QoS requirements. PIS can achieve the fairness to some extent by improving the performance of the non-real-time traffic. While, PIS should consider carefully about the transmission amount of non-real-time packets in the ONU per cycle, in case that the performance of real-time traffic might be impaired.

Reference [42] considers the OLT-level and ONU-level bandwidth allocation corresponding to the inter-ONU and intra-ONU scheduling respectively. The OLT-level allocation deploys a frame-based Weighted Round-Robin (WRR) scheduling scheme where the frame sizes are fixed. The idle capacity of the under-loaded ONUs can be shared in a weighted manner among all over-loaded ONUs. The ONU-level allocation is based on the Start-time Fair Queuing (SFQ) scheme where packets are transmitted in the increasing order of the time-stamps. It reduces the implementation complexity by only maintaining the time-stamp for the Head-Of-Line (HOL) packet in each queue. This scheme can provide the fair scheduling between different traffic flows in the same ONU, while it does not take the traffic's specific QoS requirements into consideration.

Combined with MPCP, [43] and [44] present a bandwidth allocation algorithm by using the threshold reporting method for the individual queue in ONUs. Several threshold values, which can be assigned statically or dynamically, are associated to each queue to determine the Queue Report (QR) fields in the REPORT message. Upon receiving REPORT messages from ONUs, the OLT updates the information table according to the QR values and calculates the bandwidth allocated to each ONU by the relative algorithm. The ONU deploys the Full Priority Scheduling (FPS) and Interval Priority Scheduling (IPS) to transmit traffic from different queues, which are similar to the SPQ and TSQ respectively. However, this method will result in the implementation complexity for ONUs to assign multiple threshold values when generating the REPORT message. Furthermore, the OLT should maintain the additional table to keep the information of the individual queue's request of the ONU, which will cause the extra cost.

A hierarchical DBA scheme is proposed for the OLT's scheduler and ONU's scheduler in [45]. Four classes of priority traffic are considered in the scheme. For the OLT's scheduler, bandwidth is first guaranteed for class *P0*, *P1*, and *P2*. Then it performs the dynamic allocation for class *P1*, *P2* and *P3* with the consideration of queue length and priority. The OLT only sends the aggregated bandwidth to the ONU through the control message. Then the ONU's scheduler performs the similar algorithm with that used by the OLT's scheduler, to allocate the bandwidth among different classes of traffic. This scheme leads to the expense that both the OLT and ONUs perform the algorithm to calculate the bandwidth allocation.

Using virtual time schedulers, [46] proposes the Modified Start-time Fair Queuing (M-SFQ) algorithm for intra-ONU scheduling in EPONs. Per-queue start and finish times are maintained in each ONU. Upon receiving a Grant, the ONU picks up the queue with minimum HOL start time for transmission. The local start and finish times are updated after transmission. Another algorithm, the Urgency Fair Queuing (UFQ) [47] [48], also deploys virtual time for intra-ONU scheduling. UFQ takes into account the QoS requirements of different traffic types and schedules transmission within the ONU according to packets' urgency. It has introduced an Urgency Parameter as the transmission priority, which reflects the packet's QoS requirements including the delay bound requirement, transmission time, etc. With UFQ, the QoS traffic can give some extra bandwidth to the best-effort traffic and still be provided with necessary QoS services. By combining the traffic classification with fair scheduling, UFQ can achieve fairness among different types of traffic in the same ONU.

Reference [49] proposes a fine scheduling mechanism for EPON upstream bandwidth allocation. The inter-ONU scheduling modifies the WRR algorithm deployed in [43] to prevent overloaded ONUs from getting more bandwidth than requested. Each ONU needs to send two requests in the REPORT message: maximum window size requirement and minimum window size requirement. The intra-ONU scheduling deploys a hierarchical scheduler that combines the Modified Token Bucket (M-TB) algorithm [50] and M-SFQ algorithm to achieve both inter-class and intra-class scheduling.

In [51] and [52], both inter-ONU scheduling and intra-ONU scheduling are considered as a whole and only one scheme, the Fair Queuing with Service Envelopes (FQSE), is adopted for the hierarchical scheduling. A Service Envelope (SE) represents the time slot size given to the node. The ONU generates a SE from the SEs of its users and sends to the OLT in the REPORT message. Then in a hierarchical manner, the OLT calculates the time slot start time for each ONU and the ONU decides the transmission start time for its users. FQSE can achieve the fairness among all end users, which is called the cousin-fair by the authors.

6.4.2.2.2 Schemes with OLT Decision

References [53] and [54] first propose a priority-based algorithm for the intra-ONU scheduling to overcome the light-load penalty in SPQ. Being polled by the OLT, an ONU first transmits packets arriving before it sent out the REPORT message based on their priorities. If all such packets are served and the current transmission window can carry more packets, then packets arriving after the report time can be transmitted. This is actually the SPQ scheme with the gated service. It can improve the fairness among all classes of traffic by allowing them access to the channel as reported to the OLT.

Then the authors present a dynamic bandwidth allocation scheme where the OLT assigns the guaranteed bandwidth to ONUs in proportion to the SLA requirements. The excess bandwidth is further distributed among the heavy-loaded ONUs proportional to their requests. In this scheme, the OLT should decide the allocation to all ONUs after getting REPORT messages from them, which will result in some idle time when the ONUs are waiting for the GRANT messages. In order to compensate the idle time, the authors deploy a gate-ahead mechanism where the light-loaded ONUs are scheduled for transmission early with the request size, without waiting until all REPORT messages reach the OLT. Within the allocated time slot, the ONU can request the OLT to decide the bandwidth for each class of traffic. It should report individual queue status to the OLT, and the OLT will generate multiple transmission grants to the ONU for different queues.

Reference [55] has proposed an improvement of scheduling control mechanism to address the idle period problem. The OLT maintains a tracker to record the ending time of the last scheduled ONU's timeslot and updates the tracker every time the next ONU is scheduled. It also employs an early allocation mechanism to the light-loaded ONU. Besides, the OLT accumulates the excessive bandwidth from the light-loaded ONU and schedules a heavy-loaded ONU instantaneously if two conditions are met: the tracker value is smaller than the ending time of the idle period; the next REPORT message arrives later than the tracker time. Compared to the one in [53], network performances can be improved in terms of packet delay, queue length and throughput with high traffic load.

A cyclic polling scheme is suggested in [56] and [57] that is typically a joint polling scheme. The OLT grants transmission for each of the three priority queues in ONUs at the beginning of each cycle. The high-priority traffic is first assigned the fixed bandwidth in the scheme. Then the remainder bandwidth is allocated to the medium-priority traffic flows trying to meet their requests. If there is still bandwidth left, it is distributed between all low-priority traffic flows in proportion to their requests. The OLT has the extra burden to control both the scheduling between ONUs and the scheduling within an ONU. Moreover, the ONUs with only low-priority traffic will starve for transmission at the high system load by the scheme.

Reference [58] introduces a Hybrid Slot Size/Rate (HSSR) scheme to minimize the traffic's delay variation. HSSR divides a frame into two parts, the steady part and the dynamic part, for the transmission of two classes of traffic. The steady part of the frame is used to carry the high-priority traffic, where the fixed time slot is allocated to each ONU for the high-priority traffic transmission according to the user's subscription rate. Thus the minimum delay variation can be achieved for the high-priority traffic. The dynamic part of the frame is allocated to transmit the best-effort traffic in ONUs. If the high-priority traffic cannot consume the fixed time slot, the surplus bandwidth can be assigned to the best-effort traffic in the same ONU. This is a drawback that other ONUs cannot share the surplus bandwidth. In the case that the rate of high-priority traffic exceeds its subscription rate so that the fixed time slot is not enough for transmission, the excess high-priority traffic will be redirected into the best-effort queue and be transmitted in the dynamic part of the frame. This redirection mechanism will definitely degrade the QoS received by the high-priority traffic.

Based on the similar idea with [58], authors of [59] and [60] propose a Two-Layer Bandwidth Allocation (TLBA) scheme including the class-layer and ONU-layer allocation for EPONs. The class-layer allocation divides the frame into three sub frames for three classes of traffic. The sub frame sizes are variable according to the requests of the corresponding service classes. Using the weight, a bandwidth threshold is set for each class of traffic to guarantee the minimum bandwidth. The surplus bandwidth of the light-loaded traffic can be distributed to other classes of traffic proportional to their weights. In the ONU-layer allocation, the bandwidth of each traffic class is distributed to all ONUs following the max-min fairness principle. To determine weights for the frame-level bandwidth allocation, [61] proposes a frame division method for supporting prioritized DBA schemes for three traffic classes in EPONs. This method decides weights according to the traffic load and access delay bound of each traffic class.

For HSSR and TLBA, an ONU should separately transmit different classes of traffic by multiple segments in one polling cycle. Each segment of transmission experiences one propagation time. Moreover, a guard time is required for two consecutive transmission segments from different ONUs. Consequently, HSSR and TLBA will increase the overhead of the upstream propagation time and lead to the extra guard time, resulting in low channel utilization, especially for EPON networks with many classes of traffic.

In [62], a Burst-Polling (BP) based Delta Dynamic Bandwidth Allocation (DDBA) scheme is proposed for the inter-ONU scheduling. There are three classes of traffic, Expedited Forwarding (EF), Assured Forwarding (AF) and Best Effort (BE), in the system. The OLT sends multiple GATE messages to ONUs in a burst manner after receiving REPORT messages from all ONUs. In DDBA, the requested bandwidth of EF and AF traffic can be adjusted by the request difference between two consecutive cycles. In the priority-based intra-ONU scheduling method, two different sets of bandwidth weight, for normal and high EF profile respectively, are applied to get the grant size for each class of traffic in the same ONU.

A two-step scheduling scheme is presented in [63] to reduce the scheduling complexity. It separates the process of the transmission start-time decision from the process of the GATE message generation. In the first step, the grant scheduler converges four kinds of GATE to produce the grant size for each ONU. Then in the second step, the scheme decides the grant start-time considering the grant size and RTT of each ONU. The OLT's configuration will be very complex and the extra burden is exerting on the OLT to maintain the multiple GATE queues.

Reference [64] introduces a Dynamic Credit Distribution (D-CRED) scheme in order to improve the utilization of EPONs. It uses two separate requests in one REPORT message including the buffer occupancy and the MAC frame boundary. The OLT also has two grants for one ONU per cycle: one is for data packet transmission and another is for the REPORT message transmission. According to the requests of ONUs, the OLT dynamically changes the credit value that represents the assigned bandwidth to each ONU for data transmission. In order to support QoS in EPONs, the authors apply a satisfaction-based credit allocation and a weighted-fair credit allocation method for the fair allocation based on D-CRED.

A Dynamic Bandwidth Allocation with Multiple Services (DBAM) algorithm has been presented in [65] to incorporate REPORT/GATE mechanism with class-based bandwidth allocation. It applies Priority Queuing (PQ) to enqueue different classes of traffic, and adopts Limited Bandwidth Allocation (LBA) to arbitrate bandwidth allocation among ONUs. DBAM employs a classed-based traffic prediction to estimate traffic arriving during the waiting time, and takes this part of traffic into account for bandwidth allocation. The estimation of current interval is made according to the information of the previous interval. The authors then extended this method to Limited Sharing with Traffic Prediction (LSTP) in [66]. LSTP predicts the traffic arriving during the waiting time according to the actual traffic amounts in previous L intervals. Data delay can be reduced with LSTP by allocating extra bandwidth based on the prediction.

References [67] has proposed a Class-of-service Oriented Packet Scheduling (COPS) scheme to provide global fairness for different Class-of-Services (CoSs) within an ONU and also between ONUs. It employs a credit pooling scheme that allows the combined per-CoS traffic from all ONUs add up to the amount agreed upon bandwidth partitioning process. The OLT maintains credit pools for different CoSs and for different ONUs, respectively. COPS uses a weighted-share policy that allows different ONUs to obtain a portion of the bandwidth allocated to each CoS in proportion to their weights. This algorithm includes two rounds of execution. The first round gives rise to the allocation of most of the available grants to ONUs. The second round distributes the unallocated grants from round 1 between unsatisfied ONUs. After collecting report messages from all ONUs, the OLT executes COPS at the beginning of every transmission cycle.

To eliminate the idle time of executing COPS algorithm, a Fast COPS (FCOPS) has been presented in [68]. In FCOPS, the OLT computes grants for each ONU instantaneously upon receiving its request message. The algorithm includes only one round of execution. FCOPS may introduce fairness problem that first few ONUs may use up credits of CoSs. To solve this problem, a traffic shaping mechanism should be employed for each CoS at every ONU.

Generally, in the schemes with OLT-ONU decision, the OLT's operation can be simplified by appointing ONUs to schedule transmissions among different types of traffic. The GATE and REPORT message only need to contain one grant and one request for each ONU. However, the operation of ONUs will be complicated accordingly. In addition, because the OLT may not consider the status of each queue when making the decision, the bandwidth allocation results may be unfair to the individual traffic. For the schemes with OLT decision, grants are "colored" that they specify how many bytes the ONU should transmit for each traffic. An ONU's implementation can be very simple because it only needs to transmit traffic according to the

Table 6.1 Summary of centralized TDMA-based DBA schemes in EPONs

Category			Strength	Weakness
Schemes with statistical multiplexing	Category common features		- multiple ONUs statistically share the upstream transmission link without any privilege - provide the common service to all ONUs without difference - easy to design and implement	- cannot meet different transmission requirements of end users - not suitable for the future emerging access networks
	Interleaved polling Schemes		- the downstream polling message transmission is overlapped with the upstream data transmission - will not take extra time to transmit the downstream polling messages - effectively improve the channel utilization	- the OLT is required to carefully set the time for sending out the downstream polling messages
	Non-interleaved schemes		- no interleaved transmission for downstream polling messages - the operation of the OLT is relatively simple	- the ONUs have to wait for the polling messages transmitted from the OLT - the channel utilization is low
Schemes with QoS assurances	Category common features		- individual ONU can request specific bandwidth to meet its transmission requirements - can guarantee QoS for different end users	- relatively complicated to design and implement the schemes
	Schemes with absolute QoS assurances		- can support the accurate service as the SLA contracts specify - provide the deterministic services to ONUs that can satisfy their specific transmission requirements	- may lead to the implementation complexity
		Category common features	- implementation is relatively simple	- only provide the probable QoS without quantitative guarantee

Table 6.1 (continued)

	Schemes with Relative QoS assurances	OLT-ONU decision	- both the OLT and ONUs participate to decide the bandwidth allocation - the OLT only needs to issue one grant per ONU and the ONU further assigns the bandwidth to its traffic - the OLT's operation is relatively simple - the control messages only need to contain one grant and one request for each ONU	- the operation of ONUs is complicated - the bandwidth allocation may be unfair to the individual traffic because the OLT may not consider the status of each queue
		OLT decision	- the OLT alone decides the bandwidth allocated to each ONU and to each individual traffic - the OLT will issue multiple grants to each ONU corresponding to the multiple types of traffic - the ONU's implementation is relatively simple - each type of traffic is liable to obtain the fairness	- ONUs need to report each queue's status to the OLT, and the OLT is required to issue multiple grants to an ONU - the OLT needs to decide bandwidth assignment for each individual queue

OLT's decision. Each type of traffic is more possible to obtain the fairness compared to the schemes with OLT-ONU decision. For such schemes, an ONU is required to report each queue's status to the OLT, and the OLT needs to issue multiple grants to the ONU in each polling cycle. The OLT is exerted additional burden to decide the bandwidth assignment for each individual queue in the ONU.

6.5 Summary

Firstly, this chapter has reviewed the EPON technology that is considered to be the best candidature for the next-generation access networks. Then a thorough survey has been provided on the MAC protocols for EPON upstream transmissions, among which the centralized TDMA-based DBA schemes are the most promising ones.

A classification has been presented to divide the numerous DBA schemes into corresponding groups. Besides the description and comments for each individual scheme, this chapter has also provided the common features as well as the merits and shortcomings for each category of DBA schemes. Table 6.1 gives a summary for all categories of existing centralized TDMA-based DBA schemes for the EPON networks.

References

1. G. Kramer, B. Mukherjee, and A. Maislos, "Multiprotocol over DWDM: Building the Next Generation Optical Internet", *John Wiley & Sons*, Mar. 2003.
2. G. Kramer, G. Pesavento, "Ethernet Passive Optical Network (EPON): Building a Next-Generation Optical Access Network", *IEEE Communications Magazine*, Vol. 40, Issue 2, Feb. 2002, pp. 66–73.
3. ITU-T Recommendation G.983.1, "Broadband Optical Access Systems Based on Passive Optical Networks (PON)," Oct. 1998.
4. H. Ueda, K. Okada, B. Ford, G. Mahony, S. Hornung, D. Faulkner, J. Abiven, S. Durel, R. Ballard, J. Ericson, "Deployment Status and Common Technical Specifications for a B-PON System", *IEEE Communications Magazine*, Vol. 39, Issue 12, Dec. 2001, pp. 134–141.
5. ITU-T Recommendation G.984.3, "Gigabit-capable Passive Optical Networks (GPON): Transmission Convergence Layer Specification", Oct. 2003.
6. J.D. Angelopoulos, H-C. Leligou, T. Argyriou, S. Zontos, "Efficient Transport of Packets with QoS in an FSAN-Aligned GPON", *IEEE Communications Magazine*, Vol. 42, Issue 2, Feb. 2004, pp. 92–98.
7. IEEE 802.3ah Ethernet in the First Mile Task Force: http://www.ieee802.org/3/efm/.
8. H. Frazier, G. Pesavento, "Ethernet Takes on the First Mile", *IT Professional*, Vol. 3, Issue 4, July-Aug. 2001, pp. 17–22.
9. K.H. Kwong, D. Harle, I. Andonovic, "Dynamic Bandwidth Allocation Algorithm For Differentiated Services over WDM EPONs", *Proceedings of The Ninth IEEE International Conference on Communications Systems (ICCS'2004)*, Sep. 2004, pp. 116–120.
10. G. Thomas, "Improved Performance in Ethernet Passive Optical Networks", *Proceedings of the First International Workshop on Community Networks and FTTH/P/x (CNFT)*, Oct. 2003.
11. M.P. McGarry, M. Reisslein, M. Maier, "WDM Ethernet Passive Optical Networks", *IEEE Communications Magazine*, Vol. 44, Issue 2, Feb. 2006, pp. 15–22.
12. A.R. Dhaini, C.M. Assi, M. Maier, A. Shami, "Dynamic Wavelength and Bandwidth Allocation in Hybrid TDM/WDM EPON Networks", *Journal of Lightwave Technology*, Vol. 25, Issue 1, Jan. 2007, pp. 277–286.
13. C. Chang-Joon, E. Wong, R.S. Tucker, "Optical CSMA/CD Media Access Scheme for Ethernet Over Passive Optical Network", *IEEE Photonics Technology Letters*, Vol. 14, Issue 5, May 2002, pp. 711–713.
14. E. Wong, C.-J. Chae, "CSMA/CD-based Ethernet Over Passive Optical Network for Delivery of Voice-over-IP Traffic", *Proceedings of The 16th Annual Meeting of the IEEE Lasar & Electro-Optics Society*, Oct. 2003, pp. 447–448.
15. E. Wong, C. Chang-Joon "CSMA/CD-Based Ethernet Passive Optical Network with Optical Internetworking Capability Among Users", *IEEE Photonics Technology Letters*, Vol. 16, Issue 9, Sep. 2004, pp. 2195–2197.
16. W. Yang, T.H. Cheng, "A Hybrid-Slot Decentralized Control Protocol With Quality of Service for Passive Optical Networks", *Proceedings of The Ninth IEEE International Conference on Communications Systems (ICCS'2004)*, Sep. 2004, pp. 505–507.
17. C. Foh, L. Andrew, M. Zukerman, E. Wong, "FULL-RCMA: A High Utilization EPON", *Proceedings of Optical Fiber Communications Conference (OFC'2003)*, Vol. 1, Mar. 2003, pp. 282–284.
18. C.H. Foh, L. Andrew, E. Wong, M. Zukerman, "FULL-RCMA: A High Utilization EPON", *IEEE Journal on Selected Areas in Communications*, Vol. 22, Issue 8, Oct. 2004, pp. 1514–524.
19. A. Hadjiantonis, S. Sherif, A. Khalil, T. Rahman, G. Ellinas, M. F. Arend, M.A. Ali, "A Novel Decentralized Ethernet-Based Passive Optical Network Architecture", *Proceedings of IEEE International Conference on Communications (ICC'2004)*, Vol. 3, June 2004, pp. 1781–1785.

20. S.R. Sherif, A. Hadjiantonis, G. Ellinas, C. Assi, M.A. Ali, "A Novel Decentralized Ethernet-Based PON Access Architecture for Provisioning Differentiated QoS", *Journal of Lightwave Technology*, Vol. 22, Issue 11, Nov. 2004, pp. 2483–2497
21. P.M. Micheal, M., Martin, R., Martin, "Ethernet PONs: A Survey of Dynamic Bandwidth Allocation (DBA) Algorithms", *IEEE Communications Magazine*, Vol. 42, Issue 8, Aug. 2004, pp. S8–S15.
22. G. Kramer, B. Mukherjee, G. Pesavento, "IPACT: A Dynamic Protocol for an Ethernet PON (EPON)", *IEEE Communications Magazine*, Vol. 40, Issue 2, Feb. 2002, pp. 74–80.
23. G. Kramer, B. Mukherjee, G. Pesavento, "Interleaved Polling with Adaptive Cycle Time (IPACT): A Dynamic Bandwidth Distribution Scheme in an Optical Access Network", *Photonic Network Communications*, Vol. 4, Issue 1, Jan. 2002, pp. 89–107.
24. H.-J. Byun, J.-M. Nho, J.-T. Lim, "Dynamic Bandwidth Allocation Algorithm in Ethernet Passive Optical Networks", *Electronics Letters*, Vol. 39, Issue 13, June 2003, pp. 1001–1002.
25. Y. Zhu, M. Ma, T.H. Cheng, "IPACT with Grant Estimation for EPON", *Proceedings of the 10th IEEE International Conference on Communications Systems (ICCS'2006)*, Oct. 2006, pp. 1–5
26. Y. Zhu, M. Ma, "IPACT with Grant Estimation (IPACT-GE) Scheme for Ethernet Passive Optical Networks", *Journal of Lightwave Technology*, Vol. 26, Issue 14, July 2008, pp. 2055–2063
27. K. Son, H. Ryu, S. Chong, T. Yoo, "Dynamic Bandwidth Allocation Schemes to Improve Utilization Under Non-Uniform Traffic in Ethernet Passive Optical Networks", *Proceedings of IEEE International Conference on Communications (ICC'2004)*, Vol. 3, June 2004, pp. 1766–1770.
28. T. Shan, J. Yang, C. Sheng, "EPON Upstream Multiple Access Scheme", *Proceedings of International Conferences on Info-tech and Info-net (ICII'2001)*, Vol. 2, Oct. 2001, pp. 273–278.
29. J. Zheng, H.T. Mouftah, "Adaptive Scheduling Algorithms for Ethernet Passive Optical Networks", *Communications, IEE Proceedings*, Vol. 152, Issue 5, Oct. 2005, pp. 643–647
30. C. Jihyoung, L. Taesik, H. Ikpyo, K. Sunkyoung, Batchuluun, "Modified of polling mechanism with threshold and precondition for market activation in Ethernet PON", *Proceedings of International Conference on Communication Technology (ICCT'2003)*, Vol. 1, Apr. 2003, pp. 738–742.
31. A. Gumaste, I. Chlamtac, "A Protocol to Implement Ethernet Over PON", *Proceedings of IEEE International Conference on Communications (ICC'2003)*, Vol. 2, May 2003, pp. 1345–1349.
32. M. Ma, Y. Zhu, T.H. Cheng, "A Bandwidth Guaranteed Polling MAC Protocol for Ethernet Passive Optical Networks", *Proceedings of IEEE INFOCOM'2003*, Vol. 1, Mar–Apr. 2003, pp. 22–31.
33. Y. Zhu, M. Ma, T.H. Cheng, "A Novel Multiple Access Scheme for Ethernet Passive Optical Networks", *Proceedings of IEEE GLOBECOM'2003*, Vol. 5, Dec. 2003, pp. 2649–2653.
34. M. Ma, Y. Zhu, T.H. Cheng, "A Systematic Scheme for Multiple Access in Ethernet Passive Optical Access Networks", *Journal of Lightwave Technology*, Vol. 23, Issue 11, Nov. 2005, pp. 3671–3682
35. Y. Zhu, M. Ma, T.H. Cheng, "Hierarchical Scheduling to Support Differentiated Services in Ethernet Passive Optical Networks", *Computer Networks*, Vol. 50, Issue 3, Feb. 2006, pp. 350–366
36. Y. Zhu, M. Ma, T.H. Cheng, "An Efficient Solution for Mitigating Light-Load Penalty in EPONs", *Computers & Electrical Engineering*, Vol. 32, Issue 6, Nov. 2006, pp. 426–431
37. L. Zhang, E.-S. An, C.-H. Youn, H.-G. Yeo, S. Yang, "Dual DEB-GPS Scheduler for Delay-Constraint Applications in Ethernet Passive Optical Networks", *IEICE Transactions on Communications*, Vol. E86-B, No.5, May 2003, pp. 1575–1584.
38. L. Zhang, G.-S. Poo, "Delay Constraint Dynamic Bandwidth Allocation in Ethernet Passive Optical Networks", *Proceedings of The Ninth IEEE International Conference on Communications Systems (ICCS'2004)*, Sep. 2004, pp. 126–130.

39. X. Chen, M. Yu, Y. Zhang, Y. Deng, "A Novel Upstream Dynamic Bandwidth Assignment Scheme for Ethernet PONs", *Proceedings of International Conference on Communication Technology (ICCT'2003)*, Vol. 1, Apr. 2003, pp. 748–750.
40. G. Kramer, B. Mukherjee, R. Hirth, "Supporting Differentiated Classes of Service in Ethernet Passive Optical Networks", *Journal of Optical Networking*, Vol. 1, Issue 8&9, 2002, pp. 280–298.
41. M. Ma, L. Liu, T.H. Cheng, "Adaptive Scheduling for Differentiated Services in Ethernet Passive Optical Networks", *Proceedings of The Ninth IEEE International Conference on Communications Systems (ICCS'2004)*, Sep. 2004, pp. 102–106
42. N. Ghani, A. Shami, C. Assi, M.Y.A. Raja, "Quality of Service in Ethernet Passive Optical Networks" *Proceedings of IEEE/Sarnoff Symposium on Advances in Wired and Wireless Communication*, Apr. 2004, pp. 161–165.
43. D. Nikolova, B. Van Houdt, C. Blondia, "Dynamic Bandwidth Allocation Algorithms in EPON: A Simulation Study", *Proceedings of Opticomm'2003*, pp. 369–380.
44. D. Nikolova, B. Van Houdt, C. Blondia, "QoS Issues in EPON", *Proceedings of the First International Workshop on Community Networks and FTTH/P/x*, Oct. 2003.
45. K.-H. Ahn, K.-E. Han, Y.-C. Kim, "Hierarchical Dynamic Bandwidth Allocation Algorithm for Multimedia Services over EPONs", *ETRI Journal*, Vol. 26, Issue 4, Aug. 2004, pp. 321–330.
46. N. Ghani, A. Shami, C. Assi, M.Y.A. Raja, "Intra-ONU Bandwidth Scheduling in Ethernet Passive Optical Networks", *Communications Letters, IEEE*, Vol. 8, Issue 11, Nov. 2004, pp. 683–685
47. Y. Zhu, M. Ma; T.H. Cheng, "An Urgency Fair Queuing Scheduling to Support Differentiated Services in EPONs", *Proceedings of IEEE GLOBECOM'2005*, Vol. 4, Nov. 2005, pp. 1925–1929
48. Y. Zhu, M. Ma, "Supporting Differentiated Services with Fairness by an Urgency Fair Queuing Scheduling Scheme in EPONs", *Photonic Network Communication*, Vol 12, Issue 1, 2006, pp. 99–110
49. B. Chen, J. Chen, S. He, "Efficient and Fine Scheduling Algorithm for Bandwidth Allocation in Ethernet Passive Optical Networks", *IEEE Journal of Selected Topics in Quantum Electronics*, Vol. 12, Issue 4, July–Aug. 2006, pp. 653–660
50. J. Chen, B. Chen, S. He, "A Novel Algorithm for Intra-ONU Bandwidth Allocation in Ethernet Passive Optical Networks", *Communications Letters, IEEE*, Vol. 9, Issue 9, Sep 2005, pp. 850–852
51. G. Kramer, A. Banerjee, N. Singhal, B. Mukherjee, S. Dixit, Y. Ye, "Fair Queuing with Service Envelopes (FQSE): A Cousin-Fair Hierarchical Scheduler for Ethernet PONs", *Proceedings of Optical Fiber Communications Conference (OFC'2004)*, Vol. 1, Feb. 2004. pp. 833–835.
52. G. Kramer, A. Banerjee, N. Singhal, B. Mukherjee, S. Dixit, Y. Ye, "Fair Queueing With Service Envelopes (FQSE): A Cousin-Fair Hierarchical Scheduler for Subscriber Access Networks", *IEEE Journal on Selected Areas in Communications*, Vol. 22, Issue 8, Oct. 2004, pp. 1497–1513.
53. C.M. Assi, Y. Ye, S. Dixit, M.A. Ali, "Dynamic Bandwidth Allocation for Quality-of-Service Over Ethernet PONs", *IEEE Journal on Selected Areas in Communications*, Vol. 21, Issue 9, Nov. 2003, pp. 1467–1476.
54. C. Assi, Y. Ye, S. Dixit, "Support of QoS in IP-based Ethernet-PON", *Proceedings of IEEE GLOBECOM'2003*, Vol. 7, Dec. 2003, pp. 3737–3741.
55. J. Zheng, "Efficient Bandwidth Allocation Algorithm for Ethernet Passive Optical Networks", *Communications, IEE Proceedings*, Vol. 153, Issue 3, June 2006, pp. 464–468
56. S. Choi, "Cyclic Polling-Based Dynamic Bandwidth Allocation for Differentiated Classes of Service in Ethernet Passive Optical Networks", *Photonic Network Communications*, Vol. 7, Issue 1, Jan. 2004, pp. 87–96.
57. S. Choi, J.-d. Huh, "Dynamic Bandwidth Allocation Algorithm for Multimedia Services over Ethernet PONs," *ETRI Journal*, Vol. 24, Issue 6, Dec. 2002, pp. 465–468.

58. F. An, H. Bae, Y. Hsueh, M. Rogge, L. Kazovsky, K. Kim, "A New Media Access Control Protocol Guaranteeing Fairness Among Users in Ethernet-Based Passive Optical Networks", *Proceedings of Optical Fiber Communications Conference (OFC'2003)*, Vol. 1, Mar. 2003, pp. 134–135.
59. J. Xie, S. Jiang, Y. Jiang, "A Dynamic Bandwidth Allocation Scheme for Differentiated Services in EPONs", *IEEE Communications Magazine*, Vol. 42, Issue 8, Aug. 2004, pp. S32–S39.
60. J. Xie, S. Jiang, Y. Jiang, "A Class-based Dynamic Bandwidth Allocation Scheme for EPONs", *Proceedings of The Ninth IEEE International Conference on Communications Systems (ICCS'2004)*, Sep. 2004, pp. 357–360.
61. S. Jiang, J. Xie, "A Frame Division Method for Prioritized DBA in EPON", *Selected Areas in Communications, IEEE Journal on*, Vol. 24, Issue 4, Apr. 2006, pp. 83–94
62. Y.-M. Yang, J.-M. Nho, B.-H. Ahn, "An Enhanced Burst-Polling based Delta Dynamic Bandwidth Allocation Scheme for QoS over E-PONs", Proceedings of the ACM Workshop on Next-generation Residential Broadband Challenges *(NRBC'2004)*, 2004, pp. 31–36.
63. H.-S. Lee, T.-W. Yoo, J.-H. Moon, H.-H. Lee, "A Two-Step Scheduling Algorithm to Support Dual Bandwidth Allocation Policies in an Ethernet Passive Optical Network", *ETRI Journal*, Vol. 26, Issue 2, Apr. 2004, pp. 185–187.
64. H. Miyoshi, T. Inoue, K. Yamashita, "QoS-aware Dynamic Bandwidth Allocation Scheme in Gigabit-Ethernet Passive Optical Networks", *Proceedings of IEEE International Conference on Communications (ICC'2004)*, Vol. 1, June 2004, pp. 90–94.
65. L. Yuanqiu, N. Ansari, "Bandwidth Allocation for Multiservice Access on EPONs", *Communications Magazine, IEEE*, Vol. 43, Issue 2, Feb. 2005, pp. S16–S21
66. L. Yuanqiu, N. Ansari, "Limited Sharing with Traffic Prediction for Dynamic Bandwidth Allocation and QoS Provisioning over Ethernet Passive Optical Networks", *Journal of Optical Networking*, Vol. 4, Issue 9, Sep. 2005, pp. 561–572
67. H. Naser, H.T. Mouftah, "A Joint-ONU Interval-Based Dynamic Scheduling Algorithm for Ethernet Passive Optical Networks", *Networking, IEEE/ACM Transactions on*, Vol. 14, Issue 4, Aug. 2006, pp. 889–899
68. H. Naser, H.T. Mouftah, "A Fast Class-of-Service Oriented Packet Scheduling Scheme for EPON Access Networks", *Communications Letters, IEEE*, Vol. 10, Issue 5, May 2006, pp. 396–398

Chapter 7
Scheduling Transmission of Multimedia Video Traffic on WDM Passive Optical Access Networks

Yang Qin and Maode Ma

Abstract Passive optical networks have shown to be much more attractive than traditional access networks due to their huge bandwidth provisioning. In this paper, we study the issue of QoS service of transmitting video traffic over a passive wavelength division multiplexing (WDM) optical network, which serves as an access network. We explore the possibility of providing an integrated service directly at the WDM optical layer with consideration of the unique QoS requirements from video traffic stream. We introduce a novel multimedia traffic scheduling framework for the QoS service in the WDM passive optical access networks. The simulation results show that the proposed mechanism is capable of providing QoS guarantee for MPEG video applications while achieving high overall throughput.

Keywords QoS service · WDM passive optical access network · Multimedia traffic · Scheduling

7.1 Introduction

In the new era of information technology, extensive usage of bandwidth intensive applications will create increasing much heavier traffic load over computer networks. It is evident that multimedia traffic loads are becoming much more common and applications such as video on demand are also becoming much more popular [1].

In satisfying the increasing demands for bandwidth, optical networks and Wavelength Division Multiplexing (WDM) technologies represent a unique opportunity due to their potential, almost unlimited bandwidth. The WDM optical networks have the capability to allow multiple channels transmission on a single optic fiber. This great property has led WDM technology to emerge as the predominant data communication technology for the next generation network. Not only in backbone networks, passive WDM optical networks can also play important roles in access

Y. Qin (✉)
School of Electrical and Electronic Engineering, Nanyang Technological University, Nanyang Avenue, Singapore 639798

DOI 10.1007/978-1-4020-9889-5_7,

networks to play a potential role of Fiber to the "last mile". Currently, much attention has focused on the Ethernet passive optical networks (EPON). As an alternative, passive WDM optical networks are emerging as a promising broadband access technique which will meet the ever-increasing bandwidth requirement from the end users [2–4]. Supported by the WDM technique, gigabit service to each optical network unit (ONU) is possible because one optical fiber could provide 320 Gbps data transmission rate nowadays. With the huge bandwidth of several Gbps, each ONU is able to support various multimedia applications, which require a lot of bandwidth for video and audio transmissions, such as video on demand applications. With the vast bandwidth, each ONU can manage a large number of users attached to it. It makes the fiber-to-the-building solution much more feasible and economical. Conventional EPON architectures only support both downstream and upstream transmission with a limited set of wavelength channels for communications between the optical line terminal (OLT) and each ONU. The ONUs cannot directly communicate with each other. The inter-ONU traffic must first be transmitted, via the upstream carrier, back to the OLT, where it is electronically routed [2] and modulated on the respective downstream carriers that are destined to other ONUs. This consumes much bandwidth on both the downstream and upstream carriers. Besides, the additional round-trip propagation time between the ONU and the OLT as well as the increased traffic loading to the OLT for scheduling and routing of the inter-ONU traffic, imposes extra latency to the inter-ONU traffic. Therefore, it is desirable to provide direct connections for the communication among ONUs in the optical access networks. There are some research proposals based on $(N+1) \times (N+1)$ star coupler to provide information broadcasting of the inter-ONU traffic [5, 6].

By the WDM technique with a passive star coupler, each ONU could have direct communication to any of other ONUs including the OLT, which locates in the central office to have the backbone network connections. Each ONU could use a unique wavelength as a transmission channel for the direct communication without interfering with other ONUs' transmission so that upstream bandwidth sharing problem in EPON will be gone. Furthermore, the solution supported by WDM technique to the access works can provide an extensive flexibility in the sense that there is no much difference between the ONUs and the OLT anymore. Since the operations of the passive WDM optical networks require each ONU to equip a pair of tunable transceivers, the only issue which blocks the WDM technique to be an attractive solution for access networks is the cost of the tunable transceivers. One solution to reduce the cost of the transceivers is to equip one tunable transmitter and one fixed receiver tuned to a home channel for each ONU. Then the total cost on the equipments could be sufficiently reduced. However, with the future development and wide application of the optical network products, the cost of transceivers could not be a problem to block the WDM optical networks being used as access networks.

In this paper, we consider constructing a single-hop passive star coupler based WDM optical network as an access network, in which all nodes including ONUs and the OLT will be connected to a single passive star coupler (PSC). We further study the issue of QoS service to support compressed video traffic on such an optical access network. As a solution to the problem, we propose and design an efficient and practical MAC protocol to control and schedule the transmission of the video traffic. The rest of the paper is organized as follows. In Section 7.2, we describe the proposed network configuration and the compressed video traffic. We propose our scheduling mechanism in Section 7.3. In Section 7.4, we depict the results of our simulation study on the proposed scheme. Section 7.5 concludes the paper with a summary.

7.2 Network Configuration and Traffic

Existing research has largely focused on scheduling of data traffic [7–10] over WDM optical networks. Relatively less attention has been given to the development of proper scheduling algorithm which can support multimedia applications. In [7], the authors proposed three algorithms scheduling variable-length messages in a single-hop multi-channel PSC-based optical network. The algorithms concentrate on how to reduce the penalty of large tuning overhead and schedule variable-length messages. In [8], the authors proposed a scheduling algorithm to improve the network performance when both real-time and non real-time messages are transmitted in one single-hop network topology. In [9], the authors proposed an efficient algorithm to schedule non-uniform packet traffic over a tunable transmitter and fixed receiver (TT-FR) optical network. In [10], a traffic scheduling for heterogeneous traffic has been proposed. However, less attention has been paid to the MPEG video transmission over optical networks.

In this paper, we consider constructing a, single-hop PSC-based WDM optical network with N-nodes as N ONUs. There are $W + 1$ wavelengths as $W + 1$ logical channels in the system, where W channels are used for actual data transmission, referred as data channels. One channel is used exclusively for reservation or control information exchange, called control channel. Each ONU is equipped with one transmitter and one receiver, both fixed tuned to the control channel for control message transmission and reception. Additionally, each node has one pair of transceivers tunable to all the data channels for data transmission. Data channels and control channel are assumed to be synchronized into time slots. The time on the data channel has been divided into data slots and the time on the control channel has been divided into control slots. The control slot is further divided into mini-slots with each holding a control packet from each ONU.

The traffic streams in the network are from MPEG video applications. A MPEG video bit stream is a video sequence, each of which is divided into one or more groups of pictures. Each group of picture is composed of one or more pictures of three different types of frames, I-, P- and B- frames. I-frames are coded independently, entirely without reference to other frames. P-frames obtain predictions from temporally preceding I or P frames in the sequence, whereas B-frames obtain predictions from the nearest preceding and/or upcoming I or P- frames in the sequence. Hence, an I-frame is considered to contain the most important information in one picture. Without an I-frame, the quality of MPEG picture is not acceptable. Figure 7.1 gives an illustration of the MPEG picture.

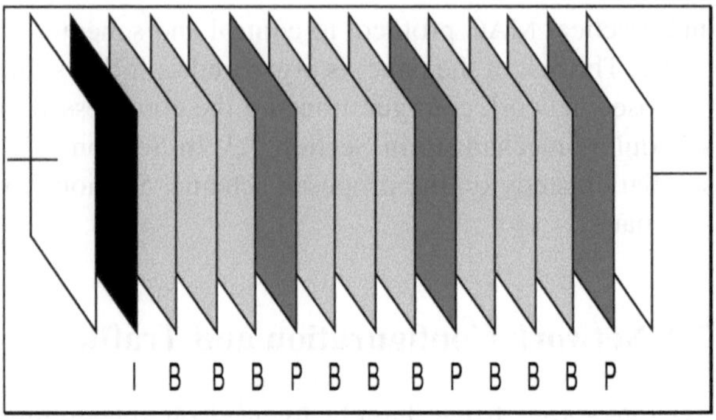

Fig. 7.1 A typical group of pictures of MPEG traffic

For the sake of simplicity and better illustration of basic concepts of multimedia scheduling framework, in this paper, we assume that there is an equal distance between any ONU and the star coupler. We consider that each ONU can support *m* multimedia application users. The multimedia video traffic will be separately buffered at each ONU according to the traffic types. We divide the traffic into 2 generic types of traffic in the network, one is for I-frame traffic, and the other is for B-frame and P-frame traffic. I-frames will be buffered separately. B-frames and P-frames will share the same buffer. We assume that I-frame traffic has the QoS requirements such as worst delay bound and packet loss ratio, while the B-frame and P-frame traffic either has minimum or no QoS requirement such as the best effort traffic. With the above considerations, we have come up with the structure of the buffers at each ONU. At each ONU, we define *2m* queues, where *m* is the number of MPEG video applications an individual ONU will accommodate. For each application at an ONU, there are 2 sub-queues. One is the queue for the traffic consisting of I frames. And the other is the queue for the traffic consisting of B and P-frames. In fact, the buffer of an ONU is essentially the frame scheduling queue in the queue system of an ONU.

7.3 Proposed Scheduling Schemes

Based on the architecture of the optical access networks and the features of the compressed video traffic, we design our scheduling scheme with two scheduling phrases. One is the frame scheduling and the other is the transmission scheduling. We notice that each frame is a variable-size message due to the features of the MPEG video traffic. We design the queuing system at each ONU to have 2 queues. One is the frame scheduling queue as mentioned in last section. The other is a queue named as transmission queue for the purpose to schedule message transmission. The frame scheduling scheme is responsible for managing the compressed video frames as messages in the frame scheduling queue to the transmission queue at an ONU. And the transmission scheduling scheme is to coordinate message transmissions among multiple ONUs. The design of 2 layers of scheduling is necessary to achieve the scalability of the scheduling to make the size of the control slot is invariant to the number of multimedia traffic flows at each ONU but only depends on the number of ONUs in the networks. As a result of such design, a fairly large number of traffic flows can be accommodated at each ONU. More importantly, this design can offer great flexibility in that the scheduling algorithms at each level can be locally optimized. Sophisticated fair queue scheduling algorithms can be readily used at the frame scheduling level while optimized transmission scheduling, independent of the frame scheduling, can be developed.

Contrary to the general belief, multiple turnaround times including propagation time and tuning delay can be justified for QoS requirements even for individual message. It implies that a re-scheduling can be initiated when the previous scheduling fails due to channel/destination conflicts. The challenge lies in the design of an effective scheduling mechanism which can guarantee QoS requirements of individual traffic stream while at the same time yielding high overall throughput.

The main complication is that the reservation over the control channel might fail due to either the destination conflict or channel conflict, in which case re-scheduling is necessary. When a reservation packet is sent from an ONU, at the end of one round-trip propagation delay, if the scheduled transmission fails, a decision has to be made by the transmission scheduler to tell whether to re-schedule the message which failed to transmit or to schedule a new message from the transmission queue. Since the MPEG traffic is a multimedia video application, it certainly makes sense to re-schedule the same frame as soon as possible. In our scheme, we will re-transmit the same reservation packet immediately in the subsequent control slots to make a new reservation. It's possible that multiple reservations at one control slot from different source nodes are targeting the same destination, due to the destination conflict, no one can succeed. The immediate re-transmission of all those in a few succeeding control slots could also collide and lead to poor channel utilization. To overcome this problem, we design to use randomized back-off algorithm to separate the rescheduling reservations.

7.3.1 Frame Scheduling Algorithm

The frame scheduling essentially selects a proper frame if there is any from multiple multimedia applications of I, B, and P-frames queues, and moves it into the transmission queue ready for transmission scheduling. Depending on the frame types, different scheduling at the frame scheduling can be used. We recognize that there might be a need for sophisticated scheduling algorithm for the frame scheduling. Since B and P frames of traffic might need some minimum level guarantee, we decide to use weighted fair queuing (WFQ) scheduling scheme to enhance the granularity of bandwidth management and fairness. We assume that the traffic streams consisting of B-frames and P-frames are best effort traffic streams and we enforce the traffic streams consisting of only I-frames having strict priority over the non-real-time or the best effort traffic streams consisting of B-frames and P-frames. For each MPEG video traffic application stream at the ONU, whenever there is a frame in its of I-frame queue, the frames in its B and P-frame queues can not be scheduled. For the scheduling among m video traffic flows, a simple round robin policy is employed without priority.

7.3.2 Transmission Scheduling Algorithm

In the transmission scheduling phrase, a control packet will be sent in the corresponding control mini-slot over the control channel. Depending on whether the transmission queue is empty or not, and the frame types, a number of scenarios are possible. The following summarizes the priority order of the reservation packet transmission.

- A failed I-frame reservation packet will be re-transmitted in the next control slot with the highest priority.
- A new non-blocked I-frame reservation packet will be transmitted with second higher priority.
- A failed B-frame and P-frame reservation packet will be transmitted with third higher priority.
- A new non-real-time traffic reservation packet will be transmitted the lowest priority.
- Within one control frame, each ONU has one control slot for sending out its reservation. Depending on the reservation type, the following is the order of transmission scheduling to avoid both channel conflict and destination conflict.
- An I-frame has higher priority than B-frame and P-frame for the transmission.
- An I-frame or a B-frame or a P-frame with an earlier arrival has higher priority than the same type of the frames with late arrival.
- Random selection with equal probability further breaks the tie.

7.4 Simulation Results and Discussion

There are a number of issues we want to study including the system performance under various traffic conditions with compressed MPEG video traffic and the impact on the system performance from a wide range of system parameters. In this section, we also discuss the results obtained from simulations to prove that our proposed 2 tier scheduling scheme is an efficient scheme applicable in the PSC-based WDM optical access networks. The following summarizes the assumptions in the simulation experiment design.

- The a frame or a message with variable-size consists of numbers of fixed-size packets
- The system is synchronized with slot time equal to a packet transmission time.
- The length of a control packet equals to the length of a data packet.
- Packet delay is given in number of time slots.
- The transmitter and receiver are tunable.
- Uniform traffic distribution has been considered.

- Traffic loading is given in terms of the number of ONUs which generates MPEG traces over the total number of ONUs in the network in one time slot, e.g. 30%, or 50%.

Frames are generated at each ONU following the trace of MPEG video traffic given in [11]. Each trace maintained in a trace file describes the characteristics of the MPEG compressed video traffic based on time axis. A typical trace file has the format: time stamp, packet size, fragment size temporal reference, frame type, and sequence number. In an MPEG trace, I-frame is normally longer than P- and B- frames. Assume a 1 Gbps transmission rate of one WDM wavelength, one data slot time, which is the transmission time of a packet, has been defined as 80 μs as the transmission time of 10000 Bytes. In a 1-Km local area environment, the propagation delay is 5 μs, the propagation delay is relatively shorter than 80 μs. While the propagation delay increases with the distance increasing, the pipelining of the propagation time with frame transmission can be adopted, which has been shown in [12, 13] to be able to leverage the impact. Tuning is another overhead potential for each frame transmission which depends on the technology used, ranging from a few nanoseconds (ns) to milliseconds (ms), but more commonly at the microseconds magnitude (μs).

7.4.1 Packet Loss Rate

In this section, we examine the packet loss rate of each of three types of the frames, respectively. We consider a network with 24 ONUs with 50% traffic loading, which means that every ONU in the network generates MPEG frames following the MPEG trace file in [11]. In our system, both I-frame and B, P-frame queues have 300 time slots as their buffer size.

We can observe from Fig. 7.2 that the loss rate of I- frame is very low compared with the B-frame and P-frame. With the number of wavelength increasing, the loss rates of B-frame and P-frame also decrease. When the number of wavelengths in the network is 12, the loss rates of I-frame, B-frames and P-frames are almost as low as 1% of the total traffic load.

Figure 7.3 depicts the relationship between the loss rate and the traffic load in the network with 8 wavelengths. It shows that when the traffic intensity has increased, the loss rates of B-frames and P-frames will get larger too. And the loss rate of P-frames is much larger than that of B- frames. However, the loss rate of I-frames has always been zero without change until the traffic load reaches 0.9.

7.4.2 Average Packet Delays

Figure 7.4 demonstrates that the average delay to I-frame can be as small as a few time slots even when the number of wavelengths in the networks is only 8. Hence the QoS of I-frame traffic transmission can be guaranteed. As we observed from the experiment, the quality is still acceptable if we play MPEG video with only I-frames. Therefore, the quality of playing the MPEG can be guaranteed.

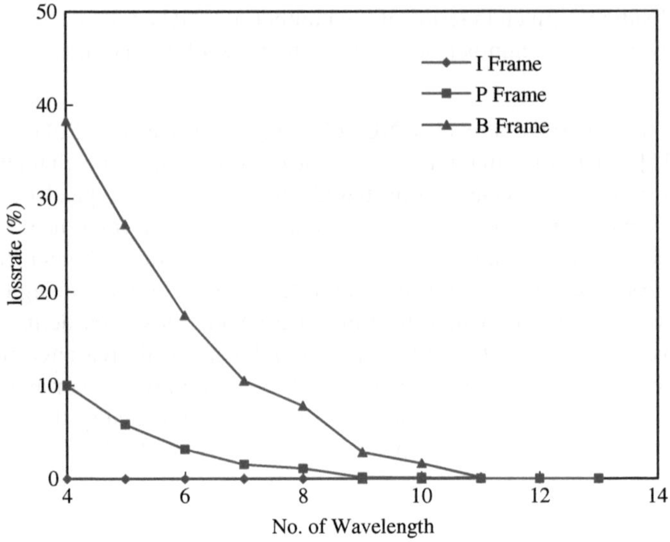

Fig. 7.2 Loss rate vs. no. of wavelengths

7.4.3 System Throughput

Figure 7.5 shows the normalized throughput of each ONU in the network. As a matter of fact, the simulation shows that the throughput of each ONU can achieve to almost 99% when the traffic load increases to 0.8. It demonstrates that the scheduling scheme works effectively to arrange traffic transmission. The network can be fully utilized when the traffic load is higher.

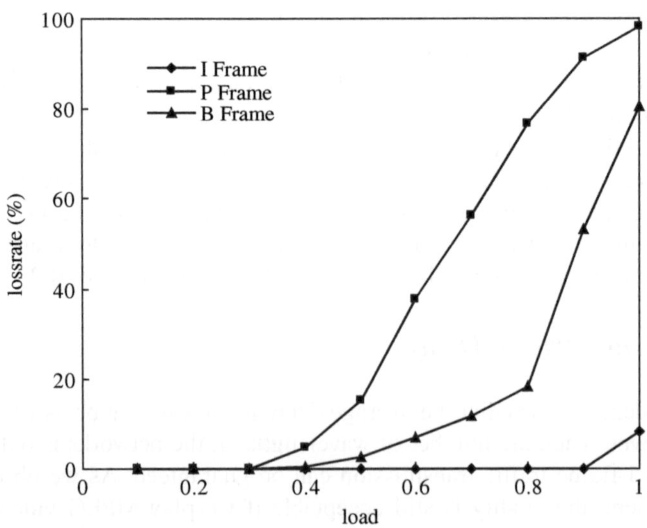

Fig. 7.3 Loss rate vs. traffic load

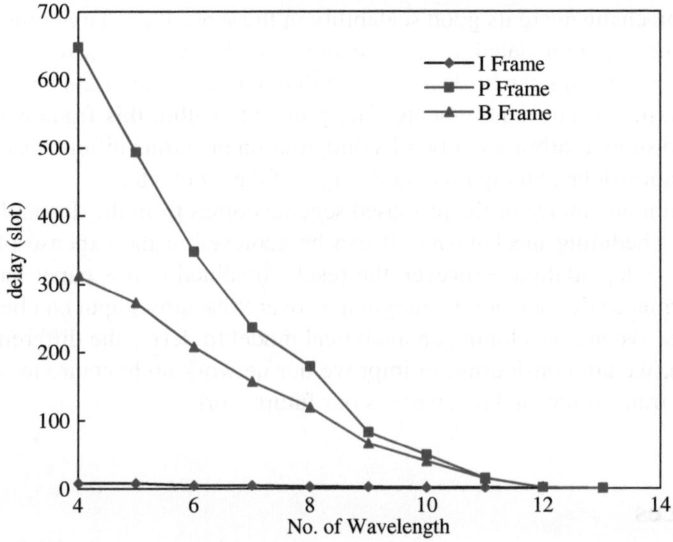

Fig. 7.4 Packet delay vs. no. of wavelengths

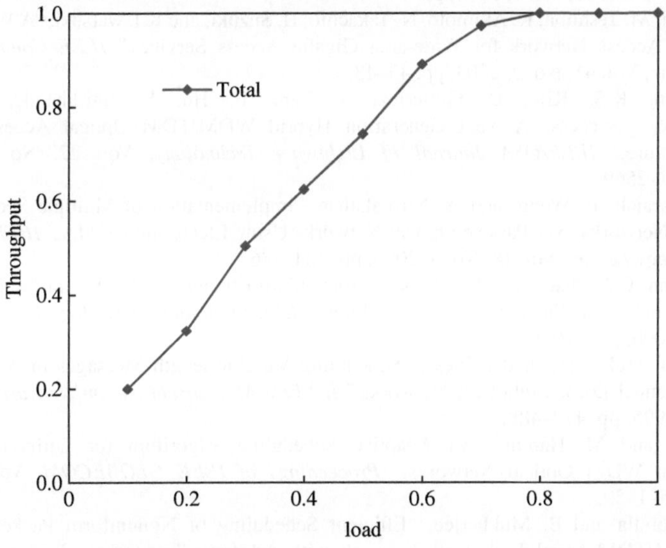

Fig. 7.5 Normalized throughput vs. traffic load

7.5 Conclusions

In this paper, we consider scheduling of multimedia traffic over a PSC-based WDM access optical network. We have proposed a hierarchical scheduling framework to support QoS service of the compressed video traffic. The main features of the proposed mechanism are its good scalability in the sense that a large number traffic flows can be accommodated, and its adaptive flexibility in the sense that optimal scheduling algorithms can be designed at different levels, which can be easily tuned to the specific system requirements. The protocols within this framework can be modified to support arbitrary network configuration including different assumptions on propagation delay, tuning time, and length of data unit, etc.

The major advantage of the proposed scheme comes from the decoupling of the two level scheduling mechanisms. It can be achieved at the expense of possible performance degradation. However, the results obtained in this paper have shown that performance degradation is marginal as over 90% throughput can be observed in all cases. We are developing an analytical model to derive the difference. In the mean time, we are considering to improve our network architecture to reduce the number of transmitters and receivers as our future work.

References

1. Telcordia Netsizer website, http://www.netsizer.com.
2. I.V. Voorde, C.M. Martin, I. Vandewege, and X.Z. Oiu, "The SuperPON Demonstrator: An Explosion of Possible Evolution Paths for Optical Access Networks," *IEEE Communications Magazine*, Vol. 38, No. 2, 2000, pp. 74–82.
3. J-I. Kani, M. Teshima, K. Akimoto, N. Takachio, H. Suzuki, and K.I. watsuki, "A WDM-Based Optical Access Network for Wide-area Gigabit Access Services," *IEEE Communications Magazine*, Vol. 41, No. 2, 2003, pp. 43–48.
4. F-T. An, K.S. Kim, D. Gutierrez, S. Yam, E. Hu, K. Shrikhande, and L.G. Kazovsky, "Success: A Next Generation Hybrid WDM/TDM Optical Access Network Architecture," *IEEE/OSA Journal of Lightwave Technology*, Vol. 22. No. 11, 2004, pp. 2557–2569.
5. N. Nadarajah, E. Wong, and A. Nirmalathas, "Implementation of Multiple Secure Virtual Private Networks over Passive Optical Networks Using Electronic CDMA," *IEEE Photonics Technology Letters*, Vol. 18, No. 3, 2006, pp. 484–486.
6. A.V. Tran, C.J. Chae, and R.S. Tucker, "Bandwidth-Efficient PON System for Broad-Band Access and Local Customer Internetworking," *IEEE Photonics Technology Letters*, Vol. 18, No. 5, 2006, pp. 670–672.
7. F. Jia, B. Mukherjee, and J. Iness, "Scheduling Variable-length Messages in A Single-Hop Multichannel Local Lightwave Networks," *IEEE/ACM Transactions on Networking*, Vol. 3, No. 4, 1995, pp. 477–488.
8. M. Ma and M. Hamdi, "An Adaptive Scheduling Algorithm for Differentiated Services on WDM Optical Networks," *Proceedings of IEEE GLOBECOM*, Vol. 3, 2001, pp. 1455–1459.
9. M.S. Borella and B. Mukherjee, "Efficient Scheduling of Nonuniform Packet Traffic in A WDM/TDM Local Lightwave Network with Arbitrary Transceiver Tuning Latencies," *Proceedings of IEEE INFOCOM*, Vol. 1, 1995, pp. 129–137.
10. B. Li and Y. Qin, "Traffic Scheduling in A Photonic Pakcet Switching System with QoS Guarantee," *IEEE/OSA Journal of Lightwave Technology*, Vol. 16, No. 12, 1998, pp. 2281–2295.
11. F.H.R. Fitzek, M. Zorzi, P. Seeling, and M. Reisslein, "Video and Audio Trace Files of Pre-Encoded Video Content for Network Performance Measurements," *Proceedings of the First IEEE International Conference of Consumer Communications and Networking 2004*, CCNC 2004, Jan. 2004, pp. 245–250.

12. M. Azizoglu, R.A. Barry, and A. Mokhtar, "Impact of Tuning Delay on the Performance of Bandwidth-Limited Optical Broadcast Uniform Traffic," *IEEE Journal on Selected Areas in Communications*, Vol. 14, No. 5, 1996, pp. 935–944.
13. B. Li, A. Ganz, and C.M. Krishna, "A Novel Transmission Coordination Scheme for Single Hop Lightwave Networks," *Proceedings of IEEE Globecom 1995*, Vol. 3, 1995, pp. 1784–1788.

Chapter 8
MAC Protocols for Single-Hop Passive-Star Coupled WDM Optical Networks

Xiaohong Huang and Maode Ma

Abstract WDM is an effective technique for utilizing the large bandwidth of an optical fiber. By allowing multiple messages to be simultaneously transmitted on a number of channels, WDM has the potential to significantly improve the performance of optical networks. A passive star coupler, equipped with tunable transmitters and tunable receivers, can be used to construct a multi-access LAN/MAN using WDM channels. It has the potential of sharing the enormous bandwidth of the optical medium among all the network users. In order to fully exploit the enormous available bandwidth of the optical fiber, efficient MAC protocols are needed to efficiently allocate and coordinate the system resources. Generally, the key requirements and features of access protocols for LANs/MANs comprise flexibility in terms of bandwidth allocation and configuration, low cost and compatibility with existing network architectures and protocols. This article presents an introduction to single-hop passive-star coupled WDM optical networks, followed by a comprehensive survey of state-the-art MAC protocols for WDM optical networks.

8.1 Passive Star-Coupled WDM Optical Networks

The passive star-coupled WDM optical network is the simplest and most popular topology for high-speed LANs/MANs. The star topology is attractive, first because of its logarithmic splitting loss in the coupler (since the splitter portion of the coupler circuit is essentially a binary tree type structure), and second because of no tapping or insertion loss (as in linear bus). Moreover, the passive star network can typically support a larger number of users than, for example a linear bus topology because power loss and tapping loss in linear buses limit the number of users that can be attached without adding broadband optical amplifiers. In addition, the passive property of the optical star coupler is important for network reliability, since no power is needed to operate the coupler.

X. Huang (✉)
Network Technology Research Institute, Beijing University of Posts and Telecommunications, Beijing, China

M. Ma. (ed.), *Current Research Progress of Optical Networks*,
DOI 10.1007/978-1-4020-9889-5_8,

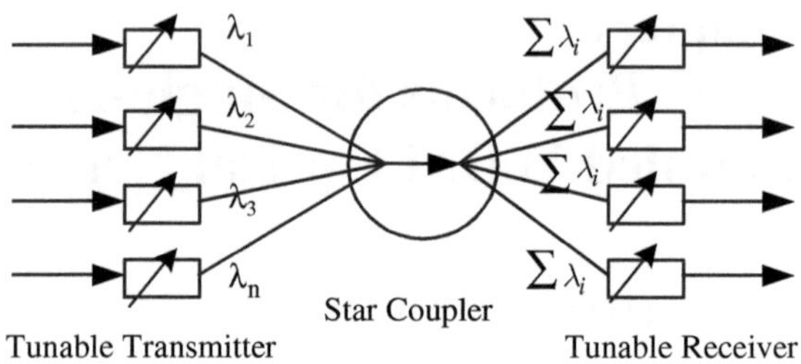

Fig. 8.1 A broadcast-and-select WDM network

The passive star coupler is a broadcast device, in which all coming signals are combined onto a single fiber and then this is split to direct $1/n$ part of the combined signal back to each station. It is the key component to configure the broadcast-and-select network, which is set up by connecting computing nodes via two-way fibers to a passive star coupler, as shown in Fig. 8.1. A fiber from each node is connected to a star coupler where the signals from all the nodes are mixed. Another set of fibers travels from the hub to each node. Communication between the source and destination nodes proceeds in one of the following two modes: single-hop, in which communication takes place directly between two nodes [1], or multi-hop, in which information from a source to a destination may be routed through the intermediate nodes of the network [2]. In our research, we will focus on the single-hop architecture.

In single-hop passive star-coupled WDM optical networks, a node sends its packets to the star coupler on one available wavelength by using a laser device which emits an optical data stream. The data streams from multiple sources are optically combined at the star coupler and the signal power of each stream is evenly split and forwarded to all of the nodes. At the receiving node, the node's receiver, typically an optical WDM filter has to be properly tuned to one of the wavelengths to receive the respective data stream. Based on whether the nodal transceivers are tunable or not, WDM systems may be classified as: Fixed Transmitter/Fixed Receiver (FT-FR) systems, Tunable Transmitter/Fixed Receiver (TT-FR) systems, Fixed Transmitter/Tunable Receiver (FT-TR) systems, and Tunable Transmitter/Tunable Receiver (TT-TR) systems. Systems with pretransmission coordination, i.e., employing a Control Channel (CC), can be formally specified by using an additional index CC such as, for instance, CC-TT-TR.

So far, some experimental testbeds have recently been built in several research laboratories. The representatives of WDM LAN/MAN prototypes based on the passive-star topology are the LAMBDANET ([3], Bellcore), Rainbow ([4, 5], IBM) and STARNET ([6], Stanford University) testbeds, which are discussed as follows.

(1) LAMBDANET

The maximum number of supported nodes (N) is 16 for LAMBDANET. The transmission speed of each node is 2 Gb/s and therefore the network throughput is 32 Gb/s. The network range is 57.5 km. In this system, the number of wavelength channels (C) accommodated has to be the same with the number of networks nodes (N), i.e., $N = C$. The LAMBDANET yields a FT-FR^N architecture, i.e., each node is equipped with one fixed-tuned transmitter and *N* fixed-tuned receivers. Each receiver is fixed to each available wavelength, a physical channel. The advantages of this prototype are its simplicity of design and architectural support of multicasting. No tunable components as well as no complicated protocols are required in this system. However, every node has to be equipped with *N* receivers so that the cost per node is proportional to the number of nodes. Moreover, this architecture is not scalable since *C* channels are needed. This design has a number of potential applications including transmission of voice and data, and distribution of video services.

(2) Rainbow

The Rainbow LAN/MAN supports 32 nodes each operating at 300 Mb/S. Therefore, the network throughput is 9.6 Gb/s. Each node is equipped with one fixed transmitter which is fixed tuned to its own unique wavelength channel (home channel) and one tunable receiver which can be tuned to any available wavelength channel, i.e. yielding a FT-TR architecture. The tuning time of the receiver may be up to 25 ms. When a receiver is idle, it scans all the incoming wavelengths in a round-robin fashion until it finds a channel with a setup request containing the receiver's address. After that, an acknowledgment is transmitted to the source node to make the actual connection setting up. This is so-called in-band receiver polling mechanism. By incorporating some higher-layer protocols, the Rainbow-II optical network [5] extends the Rainbow network in the aspect that the nodes are operating at 1 Gb/s over a distance of 10–20 km. The long setup-acknowledgment delay in this system makes it not suitable for packet-switched traffic. However, it works well for those circuit-switched applications, like telephony and video-conferencing.

(3) STARNET

STARNT is a passive star coupler based testbed constructed at the Optical Communications Research Laboratory (OCRL) of Stanford University. It offers all users two logical subnetworks: a high-speed reconfigurable packet-switched data subnetwork and a moderate-speed fix-tuned packet-switched control subnetwork. Thus STARNET supports traffic with a wide range of speed and continuity characteristics. Each node in this system contains one fixed transmitter and two tunable receivers, leading to a FT-TR^2 system. The transmitter is fixed-tuned to the node's home channel, while one receiver (main receiver) is dedicated to the high-speed data network and the second receiver (auxiliary receiver) is dedicated to the low-speed data network. The transmission speed is 2.5 GB/s for fast network and 125 MB/s FDDI compatible for slower control network. Using simulation and few experimental results, it is shown that STARNET is highly suitable for low speed applications, like email and telephony, as well as high speed multimedia network applications, like video-conferencing and HDTV.

The testbeds addressed above represent proof-of-concept systems for investigating the practicality of WDM LAN/MAN and the deployment of different WDM devices. Note that, so far, only very simple access protocols have been used in these prototypes. However, the proposals for much more sophisticated and efficient MAC protocols are needed in future systems.

In single-hop networks, a significant amount of dynamic coordination between nodes is required. For a packet transmission to occur, one of the transmitters of the sending node and one of the receivers of the destination node must be tuned to the same wavelength for the duration of the packet's transmission. It is important that transmitters and receivers tune to the same channels quickly, so that packets can be transmitted and received in quick succession. However, in comparison to packet transmission times, the tuning time of these transceivers is relatively long. Moreover, the tunable range for these transceivers is small. So the key challenge in single-hop architecture is to develop appropriate MAC protocols for efficiently coordinating the data transmissions and efficiently exploiting the potential vast bandwidth of an optical fiber to meet the increasing information transmission demand under the constraints of the network resources and the constraints imposed on the transmitted information. Note that *channel collisions* will occur when two or more signals simultaneously arrive at the star coupler on the same wavelength and *receiver collisions* will occur when two or more signals are destined to the same destination. So a proper medium access protocol has to either prevent such collisions or efficiently resolve them when they occur.

According to the network service provided to the transmitted information, the MAC protocols of WDM networks can be roughly divided into three categories as follows: MAC protocols for packet transmission, MAC protocols for variable-length message transmission, and MAC protocols with QoS concerns.

8.2 Legacy MAC Protocols for Packet Transmission

The MAC protocols dedicated for fixed-length packet transmission are so-called "legacy" protocols, which are often adopted from legacy shared medium networks. For a single-hop system to be efficient, the bandwidth allocation among the contending nodes must be dynamically managed. Such systems can be categorized into two categories: those employing pre-transmission coordination, and those not requiring any pre-transmission coordination.

8.2.1 Non Pre-Transmission Coordination Protocols

Protocols with non pre-transmission coordination do not have to reserve any channels. Arbitration of transmission rights is performed either in a pre-assigned fashion (fixed assignment protocols and partial fixed assignment protocols) or through contention-based data transmissions (random access protocols) on the regular data channels.

8.2.1.1 Fixed Assignment Protocols

A simple technique that allows single-hop communication is based on fixed assignment technique, Time Division Multiplexing (TDM) extended over a multi-channel environment [7]. Each node is equipped with one tunable transmitter and one tunable receiver; hence these systems are classified as TT-TR systems. It is predetermined that a pair of nodes is allowed to communicate with each other in the specified time slot within a cycle on the specified channel. Several extensions to the above protocol have been proposed to improve the performance. One approach, named weighted TDM, assigns different number of time slots to different transmitting nodes according to the traffic load on each node [8]. Another approach proposed a versatile time-wavelength assignment algorithm [9]. In this protocol, node i is equipped with t^i transmitters and r^i receivers, all of which are tunable over all available channels. This algorithm is designed such that, given a traffic demand matrix, it will minimize the tuning times in the schedule, while also minimizing the packet transmission duration. Based on [9], some new algorithms [10–12] study problems such as the performance of scheduling packet transmissions with an arbitrary traffic matrix and the effect of the tuning time on the performance.

8.2.1.2 Partial Fixed Assignment Protocols

The above fixed assignment protocols are too pessimistic because their main goal is to avoid both channel collisions and receiver collisions. However, alternative protocols can be defined in which the channel allocation procedures are less restrictive. A number of such protocols have also been studied in [7]. The first one is Destination allocation (DA) protocol. By using this protocol, the number of node pairs that can communicate over a slot is increased from the earlier value of N (the number of channels) to M (the number of nodes). The second one is Source Allocation (SA) protocol in which the control of access to the channels is further reduced. Similar to the SA protocol, Allocation Free (AF) protocol has been proposed, in which all source-destination pairs have full rights to transmit on any channel over any time slot duration.

8.2.1.3 Random Access Protocols

Two slotted ALOHA (SA) protocols were proposed in [13]. In the first protocol, time is slotted on all the channels, and these slots are synchronized across all channels. In the second protocol, each packet is considered to be of L minislots, and time across is synchronized across all channels over minislots. Another two similar protocols were proposed in [14].

8.2.2 Pre-Transmission Coordination Protocols

Pre-transmission coordination protocols allocate a channel as the control channel to transmit global information about the message to all the nodes in the system. These protocols can be categorized according to the ways to access the control channels into the following subgroups.

8.2.2.1 Random Access Protocols

In [15], three random access protocols such as ALOHA, slotted-ALOHA, and CSMA are proposed to access the control channel. ALOHA, CSMA, and N-server switch scheme can be the sub-protocols for the data channels. In a typical ALOHA protocol, a node transmits a control packet over the control channel at a randomly selected time, after which it immediately transmits a data packet on a data channel, which is specified by the control packet. In [16], an improved protocol named slotted-ALOHA/delayed-ALOHA has been proposed. In this protocol, transmitting node will delay transmitting data on a data channel until it gets the knowledge of that its control packet has been successfully received by the destination node. This protocol can decrease the data channel collision and improve throughput comparing with the protocols in [15]. Similarly, in [17], one set of slotted-ALOHA protocols and one set of Reservation-ALOHA protocols have been proposed to improve the performance of the protocols in [15]. And in [18], a so-called Multi-Control-Channel protocol is proposed, which aims to improve Reservation-ALOHA-based protocols.

Both the protocols in [15] and [16] ignored "receiver collisions", indicating that the probability of receiver collisions is small for large population systems and they would be taken care of by higher-level protocols. And [17] and [18] cannot prevent receiver collisions, too. A protocol, which is especially designed to avoid receiver collision, is presented in [19].

8.2.2.2 Reservation-Based Protocols

In [20], a Dynamic Time-Wavelength Division Multiaccess (DT-WDMA) protocol is proposed. In this protocol, a channel is reserved as control channel and fixed *time-division multi-access* (TDMA) is used within each slot on it. It requires two pairs of transmitters and receivers. One pair of the transceivers is fixed to the control channel and, another pair, with a fixed transmitter and a tunable receiver, is used for data channel, i.e., yielding a FT/FT-FR/TR architecture. Although this protocol cannot avoid receiver collisions, it ensures that exactly one data can be successfully accepted when more than one data packets come to the same destination node simultaneously. One proposal [21] to improve DT-WDMA algorithm introduces the concept of resolving receiver collisions by incorporating a delay line receiver to buffer the potential collided packets. And a conflict-free protocol is proposed in [22].

In [23], a new algorithm, named Hybrid Reservation Preallocation/Time Slot Assignment (HRP/TSA), is proposed. It combines the concepts of receiver preallocation and reservation access. And nodes were allowed reservations on multiple channels in the same cycle. This work aims to reduce high time complexity and schedule lengths of the existing scheduling algorithms. It is shown that the slight increased computational overhead with scheduling is justified by reduced packet latency and higher utilization, especially for client-server traffic.

In [24], another two reservation-based protocols aiming at improving the DT-WDMA algorithm are outlined. The first one is called Dynamic Allocation Scheme (DAS), which dynamically assigns slots on a packet-by-packet basis. The second protocol is named Hybrid TDM, which combines the TDM and the DAS scheme and allows both pre-assigned and dynamic slot assignment. Time on the data channels is divided into frames consisting of several slots. In a certain period of time, one slot will be opened for a transmitting node to transmit data packets to any destination receiver.

A reservation-based multi-control-channel protocol can be found in [25]. By this protocol, x channels ($1 < x < N/2$) can be reserved as control channels to transmit control information, where N is the number of channels in the network. The objective to reserve multiple control channels in the network is to decrease the overhead of control information processing time as much as possible.

Based on the above survey, we find that, among non pre-transmission coordination protocols, those taking fixed-channel assignment approach can ensure that data is successfully transmitted and received, but they are sensitive to the dynamic bandwidth requirements of the network and they are difficult to scale in terms of the number of nodes. And those taking contention-based channel assignment approach can adapt to the dynamic bandwidth requirements, but they introduce contention on data channels. As a result, either channel collision or receiver collision will occur.

On the other hand, among pre-transmission coordination protocols, the protocols with contention-based control channel assignment still have data channel collision and receiver collision because it introduces contention on control channel. However, by continuously testing the network states, some protocols proposed in [26, 27] may have the capability to avoid both collisions. The reservation-based protocols, which take fixed control channel assignment approach, can only ensure data transmission without channel collisions. However, by introducing some information to make the network nodes intelligent, it may have potential to avoid receiver collisions as well. It also has potential to accommodate application traffic composed of variable-length messages.

8.3 MAC Protocols for Variable-Length Message Transmission

The "legacy" MAC protocols are designed to schedule fixed length packets. However, in the real world, traffic streams are often characterized as bursty, and consecutive arriving packets in a burst often have the same destination. Based on this observation, we can schedule all the fixed size packets as a whole rather than schedule them on a packet-by-packet basis. Using this kind of protocols has three advantages as follows: firstly, to an application, the performance metrics of its data units have more relevant performance measures than ones specified by individual packets; secondly, it perfectly fits the current trend of carrying IP traffic over WDM networks; and lastly, message fragmentation and reassembly are not needed.

8.3.1 Basic MAC Protocols for Variable-Length Messages

The first two MAC protocols in [17] proposed for variable-length message transmission are protocols with contention-based control channel assignment. Another two Reservation-ALOHA-based protocols in [17] are presented in order to serve the long holding time traffic of variable-length messages. Data channel collisions can be avoided in the protocols presented in [17].

The protocol in [28–30] tries to improve the reservation-based DT-WDMA protocol in [20] in three aspects: the number of nodes is larger than the number of channels; the transmitted data is a variable-length message rather than a fixed length packet; and data transmission can start without delay.

The protocol introduced in [31], called FatMAC, is a hybrid approach that combines the advantage of receiver preallocation and reservation access strategies. It reserves access on preallocation-based channels through control packets. A reservation specifies the destination, the channel and the message length of next data transmission. This protocol is based on tunable transmitters and fixed receivers. LiteMAC in [32] is an extension to FatMAC. In LiteMAC protocol, each node is equipped with a tunable transmitter and a tunable receiver rather than a fixed receiver in FatMAC. LiteMAC has more flexibility than FatMAC because of its tunable receiver and special scheduling mechanism. Both these two algorithms can transmit variable-length packets without collisions. It has been proved that these two protocols have better performance than preallcation-based protocols while less transmission channels are used than reservation-based protocols.

An intelligent reservation-based protocol for scheduling variable-length message transmission has been proposed in [33]. It has the ability to avoid both channel collision and receiver collision by sharing some global information. This makes it a milestone in the development of MAC protocols for WDM optical networks. Each node is equipped with a fixed transmitter (called control transmitter) and a fixed receiver (called control receiver), both of which are tuned to the control channel. In addition, a tunable transmitter (called data transmitter) and a tunable receiver (called data receiver) are employed at each node to enable it to access the data channels. The number of channels may be much smaller than the number of nodes. A Time Division Multiple Access (TDMA) protocol is employed to access the control channel so that the collision of control packets can be avoided. In [33], three data channel assignment algorithms have been proposed. The fundamental one is named Earliest Available Time Scheduling algorithm (EATS). This algorithm schedules the transmission of a message by selecting a data channel, which is available the soonest.

Some related protocols have been proposed to improve the performance of the network based on the same system architecture of [33]. In [34], two protocols for scheduling variable-length packet transmission have been proposed. The distinction between these protocols and other WDM protocols is that message transmissions are initiated by the receipt of a control message; other schemes schedule packet transmission for some fixed point in the future. This flexibility allows these protocols to avoid "head-of-line" blocking. In [35], the authors find that significant improvement in performance can be achieved using scheduling algorithms where message sequencing and channel assignment are simultaneously taken into consideration.

8.3.2 Novel MAC Protocols for Variable-Length Messages

As an example of the general scheduling scheme in [35], a new scheduling algorithm is proposed in [36]. The algorithm, named Receiver-Oriented Earliest Available Time Scheduling (RO-EATS), is designed based on the observation that two consecutive messages with the same destination may not fully use the available channels when EATS algorithm is employed. Therefore, this algorithm has the ability to decrease message transmission blocking caused by avoiding receiver collisions. By this protocol, the management of message transmission and reception is the same as that by the protocol in [33]. The difference between the two protocols is on the scheduling algorithm for message transmission. In the RO-EATS, it first considers the earliest available receiver among all the nodes in the network and then selects a message, which is destined to this receiver from those messages. After that, a channel is selected and assigned to the selected message by the principle of EATS algorithm. The new algorithm enforces the idea of scheduling two consecutive messages away from going to the same destination node. In this way, average message delay can be shown to be quite low and channel utilization can be shown to be high.

In [37], a novel signaling protocol called the Sampling Probe Algorithm (SPA) is proposed. This protocol is based on pre-transmission coordination, distinct from all the existing approaches in that it does not require a separated out-band signaling channel (control channel), thus alleviating the stringent requirement on the number of channels required in most control channel based protocols. The reservations occur on the same channels where data packets are transmitted, i.e., in-band signaling. This proposed scheme works well for systems under moderate or heavy traffic loading.

Recently, a persistent reservation protocol for variable-length messages in WDM-based local networks using a passive star topology is proposed in [38]. With this protocol, once a node reserves a data channel, the node persistently uses the channel until its message is completely transmitted. In this protocol, the control channel is shared by all nodes or on a contention basis using the slotted SLOHA protocol. And data channel and destination collision can be avoided with this protocol. The protocol is suitable for a network in which accommodation of variable-length messages (such as circuit-switched traffic or traffic with long holding times) is required. Moreover, the protocol enables any new node to join the network at anytime without network re-initialization.

A new channel reservation protocol using a counter for detection of a source conflict in a WDM single-hop network with non-equivalent propagation delay is introduced in [39]. In this protocol, a source conflict occurs when a source node has the right to transmit more than two messages to their destination nodes using different wavelengths in the same time slot. By investigating information about the final message which has succeeded in reservation, a source node can detect a source conflict before the assignment of wavelengths. With this protocol, the mean message delay can be dramatically reduced without degrading throughput performance as the offered load becomes large.

In [40], one novel reservation-based scheme is proposed to increase the utilization of the control slots and reduce the packet delays. Different from other reservation-based schemes, the proposed scheme can dynamically adjust the length of the control frames according to the traffic patterns of the nodes. In this way, the nodes can acquire and release the control slots dynamically so that the control slots on the control channel can be used efficiently and the packet delays can be reduced.

And in [41], one novel protocol for bursty traffic is proposed. In this paper, all the channels will be used to transmit messages. Each node is equipped with one tunable transmitter and one fixed receiver, i.e., yielding a TT/FR architecture. For each wavelength λ_i, each station maintains the information of the set of stations A_i which are estimated to be active for this wavelength. The stations in set A_i are granted permission to transmit on wavelength λ_i in a round-robin fashion. The stations which are granted permission to transmit at each time slot are selected by taking into account the network feedback information. This protocol is able to allocate the bandwidth of each wavelength to the stations according to their needs so that the number of idle sots is reduced, resulting in the increase of the network throughput.

In [42], a novel scheme with fixed transceiver array and adaptive channel allocation is proposed. In this paper, fixed receiver array instead of tunable transceiver is used, which aims to reduce the tuning latency. Each node is capable to transmit and receive multiple signals on multiple channels at the same time. Meanwhile, adaptive channel allocation is proposed, by adjusting the numbers of control and data channels adaptively according to the traffic patterns. Simulation results demonstrate the advantages of the new approaches.

In [43], a predictive online collision free coordination-based MAC protocol is proposed based on one traffic prediction scheme, which aims to eliminate the possible delay added by the schedule computation. Two phases, i.e. control phase and data phase are considered. The control phase will function as a learning period in which the predictor is trained, in which the training is based on the traffic of the network. During the data phase, each station transmits its packets based on the predicted reservations. The novel mechanism tries to scan and find idle timeslots and deletes them logically. As addressed in the paper, the schedule length is less than other online protocols, and the number of idle timeslots is decreased.

In [44], an improved version of EATS, called Priority Scheme Earliest Available Time Scheduling (PS-EATS), is proposed. In this paper, the longer messages are given higher priorities regardless of the source and destination nodes. In this way, the packet delay is reduced and the idle period time is minimized.

In [45], a novel scheme with dynamic control frames is proposed to reduce the packet delay, in which variable number of control slots are used. In this scheme, each node is able to acquire and release control slots depending on theirs loads so that control slots will only be used when nodes have data packets to send. In this way, the delay is reduced. Meanwhile, a hybrid fixed and dynamic scheme, which employs

Table 8.1 Comparison of some protocols

Protocol	Architecture	Control channel access protocol	Processing complexity	Throughput
Random Access [16]	CC-TT-TR	Slotted Aloha	low	low
DT-WDMA [20]	CC-FT/FT-FR/TR	TDMA	high	moderate
DAS, HTDM [24]	CC-FT/FT-FR/TR	TDMA	very high	high
TDMA-C [30]	CC-TT-FR/TR	TDMA	high	low
RCA[19]	CC-TT-TR	Slotted Aloha	moderate	moderate
EATS [33]	CC-FT/TT-FR/TR	TDMA	high	high

the fixed control frame scheme or the dynamic control frame scheme dynamically depending on the loads of the nodes.

So far, various architectures and protocols belonging to single-hop passive star coupled WDM optical network were reviewed. According to the survey, WDM systems can be divided into several types based on whether the nodal transceivers are tunable or not. In reality, tunable transceivers are more expensive than fixed transceivers. Table 8.1 provides a simple comparison among some protocols. As the table shows, in [16] and [19], only one pair of tunable transmitter and receiver is used, which is to avoid the costly network interface unit because of the additional fixed transmitter and fixed receiver to monitor control channel. However, the throughput achieved by these two protocols is not as high as that achieved by the protocols in [24] and [33]. In order to achieve better performance, more transmitters and receivers are used in [24] and [33]. For example, in [33], one pair of fixed transmitter and receiver is used for control channel and one pair of tunable transmitter and receiver is dedicated for data channels. It becomes evident from this comparison that a tradeoff between equipment costs (related to the item architecture in Table 8.1) and the performance has to be considered when designing the MAC protocols.

Various contributions made on the research and development of MAC protocols for WDM-based local networks were discussed in this section. However, to provide real-time service to time-constrained application streams such as video or audio information become more and more important in the design of the high-speed computer networks such as WDM optical networks. It is explicit that MAC protocols are much needed to support QoS requirements. The protocols mentioned in this section can be a start point for research on this area.

8.4 MAC Protocols with QoS Concerns

One of the important issues of high-speed networks, such as WDM optical networks, is to support real-time message transmission with QoS demands and transmission service to multimedia applications with QoS requirements. The most important aspect of the former service is that a message generated at the source node must be received at the destination node within a given amount of time. This time is referred to as message deadline. Failure to meet the deadlines of these tasks may lead to catastrophic consequences. The latter service needs a certain amount of bandwidth to deliver video/audio frames in time consistent with human perception.

8.4.1 MAC Protocols for Real-Time Service

A major challenge in the design of future generation high-speed networks is the provision of real-time service to time-constraint application streams such as video or audio information. If the delay of a message in the system exceeds its time constraint, the message is considered as late. The task of the scheduling algorithms for the protocols that provide real-time service to time-constrained messages is to schedule the transmission of the messages to meet the message time constraint as much as possible. Most of the MAC protocols that provide real-time service on passive star-coupled WDM optical networks are protocols with reservation-based pre-coordination. According to the type of the real-time service provided to the transmitted messages, the MAC protocols for real-time service can be simply classified into three types: protocols with best-effort service, protocols with deterministically guaranteed service and protocols for statistically guaranteed service.

8.4.1.1 MAC Protocols for Best-Effort Real-Time Service

A real-time protocol Time-Deterministic Time and Wavelength Division Multiple Access (TD-TWDMA), based on TDM (Time Division Multiplexing), for a fiber-optic star network is presented in [46]. Services for both guarantee-seeking messages and best-effort messages are supported by using this protocol. In this protocol, the access to each channel is divided into cycles of time-slots. Each node has a number of guaranteed slots to support guarantee-seeking messages. However, if there are no guarantee-seeking messages in a node, the slots will be release for best effort messages from other nodes (or the same node) according to a predetermined scheme. Each node is equipped with one fixed-wavelength transmitter, which is always tuned to one specific wavelength channel, and tunable receivers, which can be tuned to an arbitrary wavelength channel. It is assumed that the number of wavelengths, C, equals the number of nodes, M. There are $2M$ queues in each transmitter, M for best-effort messages and M for guarantee-seeking messages. For each type of queues, one queue is for broadcast and M-1 queues for the single destination. Message transmission scheduling is based on the queue priority. Deadline guarantees are supported where the underlying deterministic bandwidth can be changed dynamically through slot reserving.

A reservation-based MAC protocol for best-effort real-time service can be found in [47]. This protocol is for the same network structure as that in [33]. Both hard real-time and soft real-time variable-length message transmissions have been considered. The scheduling algorithms of the protocol are based on the time related dynamic priority scheme, namely Minimum Laxity First (MLF) scheduling. The principle of this dynamic scheduling scheme is that the most stringent message will get the transmission service first. This work has confirmed that when real-time traffic involved in the networks, dynamic time-based priority assignment schemes as well as priority-based scheduling algorithms should be employed to improve the real-time performance of the networks as much as possible.

In [48], a novel reservation-based MAC protocol for real-time service has been proposed, which extends the function of the protocol in [47] to provide differentiated service to benefit both real-time and non real-time applications in one topology. The scheduling algorithm Minimum Laxity First with Time Tolerance Scheduling Algorithm (MLF-TTS) schedules real-time message transmission according to their time constraints. The basic MLF scheduling policy is adopted for scheduling real-time traffic. After the real-time messages have been scheduled to transmit on certain channels in certain time slots to their destination nodes, some of them could be blocked just because there may be more than two consecutive messages going to the same destination node in a very short time period. The MLF-TTS algorithm seeks and takes this waiting time period to schedule the transmission of non real-time messages under the condition that the transmission time of these messages should be less than the time that the blocked real-time messages are waiting for their destinations available. By the MLF-TTS algorithm, the average message delay for the messages without time constraints could be expected to decrease while the message loss rate for hard real-time messages or message tardy rate for soft real-time messages is kept as low as those of the simple MLF algorithms. And the channel utilization could be expected to be high.

8.4.1.2 MAC Protocols with Deterministically Guaranteed Service

It is obvious that the QoS provided by a network service to real-time applications indicates the degree of how the real-time applications can meet their time constraints. However, best-effort real-time network service cannot ensure QoS because it cannot guarantee that real-time applications can meet their time constraints in certain degree when they are transmitted. Therefore, it is necessary to develop MAC protocols with deterministically guaranteed service concerns.

In [49], a preallocation-based Wavelength Division Multiple Access (WDMA) scheme is proposed to provide deterministic timing guarantees to support time constrained communication. A scheme called Binary Splitting Scheme (BSS) is proposed to assign each message stream sufficient and well-spaced slots to fulfill its timing requirement. Given a set of real-time message streams M specified by the maximum length of each stream C_i and the relative deadline of each stream D_i, this scheme can allocate time slots over as few channels as possible in such a way that at least C_i slots are assigned to M_i in any time window of size D_i slots so that the real-time constraints of the message streams can be guaranteed.

A modified pre-allocation based MAC protocol is proposed in [50] to guarantee reserved bandwidth and constant delay bound to the integrated traffic. The access to the transmission channels is controlled by the scheduler, which works based on the concept of computing maximal weighted matching, a generalization of maximal matching on unweighted graph. Based on this concept, several scheduling algorithms have been produced to provide scheduling. A Credit-Weighted Algorithm is proposed to serve guaranteed traffic. A Bucket-Credit Weighted Algorithm is designed to serve bursty traffic. And a Validated Queue Algorithm is a modification of the Bucket-Credit Weighted Algorithm to serve bursty traffic and keep throughput guarantee at the same time. It has been proved that those scheduling algorithms can guarantee the bandwidth reservation to a certain percentages of the network capacity and ensure a small delay bound even when bursty traffic exists.

A reservation-based MAC protocol for deterministic guaranteed real-time service can be found in [51]. In [51], a systematic scheme comprised of admission control, traffic regulation, and message scheduling that provide guaranteed performance service for real-time application streams made up of variable-length messages. A traffic intensity oriented admission control policy is developed to manage flow level traffic. A g-regularity scheme based on the Max-plus algebra theory is employed to shape the traffic. An Adaptive Round-Robin and Earliest Available Time Scheduling (ARR-EATS) algorithm is proposed to schedule variable-length message transmission. All of those are integrated to ensure that the deterministic guaranteed real-time service can be achieved.

8.4.1.3 MAC Protocols for Statistically Guaranteed Service

The MAC protocols with deterministically guaranteed service can normally guarantee specific transmission delays to real-time applications. Or under certain time constraints imposed to the real-time applications, a specific percentage of real-time messages, which can meet the time constraints, can be predicted. However, MAC protocols for statistically guaranteed service cannot provide the deterministic guaranteed QoS service. Only an estimated percentage of real-time messages, which can meet their time constraints, can be evaluated statistically. Most of the MAC protocols in this category consider the issue of providing statistically guaranteed service to both real-time and non real-time applications, i.e., differentiated service. By these protocols, statistical QoS to real-time applications can be expected by sacrificing the transmission service to non real-time applications.

A novel reservation-based MAC protocol is proposed in [52] to support statistically guaranteed service in WDM networks by using a hierarchical scheduling framework. This work is developed from a similar network structure as that in [33]. The major advantage of its protocol is that it divides the scheduling issue into flow scheduling or VC scheduling and transmission scheduling. The former is responsible for considering the order of traffic streams to be transmitted. The latter is to decide the order of the packets transmission. The packets involved in the transmission scheduling are those selected from the traffic streams by the flow schedule scheme. This protocol is expected to diminish the ratio of the packets which are over their deadlines. Another good point of this protocol is that a re-scheduling scheme is employed to compensate the failure scheduling result due to either output conflict or channel conflict. If the failure is from a real-time traffic, it certainly makes sense to re-schedule the very same packet as soon as possible. And if real-time traffic has more stringent QoS requirements, the re-scheduling scheme will ignore re-scheduling the failed non real-time packet to ensure the real-time traffic to meet its time constraints.

A MAC protocol, which is based on a multi-channel ring topology, has been presented in [53]. The transmitted information is in the form of fixed size packet. A collision free slotted MAC protocol is proposed, named as Synchronous Round Robin with Reservation (SR3), to support both QoS guarantee to real-time traffic and best-effort service to non real-time traffic. It combines a packet scheduling strategy (called SRR), a fairness control algorithm (called MMR), and a reservation mechanism. SRR achieves an efficient exploitation of the available bandwidth, MMR guarantees fair throughput access to each node, and SR3, by permitting slot reservations, leads to tighter control on access delays, and can thus effectively support traffic classes with different QoS requirements. This protocol has been proved to have the capability to provide packet-mode transport to multiple information flows with differentiated QoS requirements.

In [54], a scheduling scheme for Tbits/sec, input-queued, star-coupler WDM optical networks is presented. This work focuses on packets with fixed length. In this protocol, by using Virtual Output Queueing (VOQ), arriving cells are classified at a primal stage to a queue that corresponds to their designated destination. The scheduler determines which queue is served for transmission. Upon receiving a signal from the reservation scheduler, each node performs wavelength reservation according to two primary guidelines: (a) global switch resources status, i.e., available wavelengths at the reservation instances, and (b) local considerations, i.e., the status and priorities of the node's internal queues. Using this protocol, class-differentiated low latency and extremely high throughput is achieved.

A novel scheduling scheme, namely Priority-Differentiated Scheduling (PDS), which is designed to handle real-time (high-priority) and non real-time (low-priority) packets in WDM star networks is introduced in [55]. This protocol is for the same network structure as that in [33]. PDS allows high-priority packets to preempt the prescheduled low-priority packets. By scheduling the high-priority packets first, and then having the preempted packets rescheduled, PDS guarantees that the high-priority packets can always achieve the earlier transmission than the others in order to meet the QoS requirements. In addition, it does not sacrifice the performance of low-priority packets. As a matter of fact, low-priority packets can also benefit from PDS algorithms.

In [56], a reservation-based MAC protocol enabling efficient integration of real-time traffic and data traffic is proposed. In this paper, the control channel is divided into contention-based (Slotted ALOHA) and contention-free (TDMA) fields within a control slot. Thereby, real-time transmission can be reserved by accessing one of the Slotted ALOHA minislots. These Slotted ALOHA minislots are located after each TDMA minislot in a cyclic manner. Data transmission is reserved via access to the TDMA minislots which are uniquely assigned to the respective network nodes. If the real-time control packet of one node fails to reserve the Slotted ALOHA slot, a new real-time control packet will be sent in the next TDMA minislot associated with itself, while the reservation for data traffic had to be deferred to the next control slot. So the reservation delay for real-time traffic is bounded by two round-trip delays.

8.4.2 MAC Protocols for Multimedia Application

There is a rapid growth in the number of multimedia applications recently. The transmission of the multimedia application is a kind of real-time and stream oriented communication. The quality of service required of a stream communication includes guaranteed bandwidth (throughput), delay and delay variation (jitter). Multimedia application integrates a variety of media, namely, audio, video, images, graphics, text, and data, each of which have different QoS requirements. The protocols that provide transmission service to multimedia application should support and ensure the variety of QoS requirements of different types of media. Many researchers have shown their interests on this issue. Some research results are generated from the existing protocols for real-time service. However, some protocols are completely novel or based on new network architectures dedicated to multimedia traffic.

In [57], the feasibility of several existing protocols based on the WDM bus LAN architecture to support multimedia application is studied. By the simulation study, the authors point out that several currently existing MAC protocols such as FairNet, WDMA, and nDQDB are not satisfactory for supporting multimedia traffic in the sense of that those protocols cannot guarantee that the total delay or jitter will not grow beyond the accepted value for different classes of multimedia applications.

In [58], a study on several MAC protocols to support multimedia traffic on WDM optical networks is carried out. These protocols including Distributed Queue Dual Bus (DQDB), Cyclic-Reservation Multiple-Access (CRMA), Distributed-Queue Multiple-Access (DQMA), Fair Distributed Queue (FDQ) are distributed reservation access schemes for WDM optical network based on slotted unidirectional bus structures. The performance of these four protocols is studied to simultaneously support synchronous traffic (for various real-time multimedia applications) and asynchronous traffic (for interactive terminal activities and data transfers). They have pointed out, by the extensive simulation results, that the reservation-based protocols are suitable for integrating real-time multimedia traffic with bursty data traffic in the WDM optical network when delay constraint is somewhat relaxed. And the FDQ protocol stands out to support heterogeneous traffic.

In [59], a video-on-demand system over a passive star-coupler based WDM optical network is studied. The video-on-demand (VOD) traffic is a constant bit rate (CBR) traffic because the video/audio sources of the application are processed in advance, kept on the video server, and transmitted at a regular rate. The VOD traffic is desirable to be served by isochronous transmission service by the network. A centralized medium access control scheduler is employed to schedule the isochronous and the asynchronous traffic demands. A scheduling algorithm, named as KT-MTR, is employed for scheduling the asynchronous traffic only. And a scheduling algorithm, IATSA-MTR, is presented for scheduling both isochronous and asynchronous traffic coexisted in the network. Those scheduling algorithms are proved to be efficient for serving VOD applications in the WDM optical networks.

To efficiently support multimedia traffic streams, an efficient scheme Multimedia Wavelength-Division Multiple-Access (M-WDMA) is proposed in [60]. Each node has three tunable transmitters used to serve three different classes of traffic streams according to the corresponding sub-protocols. Three types of multimedia traffic streams, including a constant–bit-rate traffic (CBR), a variable-bit-rate traffic with large burstness (VBR1), and a variable-bit-rate traffic with longer inter-arrival times (VBR2), are considered by the proposed MAC protocol. The M-WDMA protocol consists of three sub-protocols. One is the TDM sub-protocol, which is an interleaved TDMA MAC protocol. The other one is a reservation-based sub-protocol, RSV, which controls the access to the data channels by using a multiple token method. The third one is a random access sub-protocol, CNT, which works in a way similar to that of the interleaved slotted ALOHA. The outstanding point of this protocol is that a dynamic bandwidth allocation scheme is incorporated into the protocol to dynamically adjust the portions of the bandwidth occupied by the three types of traffic streams according to their QoS demands. It has been proved that the performance of the M-WDMA is good enough for WDM optical networks in serving multimedia applications.

One bandwidth guaranteed multi-access protocol is proposed in [61], in which the control channel contains two types of minislots: reservation minislots and contention minislots. There are M access nodes and one control node in the network. Nodes requiring bandwidth guarantees, called guaranteed nodes, use reservation minislots that are assigned by the control node. The remaining nodes share contention minislots using a random access mechanism. The reservation minislots can guarantee a minimum bandwidth for the guaranteed nodes. The contention minislots enable on-demand services at the optical layer and achieve good fairness for the remaining bandwidth.

A novel MAC protocol for providing guaranteed QoS service to the MPEG compressed video/audio applications in the passive star-coupled WDM optical network is proposed in [62]. The QoS of transmission of the MPEG traffic is derived based on the frame size traces from the MPEG encoded real video sequences. A frame, which is considered as the basic element with variable-size of the MPEG traffic streams, is scheduled and transmitted at one time. A systematic scheme is proposed to guarantee the deterministic delay to the transmission of the MPEG traffic. This scheme includes an admission policy, a traffic characterization mechanism, and a scheduling algorithm as a whole to ensure the QoS of the transmission of MPEG traffic. Analytical evaluation of the guaranteed deterministic delay bound for the proposed system service schemes is based on the theory of max-plus algebra. The deterministic delay bound is verified by intensive trace-driven simulations and modeled MPEG traffic simulations. It is obvious that this protocol stands out as a state-of-the-art MAC protocol among seldom MAC protocols, which support QoS of the transmission of multimedia applications in WDM optical networks.

In [63], an interesting idea on the architecture of the WDM optical network is proposed to support varieties of traffic such as data, real-time traffic, and multicast/broadcast service. The proposed architecture, named as Hybrid Optical Network (HONET), tries to combine the single-hop and multihop WDM optical network architectures into a synergy architecture. The architecture of the HONET can be considered as a network, which consists of the multihop network with an arbitrary virtual topology and a single-hop network based on a dynamically assigned T/WDMA MAC protocol. In this virtual network architecture, real-time traffic and other connection-oriented applications can be supported by single-hop network; while non real-time data traffic, which can tolerate relatively large delay is supported by multihop network. The advantage of this virtual architecture is that it is flexible to employ different topologies of the multihop network and different MAC protocols for the single-hop network to support varieties of traffic in the optical work according to the QoS of the traffic demands.

8.5 Summary

This chapter has summarized state-of-the-art medium access control protocols for the passive star-coupled WDM optical networks. Depending on the characteristics, complexity, and capabilities of these MAC protocols, we have classified them as data and message transmission MAC protocols, MAC protocols for real-time transmission service, and MAC protocols for multimedia applications. Most of these protocols focus local area environments. Architectural, qualitative and quantitative descriptions of various protocols within each category have been provided. Some important or milestone protocols have been given quite detailed explanations to present their underlined significances. This chapter can be served as a good stating point for researchers working on this area so as to give them an overview of the research efforts conducted for the past decade.

References

1. B. Mukherjee, "WDM-based Local Lightwave Networks- Part I: Single-Hop Systems," *IEEE Network*, May 1992, pp.12–27.
2. B. Mukherjee, "WDM-based Local Lightwave Networks- Part II: Multi-Hop Systems," *IEEE Network*, July 1992, pp.20–32.
3. M. S. Goodman, H. Kobrinski, M. P. Vecchi, R. M. Bulley, and J. L. Gimlett, "The LAMBDANET Multiwavelength Network: Architecture, Applications, and Demonstrations," *IEEE Journal on Selected Areas in Communications*, vol. 8, no. 6, Aug. 1990, pp. 995–1004.
4. F. J. Janniello, R. Ramaswami, and D. G. Steinberg, "A Prototype Circuit-Switched Multi-Wavelength Optical Metropolitan-Area Network", *IEEE/OSA Journal of Lightwave Technology*, vol.11, no. 5/6, May/June 1993, pp. 777–782.
5. E. Hall, J. Kravitz, R. Ramaswami, M. Halvorson, S. Tenbrick, and R. Thosmen, "The Rainbow-II Gigabit Optical Network," *IEEE Journal on Selected Areas in Communications*, vol. 14, no. 5, June 1996, pp. 814–823.

6. T. K. Chiang, S. K. Agrawal, D. T. Mayweather, D. Sadot, C. Barry, M. Hickey, and L. G. Kazovsky, "Implementation of STARNET: A WDM Computer Communications Network," *IEEE Journal on Selected Areas in Communications*, vol. 14, no. 5, June 1996, pp. 824–39.
7. I. Chlamtac and A. Ganz, "Channel Allocation Protocols in Frequency-Time Controlled High Speed Networks," *IEEE Transactions on Communications*, vol.36, no.4, April 1988, pp. 430–440.
8. G. N. Rouskas and M. H. Ammar, "Analysis and Optimization of Transmission Schedules for Single-Hop WDM Networks", *IEEE/ACM Transactions on Networking*, vol.3, no.2, April 1995, pp. 211–221.
9. A. Ganz and Y. Gao, "Time-Wavelength Assignment Algorithms for High Performance WDM Star Based Systems", *IEEE Transactions on Communications*, vol.42, no.(2-3-4), February/March/April 1994, pp. 1827–1836.
10. G. R. Pieris and G. H. Sasaki, "Scheduling Transmissions in WDM Broadcast-and-Select Networks," *IEEE/ACM Transactions on Networking*, vol.2, no.2, April 1994, pp. 105–110.
11. M. S. Borella and B. Mukherjee, "Efficient Scheduling of Nonuniform Packet Traffic in a WDM/TDM Local Lightwave Network with Arbitrary Transceiver Tuning Latencies," *IEEE Journal on Selected Areas in Communications*, vol. 14, no. 6, June 1996, pp. 923–934.
12. M. Azizoglu, R. A. Barry, and A. Mokhtar, "Impact of Tuning Delay on the Performance of Bandwidth-Limited Optical Broadcast Networks with Uniform Traffic," *IEEE Journal on Selected Areas in Communications*, vol. 14, no. 6, June 1996, pp. 935–944.
13. P. W. Dowd, "Random Access Protocols for High Speed Interprocessor Communication Based on An Optical Passive Star Topology", *IEEE/OSA Journal of Lightwave Technology*, vol. 9, no. 6, June 1991, pp.799–808.
14. A. Ganz and Z. Koren, "Performance and Design Evaluation of WDM Stars", *IEEE/OSA Journal of Lightwave Technology*, vol. 11, no. 2, February 1993, pp.358–366.
15. I. M. I. Habbab, M. Kavehrad, and C.-E. W. Sundberg, "Protocols for Very High Speed Optical Fiber Local Area Networks Using A Passive Star Topology", *IEEE/OSA Journal of Lightwave Technology*, vol. 5, no. 12, December 1987, pp.1782–1794.
16. N. Mehravari, "Performance and Protocol Improvements for Very High-Speed Optical Fiber Local Area Networks Using a Passive Star Topology", *IEEE/OSA Journal of Lightwave Technology*, vol. 8, no. 4, April 1990, pp.520–530.
17. G. N. M. Sudhakar, M. Kavehrad, and N. Georganas, "Slotted ALOHA and Reservation ALOHA Protocols for Very High-Speed Optical Fiber Local Area Networks Using Passive Star Topology," *IEEE/OSA Journal of Lightwave Technology*, vol. 9, no. 10, October 1991, pp. 1411–1422.
18. G. N. M. Sudhakar, N. Georganas, and M. Kavehrad, "Multi-Control Channel Very High-Speed Optical Fiber Local Area Networks and Their Interconnections Using a Passive Star Topology," *Proceedings of IEEE GLOBECOM'91*, Phoenix, AZ, December 1991, pp. 624–628.
19. F. Jia and B. Mukherjee, "The Receiver Collision Avoidance (RCA) Protocol for A Single-Hop Lightwave Network", *IEEE/OSA Journal of Lightwave Technology*, vol.11, no. 5–6, May/June 1993, pp.1052–1065.
20. M.-S. Chen, N. R. Dono, and R. Ramaswami, "A Media-Access Protocol for Packet-Switched Wavelength Division Multiaccess Metropolitan Area Networks", *IEEE Journal on Selected Areas in Communications*, vol. 8, no. 6, August 1990, pp. 1048–1057.
21. I. Chlamtac and A. Fumagalli, "Quadro-Star: A High Performance Optical WDM Star Network", *IEEE Transaction on Communications*, vol. 42, no. 8, August 1994, pp. 2582–2591.
22. M. Chen and T.-S. Yum, "A Conflict-Free Protocol for Optical WDM Networks", *Proceedings of IEEE GLOBECOM'91*, December 1991, pp.1276–1291.
23. K. M. Sivalingam and J. Wang, "Media Access Protocols for WDM Networks with On-Line Scheduling", *IEEE/OSA Journal of Lightwave Technology*, vol. 14, no. 6, June 1996, pp. 1278–1286.

24. R. Chipalkatti, Z. Zhang, and A. S. Acampora, "Protocols for Optical Star-Coupler Network Using WDM: Performance and Complexity Study," *IEEE Journal on Selected Areas in Communications*, vol. 11, no. 4, May 1993, pp. 579–589.
25. P. A. Humblet, R. Ramaswami, and K. N. Sivarajan, "An Efficient Communication Protocols for High-Speed Packet-Switched Multichannel Networks," *IEEE Journal on Selected Areas in Communications*, vol. 11, no. 4, May 1993, pp. 568–578.
26. H. Jeon and C. Un, "Contention-based Reservation Protocols in Multiwavelength Optical Networks with a Passive Star Topology," *Proceedings, IEEE ICC*, June 1992, pp. 1473–1477.
27. J. H. Lee and C. K. Un, "Dynamic Scheduling Protocol for Variable-Sized Messages in a WDM-Based Local Network," *IEEE/OSA Journal of Lightwave Technology*, vol. 14, no. 7, July 1996, pp. 1595–1600.
28. K. Bogineni and P. W. Dowd, "A Collisionless Media Access Protocols for High Speed Communication in Optically Interconnected Parallel Computers," *Proceedings of SPIE*, vol. 1577, September 1991, pp. 276–287.
29. P. W. Dowd and K. Bogineni, "Simulation Analysis of A Collisonless Multiple Access Protocol for A Wavelength Division Multiplexed Star-Coupled Configuration," *Proceedings of 25th Annual Simulation Symposium*, Orlando, FL, April 1992.
30. K. Bogineni and P. W. Dowd, "A Collisionless Multiple Access Protocol for A Wavelength Division Multiplexed Star-Coupled Configuration: Architecture and Performance Analysis," *IEEE/OSA Journal of Lightwave Technology*, vol. 10, no. 11, November 1992, pp. 1688–1699.
31. K. M. Silvalingam and P. W. Dowd, "A Multilevel WDM Access Protocol for an Optically Interconnected Multiprocessor System," *IEEE/OAS Journal of Lightwave Technology*, vol. 13, no. 11, November 1995, pp. 2152–2167.
32. K. M. Sivalingam and P. W. Dowd, "A Lightweight Media Access Protocol for a WDM-Based Distributed Shared Memory System," *Proceedings of IEEE INFOCOM' 96*, 1996, pp. 946–953.
33. F. Jia, B. Mukherjee, and J. Iness, "Scheduling Variable-Length Messages in a Single-Hop Multichannel Local Lightwave Network", *IEEE/ACM Transactions on Networking*, vol. 3, no. 4, August 1995, pp.477–487.
34. A. Muir and J. J. Garcia-Luna-Aceves, "Distributed Queue Packet Scheduling Algorithms for WDM-Based Networks", *Proceedings of IEEE INFOCOM' 96*, 1996, pp. 938–945.
35. B. Hamidzadeh, M. Ma, and M. Hamdi, "Message Sequencing Techniques for On-Line Scheduling in WDM Networks", *IEEE/OSA Journal of Lightwave Technology*, vol. 17, no. 8, August 1999, pp. 1309–1319.
36. M. Ma, B. Hamidzadeh, and M. Hamdi, "A Receiver-Oriented Message Scheduling Algorithm for WDM Lightwave Networks", *Computer Networks*, vol. 31, no. 20, September 1999, pp.2139–2152.
37. L. Bo, A. Ganz, C. M. Krishna, "An In-Band Signaling Protocol for Optical Packet Switching Networks" *IEEE Journal on Selected Areas in Communications*, vol. 18, no. 10, October 2000, pp. 1876–1884.
38. J. H. Lee, "Persistent Reservation Protocol for Variable-Length Messages in WDM-based Local Networks", *Proceedings of IEEE Communications*, vol. 148, no. 2, April 2001, pp. 81–85.
39. M. Sakuta, Y. Nishino, and I. Sasase, "Channel Reservation Protocol Using a Counter for Detecting a Source Conflict in WDM Single-Hop Optical Network with Non-Equivalent Distance", *Proceedings of IEEE International Conference on Communications*, vol. 3, 2001, pp.707–711.
40. L. Hwa-Chun, and L. Pei-Shin, "Dynamic Control Frames in Reservation-Based Packet Scheduling for Single-Hop WDM Networks", *Proceedings of IEEE International Conference on Parallel Processing*, 2003, pp. 87–95.
41. G. I. Papadimitriou, M. S. Obaidat, and A. S. Pomportsis, "Adaptive Protocols for Single-Hop Photonic Networks with Bursty Traffic", *Proceedings of IEEE International Conference on Parallel Processing*, pp. 227–231, 2001.

42. L. Hwa-Chun and L. Pei-Shin, "Reducing Packet Delay in Single-Hop WDM Networks Using Fixed Transceiver Array and Adaptive Channel Allocation", *Journal of Lightwave Technolgy*, vol. 24, no. 12, Dec. 2006, pp. 4925–4935.
43. P. G. Sarigiannidis, G. I. Papadimitriou and A. S. Pomportsis, "A High-Throughput Scheduling Technique, With Idle Timeslot Elimination Mechanism", *Journal of Lightwave Technology*, pp. 4811–4827, Dec. 2006.
44. C. Papazoglou, P. G. Sarigiannidis, G. I. Papadimitriou, and A. S. Pomportsis, "A New Priority Scheme for WDM Star Networks", *2007 14th IEEE Symposium on Communications and Vehicular Technology in the Benelux*, Nov. 2007, pp. 1–3.
45. L. Hwa-Chun and L. Pei-Shin, "Using Dynamic Control Frames to Reduce Packet Delays in Transceiver Array-Based Single-Hop WDN Networks", *Journal Of Lightwave Technology*, vol. 27, no. 7, April 2008, pp. 742–755.
46. M. Jonsson, K. Borjesson, and M. Legardt, "Dynamic Time-Deterministic Traffic in a Fiber-Optic WDM Star Network", *Proceedings of Ninth Euromicro Workshop on Real Time Systems*, June 1997, pp.25–33.
47. M. Ma, B. Hamidzadeh, and M. Hamdi, "Efficient Scheduling Algorithms for Real-Time Service on WDM Optical Networks", *Photonic Network Communications*, vol. 1, no. 2, August 1999, pp.161–178.
48. M. Ma and M. Hamdi, "A Scheduling Algorithm for Differentiated Service on WDM Passive Optical Access Networks", *Journal of Optical Networking*, vol. 1, no. 11, November, 2002, pp. 386–396.
49. H.-Y. Tyan, J. C. Hou, B. Wang, and C.-C. Han, "On Supporting Temporal Quality of Service in WDMA-Based Star-Coupled Optical Networks", *IEEE Transactions on Computers*, vol. 50, no. 3, March 2001, pp. 197–214.
50. A. C. Kam, K.-Y. Siu, R. A. Barry, and E. A. Swanson, "A Cell Switching WDM Broadcast LAN with Bandwidth Guarantee and Fair Access", *IEEE/OSA Journal of Lightwave Technology*, vol. 16, no. 12, December 1998, pp. 2265–2280.
51. M. Ma and M. Hamdi, "Providing Deterministic Quality-of-Service Guarantees on WDM Optical Networks", *IEEE Journal on Selected Areas in Communications*, vol. 18, no. 10, October 2000, pp. 2072–2083.
52. B. Li and Y. Qin, "Traffic Scheduling in A Photonic Packet Switching System with QoS Guarantee", *IEEE/OSA Journal of Lightwave Technology*, vol. 16, no. 12, December 1998, pp. 2281–2295.
53. M. A. Marsan, A. Bianco, E. Leonardi, A. Morabito, and F. Neri, "All-Optical WDM Multi-Rings with Differentiated QoS", *IEEE Communications Magazine*, February 1999, pp. 58–66.
54. L. Elhanany and D. Sadot, "A Prioritized Packet Scheduling Architecture for Provision of Quality-of-Service in Tbits/sec WDM Networks", *Proceedings of IEEE International Conference on Communications*, New-Orleans, June 2000, vol. 2, pp. 695–700.
55. J. Diao and P. L. Chu, "Packet Rescheduling in WDM Star Networks with Real-Time Service Differentiation", *IEEE Journal of Lightwave Technology*, vol. 19, no. 12, December 2001, pp. 1818–1828.
56. K. Bengi and H. R. van As, "CONRAD: A Novel Medium Access Control Protocol for WDM Local Lightwave Networks Enabling Efficient Convergence of Real-Time and Data Services", *Proceedings of IEEE LCN'01*, 2001, pp. 468–476.
57. J. Indulska and J. Richards, "A Comparative Simulation Study of Protocols for A Bus WDM Architecture", *Proceedings of International Conference on Networks*, 1995, pp. 251–255.
58. W. M. Moh, T.-S. Moh, Y.-J. Chien, J. Wang, R.-J. Wang, and Y.-W. Wang, "The Support of Optical Network Protocols for Multimedia ATM Traffic", *Proceedings. International Conference on Networks*, 1995, pp. 1–5.
59. N.-F. Huang and H.-I. Liu, "Wavelength Division Multiplexing-Based Video-on-Demand Systems", *IEEE/OSA Journal of Lightwave Technology*, vol. 17, no. 2, Februray 1999, pp.155–164.

60. L. Wang, M. Ma, and H., M., "Efficient Protocols for Multimedia Streams on WDMA Networks", *IEEE/OSA Journal of Lightwave Technology*, vol. 21, no. 10, October 2003, pp. 2123–2144.
61. J. S. Choi, N. Golmie and D. Su, "A Bandwidth Guaranteed Multi-Access Protocol for WDM Local Networks," *Proceedings of IEE ICC'00*, 2000, vol. 3, pp. 1270–1276.
62. M. Ma and M. Hamdi, "Providing Deterministic Quality-of-Service Guarantees on WDM Optical Networks", *IEEE Journal on Selected Areas in Communications*, vol. 18, no. 10. October 2000, pp. 2072–2083.
63. M. Kovacevic and M. Gerla, "HONET: An Integrated Services Wavelength Division Optical Network", Proceedings. *IEEE ICC'94*, 1994, pp. 1669–1674.

Chapter 9
Efficient Traffic Grooming Scheme for WDM Network

Y. Aneja, A. Jaekel, S. Bandyopadhyay and Y. Lu

Abstract One major objective in WDM network design is to develop a logical topology and a routing scheme over the topology that minimizes the congestion of the network. This combined topology design and traffic grooming problem becomes computationally intractable, even for moderate sized networks. One standard approach is to decouple the problem of logical topology design and the problem of routing over the logical topology. Heuristics for finding the logical topology exist and a straight-forward linear program (LP), based on the node-arc formulation is normally used to solve the routing problem over a given logical topology. However, the traffic grooming problem is in itself an inherently difficult problem, and standard LP formulations are not able to solve this problem for large networks. In this paper, we have introduced a novel approach for traffic grooming, over a given logical topology, using the concept of approximation algorithms. This technique allows us to efficiently route traffic for practical sized networks and obtain solutions, which are guaranteed to be within a specified bound of the optimal solution. Simulation results from different networks demonstrate that approximation algorithms can be used to quickly generate "near-optimal" solutions to the traffic routing problem in WDM networks.

Keywords WDM networks · Congestion · Multi-commodity flow

9.1 Introduction

With the rapid development of internet technology and other applications using computer networks in recent years, the demand for high speed networks has been increasing dramatically. Optical networks using Wavelength-Division Multiplexing (WDM) are ideal candidates to meet the rapidly increasing user requirements for increased communication. In a WDM optical network, the transmissions from many different end-users can be combined and transmitted on the same fiber [1]. By

Y. Aneja (✉)
University of Windsor, Windsor, Ontario Canada N9B 3P4

M. Ma. (ed.), *Current Research Progress of Optical Networks*,
DOI 10.1007/978-1-4020-9889-5_9,

transmitting data using a number of separate optical carriers on each fiber, where each optical carrier has a distinct wavelength, the WDM technology can exploit the huge bandwidth of optical networks [2, 3].

A *lightpath* [4] in a WDM network is an end-to-end all-optical communication channel. The lightpaths determine which nodes can directly communicate with each other. It is convenient to view the lightpaths as the edges of a directed graph G (V, E) where the nodes of G (V, E) are the end-nodes or router nodes of the physical topology. Such a graph is called the *logical topology* of an optical network and the edges of such a graph are called *logical edges* [4]. Once the lightpaths are set up, the physical topology is irrelevant for determining a strategy for data communication to handle the traffic between all the nodes.

Individual requests for connections are typically for data streams at a much lower data communication rates, compared to the huge bandwidth of a lightpath. The standard *traffic grooming* problem [5, 6] is to combine the low-speed user data streams in a way that makes effective use of the bandwidth of each lightpath, so that

1. The traffic demands between all node pairs are satisfied and
2. The total traffic on a lightpath does not exceed the capacity of the lightpath.

The *routing strategy* for traffic grooming is used to determine the logical path(s) selected to route the traffic from a source node s to a destination node t. A logical path from s to t is a sequence of edges, of the form $s = v_0 \rightarrow v_1 \rightarrow \ldots \rightarrow v_p = t$, defined over the logical topology. This means there is a lightpath from $s = v_0$ to v_1, a lightpath from v_1 to v_2, ..., a lightpath from v_{p-1} to v_p. In general, there are many logical paths from s to t.

In the literature, two approaches have been adopted for determining the routing strategy for traffic grooming. In the *bifurcated model* [7], a data stream from an individual user is allowed to be split into multiple data streams, to be carried by different logical paths from the source of the data stream to its destination [8]. In the *non-bifurcated model*, each data stream is communicated using a *single* logical path, P, from the source of the data stream to its destination. The data stream becomes part of the payload of each lightpath in the logical path P. This model has been adopted in [9–13]. In this paper, we consider the *bifurcated* model.

As mentioned in [14], for a given logical topology, the traffic grooming problem is in itself an inherently difficult problem. For the bifurcated model, it has been suggested that a standard linear program (LP) may be defined to optimally route the traffic over the logical topology [15]. A LP of the type given in [15] is called a node-arc formulation [16] and requires O (n^3) constraints, where n is the number of nodes in the network. The size of such a LP increases rapidly with the number of nodes in the network and we have found that this approach is not feasible for networks with more than 25 end-nodes. Heuristic approaches can be used to generate a solution, but it is difficult to evaluate the performance of such heuristics as they provide no guarantees on the quality of the solutions.

In this paper, we address the problem of traffic grooming, under the bifurcated traffic model, for a given logical topology. We assume that the logical topology (i.e., the set of lightpaths) has already been determined, using existing techniques [17]. The topology design problem takes into consideration different WDM network constraints such as fiber capacity, the number transceivers per node, o-e-o conversion capabilities and wavelength conversion capabilities at each node. Our objective in this paper is to find a traffic routing that minimizes the *congestion* of the network, where congestion is defined as the traffic load on the edge (lightpath) carrying the maximum amount of traffic. This objective has also been used in [15, 18, 19]. Informally, we wish to distribute the traffic as evenly as possible over all the edges. This ensures that, if the traffic requirements increase in the future, the scheme does not have to be modified until the load on the edge carrying maximum traffic exceeds the capacity of a lightpath.

The main contributions of this paper are:

1. We formulate the congestion minimization problem for WDM networks as a multi-commodity network flow (MCNF) problem and show that, for larger networks, the use of standard LP formulations become computationally intractable, and a different approach is needed.
2. We introduce a novel approach based on *approximation algorithms* to generate solutions, which are guaranteed to be within a specified bound of the optimal solutions.
3. We show that our approximation algorithm based approach can generate solutions, in a reasonable time, where standard LP formulations fail.

The remainder of the paper is organized as follows. In Section 9.2, we briefly review existing techniques for the logical topology design and traffic grooming problems, as well as the multi-commodity network flow (MCNF) problem. In Section 9.3, we present our approximation algorithm based approach for minimizing congestion. We discuss our experimental results in Section 9.4, and conclude with a critical summary in Section 9.5.

9.2 Background Review

In this section, we review some relevant background material including basic concepts of multi-commodity network flows, and existing linear program-based solutions for routing in WDM networks.

9.2.1 Logical Topology Design and Traffic Grooming

Traffic Grooming in WDM can be defined as a family of techniques for combining a number of low-speed data streams from users so that the high capacity of each lightpath may be used as efficiently as possible. There are two basic approaches in traffic grooming:

For a given set of traffic requests, minimize the total network cost, with the condition that all traffic requests are satisfied.

For given resource limitations and traffic demands, maximize the network throughput, measured by the total amount of traffic that is successfully carried by the network.

The complete traffic grooming problem has been addressed in a number of recent papers. In [18, 19] the authors propose ILP formulations and heuristics for logical topology design, with the objective of minimizing congestion. In [10] the objective is to minimize the number of lightpaths in the network. This objective function is directly related to the cost of the transceivers. In [20] the authors minimize the total weighted hop count corresponding to the logical paths used to route each traffic request. The heuristic proposed in [13], where the single hop communication was favored over multi-hop communication, captures the same idea. Finally, [13] also presents a formulation that maximizes the weighted sum of requests that may be handled by the network, for a specified set of network resources. Here, the *weight* of a request is the required data communication rate, in OC-n notations. This objective was also considered in [21].

The combined design problem quickly becomes intractable, even for moderate sized networks. One widely used heuristic [15, 17, 18] is to decompose the combined problem into 3 separate sub-problems, as given below, and solve each one independently:

Logical topology design: Determine the set of lightpaths (logical edges) capable of accommodating a given traffic demand. An excellent survey of existing techniques for topology design is given in [17].

Routing and wavelength assignment (RWA): Determine a suitable route, over the physical topology, and select a single WDM channel for each lightpath in the logical topology [22]. Both ILP formulations [23] as well as heuristic solutions based on genetic algorithms [24], simulated annealing [25] and tabu search [26] have been proposed for this problem.

Traffic grooming: Given a logical topology with a feasible RWA, route the user traffic over the logical edges. For the bifurcated traffic model, one common objective of traffic grooming is to minimize the congestion of the network.

The above decomposition may be sub-optimal, and may even be infeasible in cases where a later sub-problem does not have a solution, based on the solutions used for the earlier sub-problems.

Considerable work has been done in the literature on the first two sub-problems [17, 22]. In this paper, we focus on the third sub-problem, i.e. traffic grooming, on a given logical topology.

9.2.2 WDM Routing as Multi-Commodity Network Flow (MCNF) Problem

In a multi-commodity network flow (MCNF) problem [27–29] the network is viewed as a graph where the nodes represent sources and sinks of *commodities* to be transported over the network and edges represent a way to transport a commodity

Table 9.1 Increase of number of constraints with network size

nodes	arcs	commodities	constraints
10	30	90	930
20	60	380	7660
40	120	1560	62520
100	300	9900	990300

from one node to another. Typically the commodities must be routed in a way that optimizes certain objectives, such as the total cost of transportation, or the amount of spare capacity to accommodate future growth. The problem of routing traffic over the logical topology of a WDM network can be viewed as a MCNF problem. If there is a traffic $d_j(d_j > 0)$ from source s_j to destination t_j, from a multi-commodity network flow perspective, there is a *commodity* from s_j to destination t_j. Since, in a WDM network, most nodes will be required to communicate with each other, the number of commodities for a network with n nodes is $O\ (n^2)$.

Multi-commodity flow problems reported in the operations research literature typically consider a very limited number of commodities [27]. The fact that a standard LP approach, as given in [15], for solving the routing problem in WDM networks require $O(n^2)$ commodities and $O(n^3)$ constraints means that such formulations are not useful for practical sized networks. Table 9.1 shows how quickly the number of constraints and commodities increases with the number of network nodes.

For instance, in a network with 40 nodes network, where the average node degree is 3, the number of edges is 120, the number of commodities is over 1500 and the number of constraints is over 62000. In other words, standard LP packages guarantee an optimal solution, but cannot handle even moderate size networks. Heuristics can give us fast solutions, but it is difficult to evaluate the quality of the solutions. We are not aware of any heuristic that provides a guarantee on the quality of the solution.

9.2.3 Primal-Dual Formulations

For any linear programming (LP) formulation of a problem (the *primal* formulation) there exists a corresponding *dual* formulation [30]. Some well-known relationships between a primal and its corresponding dual are as follows:

If the primal is a maximization problem, then the dual is a minimization problem and vice versa.

For the *optimal* solution, assuming both the primal and the dual are feasible, objective values for the primal and the objective value for the dual are the same.

The number of constraints in the primal is equal to the number of variables in the dual and vice versa.

The *right-hand side* (RHS) coefficients of the primal become the objective function coefficients in the dual and vice versa.

A multi-commodity flow problem is defined on a directed network $G = (V, E)$ with capacities $u : E \to R$ having K commodities. The jth commodity has an associated source s_j and a destination tj, $1 \le j \le K$. The problem is to find flows for the jth commodity, from s_j to t_j, that satisfy node conservation constraints and meet some objective function criteria so that the sum of the flows on any edge does not exceed the capacity of the edge, for all j, $1 \le j \le K$.

Let P_j denote the set of paths from s_j to t_j, and let $P = \cup_j P_j$. Variable $x(p)$ equals the amount of flow sent along path p. For the maximum multi-commodity flow problem, the objective is to maximize the sum of the flows. The constraint on each edge states that the total flow on edge e must be less than or equal to the capacity $u(e)$ of the edge. The capacity $u(e)$ of each edge e is a constant, which is specified as input to the linear program. The corresponding linear programming formulation is as follows:

$$\text{Max} \quad \sum_{p \in P} x(p)$$
$$\forall e : \quad \sum_{p : e \in P} x(p) : u(e)$$
$$\forall p : x(p) \ge 0$$

The dual to this linear program corresponds to the problem of assigning lengths to the edges of the graph so that the length of the shortest path from s_j to t_j is at least 1 for all commodities j. The length of an edge represents the marginal cost of using an additional unit of capacity of the edge. The objective function remains *linear*, since $u(e)$ is assumed to be a constant.

$$\text{Min} \sum_{e} u(e) l(e)$$
$$\forall p : \sum_{e \in P} l(e) \ge 1$$
$$\forall e : l(e) \ge 0$$

For the maximum concurrent flow problem, there are demands d_j associated with each commodity j, and the objective is to satisfy the maximum possible proportion of all demands. We will discuss such formulations in more detail in Section 9.3.

9.2.4 Approximation Algorithms

In the operations research community, approximate algorithms have been proposed to solve network flow problems in large networks. Garg and Konemann [31] give a simple, deterministic algorithm to solve the maximum flow problem. In a way similar to that in [32], this algorithm augments the flows in the network using shortest paths. They obtained an improvement in run time for the flow problems. Their main contribution is to provide an analysis for the correctness of their algorithms. Lisa K.Fleischer [33] has proposed a faster approximation algorithm for maximum multicommodity concurrent flow problem based on Garg and Konemann [31]. It is faster than Garg and Konemann's algorithm when the graph is sparse or there are a large number of commodities.

9.3 Approximation Algorithms for WDM Network

The number of variables and constraints in the MCNF formulations, for grooming in WDM networks, grow very rapidly with network size, particularly when the number of commodities is large. Therefore, it is very difficult to obtain optimal solutions, even for networks of moderate size. In this context, approximation algorithms can be an important tool for obtaining "good" solutions, in a reasonable amount of time. These algorithms do not necessarily generate the best solution, but can guarantee that their solution will be within a specified bound of the optimal solution [31] and [33].

In this section, we will present our approach, based on the concept of approximation algorithms [33], for solving the multi-commodity flow problem to minimize the congestion in WDM optical networks. The algorithms presented in [33] cannot be used directly for the traffic grooming problem in WDM networks. Therefore, we first specify the congestion minimization problem using the arc-chain formulation (Equations 9.1–9.3 and 9.4). Then we apply some transformations (Equations 9.5–9.10) to convert it to a maximization problem, which can be solved using approximation algorithms. Next, we obtain the corresponding dual formulation (Equations 9.11–9.14). Finally, we prove (Theorem 1 in Section 9.3.2) that there exists a feasible solution to the dual formulation corresponding to any feasible primary solution. We also show how to obtain the feasible dual solution and corresponding value of the objective function from the feasible primary solution. Once we have the primary and dual objective values, we can compare them to check if the solution is within the specified bound of the optimal solution. If so, the stopping criterion is met and the algorithm terminates, otherwise the next iteration is started.

Based on the relationships between the primal and the dual solutions summarised in Section 9.2.3, we can make the following observation:

Let *P1 (D1)* be a *feasible* (not necessarily optimal) solution of the primal (dual) formulation and let obj_p (obj_d) be the objective function value corresponding to *P1 (D1)*. Then, if $obj_p = obj_d$, we know (based on relationship ii of Section 9.2.1) that this is the optimal value. For all other cases, if the primal is a maximization (minimization) problem, then $obj_p < optimal\_value < obj_d$ (respectively $obj_d < optimal\_value < obj_p$). In this case, by comparing the values of obj_p and obj_d, we can determine how close we are to the optimal value. For example, if $obj_p = 100$ and $obj_d = 110$, we can say that our solution is within 10% of the optimal solution, even though we do not know exactly what the optimal solution is. This is the main concept behind the approximation algorithm presented in this section.

9.3.1 Formulation of the Minimum Congestion Problem

In this section, we consider the arc-chain formulation [16] of the minimum congestion routing problem for WDM networks. It is well-known that arc-chain formulations for minimum-cost MCNF require O (n^2) constraints [16]. This represents significant savings over the node-arc formulation requiring O(n^3) constraints.

We consider the logical topology of a WDM network viewed as a $G = (V,\ E)$, where V is the set of the n end-nodes of the WDM network and E is the set of m logical edges. We denote a logical edge from node x to y by $x \to y$. Each non-zero traffic demand d_j, between a source-destination node pair $(s_j,\ t_j)$ constitutes one commodity. If there are K node pairs with non-zero traffic between them, there will be exactly K commodities flowing over the network.

In our formulation, p_e denotes the set of paths which pass through the logical edge e and P_j denotes the set of paths (from s_j to t_j) for commodity j, *so* that $P = \bigcup_{j=1}^{K} P_j$ is the set of all valid paths (for all commodities).

As mentioned, d_j is the traffic demand for commodity j. This means that d_j units of traffic must be sent over the network, from s_j to $t_j 1 \le j \le K$. We will use $x(p)$ to denote the traffic flow on path p. The objective of the optimization is to route the traffic in such a way that the demands for all commodities are met and the congestion (μ) of the network is minimized.

The linear programming formulation for minimizing the congestion is given below.

$$\text{Objective: Min } \mu \tag{9.1}$$

$$\sum_{p \in p_e} x(p) \le \mu \qquad \forall e \in E \tag{9.2}$$

$$\sum_{p \in P_j} x(p) = d_j,\ j = 1, 2, 3, \ldots, K \tag{9.3}$$

$$x(p) \ge 0, \forall p \in P \tag{9.4}$$

(Equation 9.1) Gives the objective function, which states that the congestion should be minimized. Constraint (9.2) is actually a composite constraint. It corresponds to m individual constraints, one for each edge in the network. For each edge $e \in E$, $\sum_{p \in p_e} x(p)$ gives the total traffic flow on edge e summed over all paths containing the edge e, for all commodities in the network. Constraint (9.2) then states that the total traffic flow on edge e must be less than or equal to the congestion μ of the network. This constraint must be satisfied since congestion is, by definition, the maximum traffic flow on any edge.

Constraint (9.3) is another composite constraint. It corresponds to K individual constraints, one for each commodity in the network. For each commodity j, $1 \le j \le K$, constraint (9.3) states that the total traffic flow for commodity j, over all paths in P_j, must be greater than or equal to d_j. In other words, constraint (9.3) ensures that the traffic demands for each commodity are satisfied. Constraint (9.4) simply states that all traffic flows must be positive.

9.3.2 Primal-Dual Formulation for Congestion Minimization

We now consider the basic formulation we described in Equations (9.1–9.4), apply some transformations for convenience to give us a primal formulation that we will use in our algorithm. Then we will construct the corresponding dual formulation. Our transformations are such that

- it is easy to obtain initial feasible solutions for both the primal and corresponding dual formulations and
- it is possible to improve the primal and dual solutions iteratively, without using the time consuming standard solution methods (e.g. revised simplex method [30]) for linear programs.

$$\begin{aligned} &\text{Let } \mu = \mathrm{d}_{\max}/\lambda, \\ &\text{where } d_{\max} = max\{d_j : j = 1,\ 2,\ 3,\ \ldots,\ K\} \end{aligned} \tag{9.5}$$

Substituting $\mu = d_{max}/\lambda$ in (9.1) we get a new objective function – minimize d_{max}/λ. This is equivalent to maximizing λ, since d_{max} is a constant.

Substituting (9.5) into (9.2), we get:

$$\lambda \sum_{p \in p_e} x(p) \leq d_{\max},\ \forall e \in E \tag{9.6}$$

Let $y(p) = \lambda x(p)$. Equations 9.1–9.4 may be restated as follows:

$$\text{Maximize } \lambda \tag{9.7}$$

$$\sum_{\substack{e \\ p \in P}} y(p) \leq d_{\max}, \forall e \in E \tag{9.8}$$

$$\sum_{p \in P_j} y(p) \geq \lambda d_j, \forall j,\ j = 1, 2, 3, \ldots, K \tag{9.9}$$

$$y(p) \geq 0, \forall p \in P \tag{9.10}$$

This is a special case of the concurrent flow problem [27]. We have m constraints in (9.8) – one for each edge. We associate dual variable $l(e)$ with edge e, one dual variable for each of the m constraints of (9.8). We have K constraints in (9.9) – one for each commodity. We associate dual variable z_j for the jth commodity $1 \leq j \leq$

K in (9.9). Then, the dual formulation for (9.7, 9.8, 9.9 and 9.10) is obtained as follows:

$$\text{Minimize } d_{max} \sum_{e \in E} l(e) \tag{9.11}$$

$$\sum_{e \in P} l(e) - z_j \geq 0 \; p \in P_j, \; j = 1, 2, 3, \ldots, K \tag{9.12}$$

$$\sum_{j=1}^{K} d_j z_j \geq 1 \tag{9.13}$$

$$l(e) \geq 0, \forall e \in E, \; and \; z_j \geq 0 \; j = 1, 2, 3, \ldots, K \tag{9.14}$$

Suppose, at any given time, we have positive values for the dual variables so that the above constraints (9.12, 9.13) are satisfied. This means $l(e) > 0$, for all edge $e \in E$. We now discuss how we get, in the next iteration, a new value for each of the dual variables satisfying the constraints (9.12, 9.13).

We will use $l(p_j^*)$ to denote the length of the shortest path for commodity j. We now define D^* and z_j^* and $\hat{l}(e)$ as follows:

$$D^* = \sum_{j:1 \leq j \leq K} d_j l(p_j^*) \tag{9.15}$$

$$z_j^* = l(p_j^*)/D^* \tag{9.16}$$

$$\hat{l}(e) = l(e)/D^*, \quad e \in E \tag{9.17}$$

Theorem 1. *If the flows in the network are assigned, based on arc lengths l(e), there is a feasible solution of the dual formulation with arc lengths* $\hat{l}(e) = l(e)/D^*$, $\forall e \in E$.

Proof. $\sum_{j \in k} d_j z_j^* = \sum d_j l(p_j^*)/D^* = 1$. So, if we substitute z_j by z_j^*, then constraint (9.13) is always satisfied.

The length of the pth path for commodity j is $\sum_{e \in P} l(e)$. Since $l(p_j^*)$ is the length of the *shortest* path for commodity j, the length of any other path for commodity j must be greater than or equal to $l(p_j^*)$.

Therefore $\sum_{e \in p} l(e) \geq l(p_j^*)$ and relationship (9.18) follows.

$$\sum_{e \in P} l(e)/D^* \geq l(p_j^*)/D^* \tag{9.18}$$

By using the value of z_j^* from (9.16), we get:

$$\sum_{e \in p} l(e)/D^* \geq z_j^* \; p \in P_j, 1 \leq j \leq K \tag{9.19}$$

Then, using (9.16) and (9.17), we can write (9.19) as:

$$\sum_{e \in p} \hat{l}(e) - z_j^* \geq 0, \; p \in P_j, 1 \leq j \leq K \tag{9.20}$$

which satisfies constraint (9.12).

9.3.3 The Approximation Algorithm

In this section, we describe the approximation algorithm itself (Fig. 9.1). This is an adaptation of the algorithms presented in [30, 33]. We show how the primal and dual objective values are calculated and how the stopping criterion is applied.

1. Choose values for δ and ε ($\delta > 0$ and $\varepsilon > 0$)
2. $l(e) := \delta/d_{\max}$, $x(e) = 0$, $\forall\, e \in E$
3. $r = 0$
4. Do
   a. $r = r+1$
   b. for $j = 1$ to K
      i) $l(p_j)$:= shortest path distance for Commodity j, $j = 1,2,3\ldots\ldots K$
      ii) $x(p_j) := x(p_j) - d_j$
      iii) $l(e) := l(e)[1 - \varepsilon d_j/d_{\max}]$ $e \in p_j$
   End for
   c. primalSolution = $r * d_{\max} / \max\{x(e) : e \in E\}$
   d. dualSolution = $\sum_{e \in E} l(e) / \sum_{j \in k} d_j l(p_j)$
   e. $\mu = \max\{x(e) : e \in E\}/\mathrm{r}$
   while ((dualSolution/primalSolution) < $(1 - \varepsilon)$)

Fig. 9.1 Overview of approximation algorithm for minimizing congestion

The first step in the algorithm is to set appropriate values for δ and ε. These determine the accuracy of the final solution and the speed of convergence of the algorithm. For example, if we want the final solution to be within 10% of the optimal value, we should choose $\varepsilon = 0.1$. In our experiments we have used $\delta = 0.001$ and $\varepsilon = 0.1$.

In step 2, we set the initial values for the dual variables $l(e)$. We start with very small non-zero values (δ/d_{max}) for each arc length. In step 3, we initialize the iteration counter r.

Step 4 is the main iterative step of the algorithm and is repeated until the final solution is found. In this step, we first update the iteration counter r (step 4a). Then we consider each commodity in turn (step 4b) and for each commodity j, we take the following actions:

We calculate the shortest path for each commodity, based on the arc lengths $l(e)$. We have three different implementations of the approximation algorithm, based on how the shortest path is calculated. These variations will be discussed in Section 9.3.4.

Send d_j units of flow along the shortest path p for commodity j, and update the flow on each edge $e \in p$, by d_j.

Update the length of each edge $e \in p$ as follows:

$$l(e) := l(e)[1 - \epsilon d_j / d_{\max}]$$

In step 4c (after we have considered all the commodities in 4b), we calculate the primal objective value (L) and the dual objective value (U), using Equations (9.19) and (9.20). Finally, we compare L and U and if they are close enough ($U/L < 1 + \varepsilon$), we can stop. Otherwise, we go back to step 4 and start the next iteration.

The objective value of the primal can be obtained from the above approximation algorithm as follows:

Suppose we are looking at the solution at the end of r iterations. After r iterations, all demands have been sent over the network r times. Then the amount of flow sent for commodity j is

$$r * d_j, \quad j = 1, 2, 3 \ldots \ldots K.$$

Assume $\{f_r(e) : e \in E\}$ is the flow on edge e at the end of r iterations for all edges. Scaling all arc flows $\{f_r(e) : e \in E\}$ by r would provide a solution to the primal. We know that the congestion (μ) is the maximum flow on any edge in the network. Therefore,

$$\mu = \frac{1}{r} Max\{f_r(e) : e \in E\} = \frac{1}{r} f_r^*$$

The objective value to be maximized, in the primal formulation, is λ as given by (9.7). Therefore, using Equation (9.5),

$\lambda = d_{\max}/\mu = r * d_{\max}/Max\{f_r(e) : e \in E\}$ Hence primal objective value is

$$\lambda = \frac{d_{\max}}{\mu} = r d_{\max}/f_r^* \tag{9.21}$$

Using the values of $l(e)$ generated by the approximation algorithm (in step 4bii) the objective value for a feasible dual solution can be obtained, as follows.

Equation (9.20) shows that $\hat{l}(e)$ satisfies constraint (9.12) of the dual formulation. Putting this value of $l(e)$ in the dual objective function (9.11), the dual objective value for a feasible solution (satisfying constraints (9.12) and (9.13)) is:

$$d_{\max}{}^{*}\sum_{e\in p}\hat{l}(e) = d_{max}{}^{*}\sum_{e\in p} l(e)/\mathrm{D}^{*}$$
$$= d_{\max}\sum_{e\in E} l(e)/\sum_{j\in k} d_j l(p_j^{*}) \tag{9.22}$$

Therefore, by using (9.21) and (9.22), we can calculate the primal and dual objective values. When these values are close enough (based on some predetermined limit), we can say we have obtained a "good enough" solution.

9.3.4 Shortest Path Algorithms

One important operation in our approximation algorithm is to find the shortest path for a particular commodity. This operation must be repeated for each commodity, in each iteration. Therefore, it is extremely important to find the shortest path as efficiently as possible. We have used three different shortest path algorithms in our implementations – Floyd-Warshall algorithm (WA) [34], Dijkstra algorithm (D1) [35], and a modified version of Dijkstra's algorithm, which we call the "Efficient Dijkstra" algorithm (D2).

Efficient Dijkstra algorithm is a variation of the Djikstra's algorithm that is faster to implement for this particular problem. Normally when the lengths of some edges in a graph change, the Djikstra's algorithm has to be used from the beginning. In this application, only the lengths of edges that lie on the shortest path change in one iteration, affecting, in general a small number of paths. Our efficient Djikstra's algorithm only recalculates the lengths of those paths which are affected by the changes in edge lengths in the previous iteration. Details of these implementations are available in [36].

9.4 Experimental Results

In this section, we present experimental results on the performance of the approximation algorithm and compare these results with the optimal solutions obtained by standard LP techniques. We have carried out our experiments on a large number of networks, ranging in size from 5 nodes to 50 nodes.

We have used an existing program [37], to generate logical topologies of different sizes (up to 50 nodes). This program takes in as input the underlying physical topology, including the number of available wavelengths per fiber and the number of transceivers per node and a traffic matrix (t_{ij}) that specifies the demand for the commodity corresponding to the traffic from node i to node j, in the network. The traffic matrices used to create the logical topology were generated randomly. The entries in the traffic matrix are expressed as a percentage of the total capacity of a lightpath. For example, assume the capacity of a lightpath is 10 GB/s. Then the entry $t_{13} = 12$, in the traffic matrix, indicates that the traffic demand from source node 1 to destination node 3 is 12% of the capacity of a single lightpath or 1.2 Gb/s. Based on this information, a logical topology is generated which is guaranteed to be able to accommodate the entire traffic demand. Although, we have selected a specific method for designing the logical topology, *any* suitable heuristic for topology design [17] may be used. Our traffic grooming algorithm can be used with any topology, irrespective of how it is generated.

Given the logical topology, our objective is to route the traffic over the topology in the most efficient manner. In order to evaluate the performance of our approximation algorithm based approach, we are interested in the following parameters:

- The quality of the solutions, i.e. how close they are to the *optimal* solution and
- The speed of the algorithm, i.e. how quickly it can generate "near-optimal" solutions.

Tables 9.2 and 9.3 summarize the results of our experiments. For each physical network we considered a number of different logical topologies and traffic demands. The results presented in the following tables are averages over the different simulation runs. Details of individual experiments are available in [36]. As a benchmark for the comparisons, we generate the optimal solutions, based on standard LP formulations, using the well-known optimization tool CPLEX [38]. The LP formulation directly minimizes congestion and the constraints are generated using the node-arc representation. The remaining three columns represent the results from our approximation algorithm, using three different implementations of the shortest path algorithm - Djikstra's shortest path algorithm (D1), efficient Djikstra (D2) and Floyd-Warshal algorithm (WA) respectively. In our experiments, we have set $\delta = 0.001$, and $\varepsilon = 0.1$, for the approximation algorithm.

Table 9.2 shows the average time required to generate a solution, using CPLEX and the approximation algorithms, for different network sizes. From Table 9.2, we can see that traditional LP techniques are only feasible for small to moderate sized networks. The solution times, obtained using CPLEX, increase rapidly with the number of nodes and it fails to find a solution for networks of over 25 nodes. This is indicated by placing a "–" in the corresponding entries in Table 9.2. The running

Table 9.2 Average of running time for different networks

	Average of Running Time (in sec.) Using			
# of Nodes	Standard LP	D1	D2	WA.
5	0.075	0.177	0.193	0.185
10	3.435	2.614	2.264	3.736
14	18.318	5.373	4.334	9.799
20	432.14	35.990	22.255	73.512
25	3586.5	186.390	112.000	375.200
30	–	351.090	184.857	730.841
50	–	7795.567	5314.582	–

Table 9.3 Average of Minimum congestion for different networks

	Average of Minimum Congestion Using			
# of Nodes	Standard LP	D1	D2	WA.
5	43.88	44.46	45.10	44.56
10	77.11	82.67	79.52	82.53
14	63.24	66.87	67.35	67.00
20	58.80	63.51	63.47	63.48
25	56.37	60.16	60.02	60.21
30		55.87	56.60	55.88
50		48.56	50.27	

time of the approximation algorithm is *significantly lower*, compared to standard LP techniques, when the network size or the number of commodities is large. For example, for a 25-node network, Efficient Dijkstra (D2) requires only 3% of the time required by CPLEX.

Of the three versions of approximation algorithms that we presented, the one based on efficient Djikstra algorithm (D2) performed the best, followed by Djikstra algorithm (D1) and Floyd-Warshall (WA) algorithm. Figure 9.2 shows the average reduction in solution times compared to standard LP techniques, for each of the three approaches. For very small networks (5 nodes), CPLEX actually performs better and requires less time. However, as the network size increases, the performance of the approximation algorithm based approaches improves rapidly, with respect to CPLEX. For large networks (over 25 nodes), standard LP techniques are unable

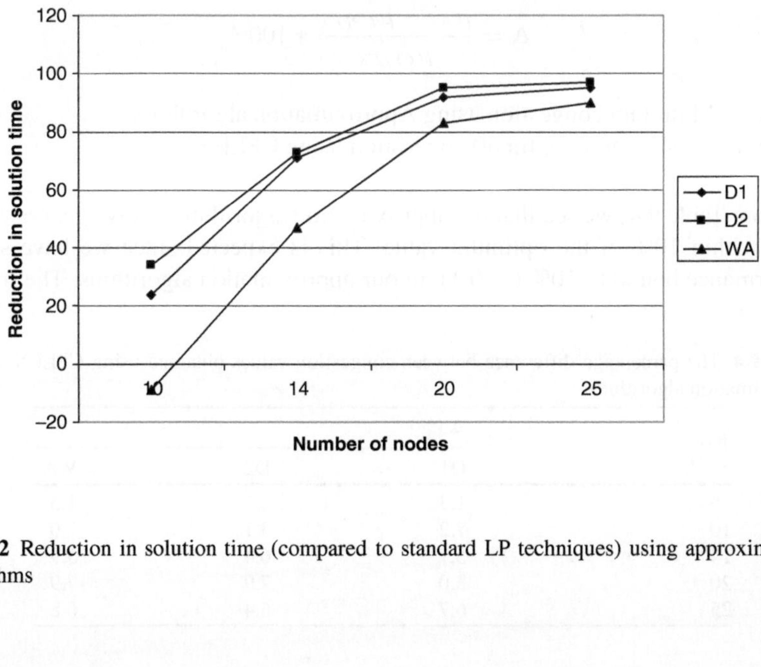

Fig. 9.2 Reduction in solution time (compared to standard LP techniques) using approximation algorithms

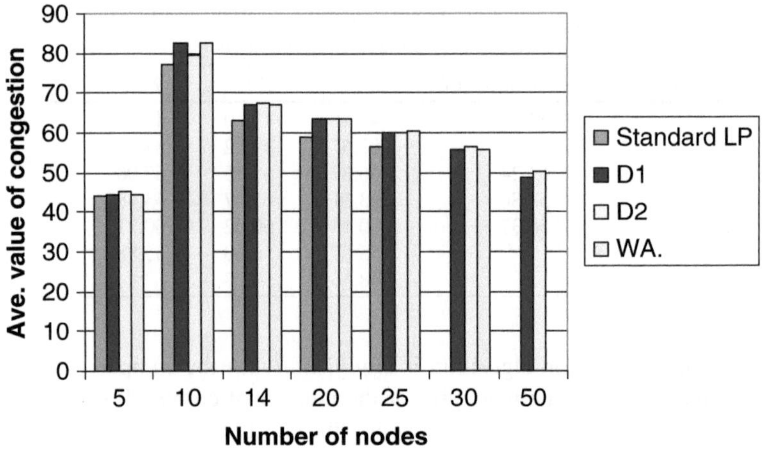

Fig. 9.3 Comparison of average congestion values

to generate optimal solutions, but our approach can still be used to generate good, near-optimal solutions.

Table 9.3 shows the congestion values obtained using CPLEX and the approximation algorithms. In all cases, the approximation algorithms produce solutions which are quite close (within 8%) to the optimal solution generated using the LP formulation. A comparison of the average congestion values is shown in Fig. 9.3. In Table 9.4, we have defined Δ, which measures the performance of the approximation algorithms in terms of the quality of the solutions.

$$\Delta = \frac{\mu_{Ap} - \mu_{CPLEX}}{\mu_{CPLEX}} * 100\%$$

μ_{Ap}: Minimum congestion using Approximation algorithms.
μ_{CPLEX}: Minimum (optimal) congestion using CPLEX.

From Table 9.4, we see that the approximation algorithms always generate solutions within 10% of the optimum value. This is expected since we have set the performance bound to 10% (= 0.1), in our approximation algorithms. The relative

Table 9.4 The percentage difference between congestion values obtained using CPLEX and the approximation algorithms

# of Nodes	Δ (%)		
	D1	D2	WA.
5	1.3	2.7	1.5
10	7.2	3.1	7.0
14	5.7	6.4	5.9
20	8.0	7.9	7.9
25	6.7	6.4	6.8

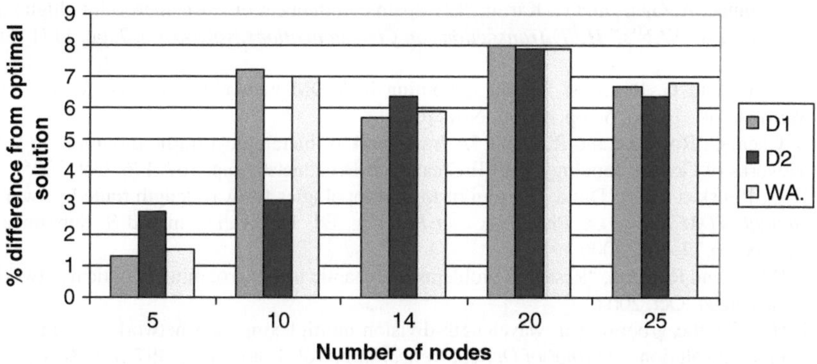

Fig. 9.4 Performance of approximation algorithms with respect to the optimal solutions

performance of the different algorithms with respect to the optimal solution is shown in Fig. 9.4.

It is well known that the "tail behavior" of column generation schemes for solving LPs is poor [39] meaning that, as a LP converges to a solution, successive iterations give relatively smaller improvements. Our approximation algorithm stops with a "reasonable" answer when we know that the actual value of the congestion is within a specified bound, (e.g. 5% or 10%) of the optimal value. As a result, our approach can generate very quick and efficient solutions, which are guaranteed to be within a specified bound of the optimal solution.

9.5 Conclusions

In this paper, we have presented a practical and efficient method for traffic grooming in WDM optical networks, under the bifurcated traffic model. We formulated the congestion minimization problem for WDM networks as a MCNF problem and then used an approximation algorithm to solve this problem. This allowed us to efficiently route traffic for practical sized networks and obtain near-optimal solutions.

We have shown that our approximation algorithm based approach is able to generate near-optimal solutions, within a pre-determined bound of the optimal. Our approach significantly reduces the solution time (by over 90%) for larger networks, and can also generate good solutions in many cases, where standard LP techniques become computationally intractable.

References

1. B. Mukherjee, "Optical Communication Networks", McGraw-Hill, 1997.
2. R. Ramaswami and K. Sivarajan, "Optical Networks: A Practical Perspective", Morgan-Kaufmann, 2002.
3. T. Stern and K. Bala, "Multiwavelength Optical Networks: A Layered Approach," Addison-Wesley, 1999.

4. I. Chlamtac, A. Ganz, and G. Karmi, "Lightpath communications: An approach to high bandwidth optical WAN's," *IEEE Transactions on Communications*, vol. 40, no. 7, pp. 1171–1182, July 1992.
5. R. Dutta and G. Rouskas, "Traffic grooming in WDM networks: past and future," *IEEE Network*, vol. 16, no. 6., pp. 46–56, Nov. 2002.
6. B. Chen, G. Rouskas, and R. Dutta, "A framework of hierarchical traffic grooming in WDM networks of General topology", IEEE/Create-Net Broadnets05, pp. 167–176, Oct. 2005.
7. G. N. Rouskas and R. Dutta, "Design of logical topologies for wavelength routed networks", *Optical WDM Networks: Principles and Practice*, Ed: K. Sivalingam and S. Subramanian, Kluwer, pp.79–102, 2000.
8. S. Huang and R. Dutta, "Research problems in dynamic traffic grooming in optical networks", Broadnets04, Oct. 2004.
9. J. Hu, "Traffic grooming in wavelength-division-multiplexing ring networks: A linear programming solution", *Journal of Optical Networking*, Vol. 1, no. 11, pp.397–408, Nov. 2002.
10. J. Q. Hu and B. Leida, "Traffic grooming, routing, and wavelength assignment in optical WDM mesh networks", IEEE INFOCOM, pp. 495–501, Mar. 2004.
11. J. Fang and A.K. Somani, "Enabling subwavelength level traffic grooming in survivable WDM optical network design", IEEE Globecom, pp. 2761–2766, Dec. 2003.
12. J-Q. Hu and E. Modiano, "Exact ILP solution for the Grooming problem in WDM ring networks," *Optical WDM Networks: Principles and Practice*, vol. II, Kluwer, 2004.
13. K. Zhu and B. Mukherjee, "Traffic grooming in an optical WDM mesh network," *IEEE JSAC*, vol. 20, no. 1, pp.122–133, Jan. 2002.
14. R. Dutta, S. Huang, and G. Rouskas, "On optimal traffic grooming in elemental network topologies", Opticomm, pp. 13–24, Oct. 2003.
15. R. Ramaswami and K.N. Sjvarajan, "Design of logical topologies for wavelength-routed optical networks", *IEEE Journal on on Selected Areas in Communications*, vol. 14, no. 5, pp. 840–851, June 1996.
16. J. Tomlin, "Minimum-cost multicommodity network flows", *Operations Research*, vol. 14, no. 1, pp. 45–51, Jan. 1966.
17. R. Dutta, and G. N. Rouskas, "A survey of virtual topology design algorithms for wavelength routed optical networks", *Optical Networks Magazine*, vol. 1, no. 1, pp. 73–89, Jan. 2000.
18. R. Krishnaswamy and K. Sivarajan, "Design of logical topologies: a linear formulation for wavelength routed optical networks with no wavelength changers", *IEEE/ACM Trans. on Networking*, vol. 9, no. 2, pp. 186–198, Apr. 2001.
19. E. Leonadi, M. Mellia, and M. A. Marsan, "Algorithms for the logical topology design in WDM all-optical networks", *Optical Networks Magazine*, pp. 35–46, Jan. 2000.
20. K. Lee and M. A. Shayman, "Optical network design with optical constraints in IP over WDM Networks," ICCCN 2004.
21. K. Lee and M. A. Shayman, "Rollout algorithms for logical topology design and traffic grooming in multihop WDM networks," IEEE Globecom 05, pp. 2113–2117, 2005.
22. H. Zang, J. P. Jue, and B. Mukherjee, "A review of routing and wavelength assignment approaches for wavelength- routed optical WDM networks", *Optical Networks Magazine*, pp. 47–60, Jan. 2000.
23. B. Jaumard, C. Meyer, and B. Thiongane, X. Yu, "ILP Formulations and Optimal Solutions for the RWA Problem", IEEE Globecom, pp. 1918–1924, 2004.
24. H. Qin, Z. Liu, S. Zhang, and A. Wen, "Outing and wavelength ssignment based on genetic algorithm," *IEEE Communications Letters*, vol. 6, no. 10, pp 455–457, Oct. 2002.
25. A. Katangur, Y. Pan, and M. Fraser, "Simulated annealing routing and wavelength lower bounds estimation on wavelength-division multiplexing optical multistage networks", *Optical Engineering*, vol. 43, no. 5, pp. 1080–1091, May 2004.
26. C. Dzongang, P. Galinier, and S. Pierre, "A tabu search heuristic for the routing and wavelength assignment problem in optical networks", *IEEE Communications Letters*, vol. 9, no. 5, pp. 426–428, May 2005.

27. Revindra K. Ahuja, Thomas L. Magnanti, and James B. OrLin, "Network flows", Prentice-Hall, Inc., 1993.
28. A. Ali, D. Barnett, and K. Farhangian, "Multicommodity network problems: Applications and computations," *IIE Transactions*, vol. 16, no. 2, pp. 127–134, 1984.
29. B. Awerbuch and T. Leighton, "Improved approximation algorithms for the multi-commodity flow problem and local competitive routing in dynamic networks", Proceedings of the Twenty-sixth Annual ACM Symposium on Theory of Computing, May 1994.
30. F. Hillier and G. Lieberman, "Introduction to operations research", McGraw-Hill, 2001.
31. N. Garg and J. Konemann, "Faster and simpler algorithms for multicommodity flow and other fractional packing problems," In 39th Annual IEEE Symnposium on Foundations of Computer Science, pp. 300–309, 1998.
32. N. Young, "Randomized rounding without solving the linear program. In ACM/SIAM[1], pp. 170–178.
33. L. K. Fleischer, "Approximating fractional multicommodity flow independent of the number of commodities", Proceedings of the 40th Annual Symposium on Foundations of Computer Science, 1999.
34. R. Floyd, "Algorithm 97, shortest path," *Communications ACM*, vol. 5, p. 345, 1962.
35. R. K. Ahuja, O. K. Mehlhorn, and R. J. Tarjan, "Faster algorithms for the shortest path problem," *Journal of ACM*, vol. 37, no. 2, pp. 213–223, 1990.
36. Y. Lu, "Approximation algorithms for optimal routing in wavelength routed WDM networks," Computer Science, University of Windsor, Masters Thesis, 2004.
37. M. Hou, "Heuristics for WDM path protection", Computer Science, University of Windsor, Master thesis, 2003.
38. http://www.ilog.com.
39. J. M. Valero de Carvalho, "Using extra dual cuts to accelerate convergence in column generation", *INFORMS Journal of Computing*, vol. 17, no. 2, pp. 175–182, Spring 2005.

Chapter 10
Current Progress in Optical Traffic Grooming: Towards Distributed Aggregation in All-Optical WDM Networks

Nizar Bouabdallah

Abstract Current trends in optical networking include switching packets directly in the optical domain, as this can take advantage of both packet flexibility and optical transparency. This enables also to improve the optical resources utilization by grooming efficiently the low-speed connections onto high capacity lightpaths. However, optical packet switching is hampered by major technological bottlenecks. In this chapter, we propose a novel solution for high-speed optical networks which reconciles packet switching and optical transparency requirements while avoiding actual technology bottlenecks. Specifically, we suggest a new concept of traffic aggregation in mesh networks that aims to eliminate both the bandwidth underutilization and scalability problems existing in all-optical wavelength routed networks. Our objective is to improve the network throughput while preserving the benefits of all-optical wavelength routed networks.

The proposed solution consists in distributing the aggregation process. So, instead of limiting the utilization of lightpaths capacity to the ingress node, each node along the path is allowed to fill on the fly the optical resource according to its availability. Therefore, the lightpath will be shared by several connections traveling from multiple ingress nodes to a single egress node. This technique combines the benefits of both the optical bypass and the statistical multiplexing gain. The feasibility of our scheme and the gain that it introduces over existing solutions are analyzed in this chapter. This is achieved through a linear integer programming formulation of the problem and by means of heuristic algorithms. The results show that our distributed aggregation technique can improve significantly the network throughput and reduce the network cost.

Keywords All-optical networks · Traffic grooming

N. Bouabdallah (✉)
INRIA, Campus de Beaulieu, F-35042 Rennes, France

M. Ma. (ed.), *Current Research Progress of Optical Networks*,
DOI 10.1007/978-1-4020-9889-5_10, © Springer Science+Business Media B.V. 2009

10.1 Introduction

The last decade has witnessed a continuous growth in data traffic. This growth, driven primarily by the proliferation of Internet access, has created a rising demand for more robust networks, with increasingly high-link capacity and node throughput. In this perspective, operators have been deploying optical networks taking advantage of the tremendous transmission capacity offered by the optical technology. In such networks, a significant portion of the network cost is due to the equipment used to convert signals from the electrical to the optical domain. In view of this, the optical layer is migrating from an opaque network, consisting of WDM links with electrical processing at the ends of the links, to an all-optical network, where traffic is switched at intermediate nodes in the optical domain. The optical layer here provides circuit-switched lightpaths to the higher layer equipment such as SONET and IP boxes.

Realizing connections in an all-optical (transparent) wavelength routed network involves the establishment of point-to-point (P-to-P) lightpaths between every edge nodes pair. These lightpaths may span multiple fiber links. All-optical cross connects (OXCs) are used at intermediate nodes to switch an incoming optical signal on a wavelength channel from an input port to an output port. This way a connection (lightpath) is routed from its source to its destination in the optical domain, optically bypassing all intermediate nodes. In view of this, the all-optical wavelength routing approach, also called optical circuit-switched approach, presents two obvious advantages. The first advantage stems from the fact that the optical bypass eliminates the need for Optical-Electrical-Optical (OEO) conversion at intermediate nodes. As a result, the node's cost decreases significantly, since in this case the number of required expensive high-speed electronics, laser transmitters and receivers is reduced. The second advantage is due to all-optical routing which is transparent with regard to the bit rate and the format of the optical signal.

In spite of the aforementioned advantages, all-optical wavelength routing still presents two drawbacks. The first one is related to the great number of wavelengths required within a large network when routing is performed at the wavelength granularity. For full connectivity, an N node all-optical network suffers from the N-squared problem, since each node requires N-1 P-to-P lightpaths, which raises a scalability issue with respect to the number of required wavelengths. The second, drawback of wavelength routing is the rigidity of the routing granularity. Such a large granularity can indeed lead to severe bandwidth waste especially when only a portion of wavelength capacity is used. Efficient use of network resources is always a concern from the operator's point of view. In wavelength routed networks, this efficiency is possible only when there is enough traffic between pair nodes to fill the entire capacity of wavelengths.

In contrast, an opaque network has the advantage of being able to use efficiently the link bandwidth since lightwave channels are detected at each node, then electronically processed, switched and reassigned to a new outgoing wavelength when needed. Nonetheless, this results in a maximum transceiver cost since nodes do not have optical bypass. The major advantage of electronic packet switching is its bandwidth efficiency achieved through statistical multiplexing. Therefore, many researchers are now focusing on bringing the packet switching concept into the optical domain. The ultimate aim is to benefit from both optical transparency and sub-wavelength multiplexing gain. However, optical packet switching (OPS) is not available yet and has been hampered by technological limitations mostly related to the fine switching granularity (optical packet) at high bit rate [1]. Currently, OPS is regarded as a solution for the long term future.

To alleviate the aforementioned shortcomings, we propose a new technique, which combines the advantages of optical bypass in transparent wavelength routed networks and statistical multiplexing. In this technique, a lightpath, which remains entirely in the optical domain, is shared by the source node and all the intermediate nodes up to the destination. So, in essence, a single lightpath is used to establish a multipoint-to-point (MP-to-P) connection. We refer to this technique as the distributed aggregation (DA) scheme [2, 3]. In this chapter, we provide a typical design of all-optical networks that function according to the DA scheme. Moreover, we assess the gain introduced by our proposal compared to existing solutions in terms of network throughput (or blocking probability) and cost savings. To achieve this, the network throughputs and costs entailed by the various approaches are evaluated.

The rest of the chapter is organized as follows. Section 10.2 discusses prior research related to this work. A detailed description of our proposed DA scheme is given in Section 10.3. We first investigate the node architecture needed to support such traffic-aggregation feature within WDM optical networks. Moreover, we emphasize the MAC (Medium Access Control) context including a description of the associated fairness control mechanism. Then, we demonstrate, through simulations, how the proposed control mechanisms achieve efficient traffic grooming on the shared lightpaths. In Section 10.4, we assess the benefits introduced by our proposal with respect to existing solutions in term of blocking probability. To achieve this, we formulate the problem as an Integer Linear Programming (ILP). Then, based on a sample small network, the network blocking probabilities of all representative approaches are compared. In Section 10.5, the comparison study is extended to large and arbitrary mesh networks by using heuristic algorithms. In addition, a cost comparison between our proposal and existing solutions is conducted. Finally, we conclude this chapter in Section 10.6.

10.2 Related Work

As mentioned before, both opaque and P-to-P all-optical networks are not consistent with the packet switching philosophy of the Internet. In next-generation networks, packet-based data traffic of bursty nature will become prevalent. Hence, the lack of packet switching in current all-optical wavelength routed networks (i.e., circuit switched networks) may lead to underutilization of critical resources. Consequently, two major enabling factors are identified as crucial for the evolution process of next-generation network architectures: packet switching and optical transparency. The trend is therefore towards switching packets directly in the optical domain, as this can take advantage of both packet flexibility and optical transparency.

A significant amount of is currently focusing on the implementation of packet switching in the optical domain. However, OPS is hampered by major technological bottlenecks, such as the lack of optical processing logic, optical memories, and cost-effective fast switching and synchronization technologies. Two promising solutions have been identified that by-pass some of these technological problems, namely, Photonic Slot Routing (PSR) [4] and Optical Burst Switching (OBS) [5]. In view of such advances, OPS is a solution that may become feasible in the future. Meanwhile, the trend is to improve the efficiency of existing and mature all-optical networks. In this area, much of the recent emphasis has been on circuit switched all-optical networks, where the goal is shifted more towards the improvement of optical resource utilization by means of new traffic aggregation schemes, rather than attempting to realize optical packet switching.

In light of the above, many interesting solutions have been proposed in the literature, see [6–10]. In what follows, we review these new solutions emphasizing how they reconcile optical transparency and sub-wavelength grooming.

10.2.1 The Multi-Hop Approach

The key idea behind multi-hop (MH) networks is to allow electronic processing at some intermediate nodes of the all-optical circuit switched network in order to increase its grooming capacity [6]. Accordingly, a packet may undergo electronic processing at some intermediate nodes before reaching its final destination. Hence, lightpaths can be seen as chains of physical channels through which packets are moved from a router to another toward their destinations. At intermediate nodes, the transit lightpaths are switched transparently through an OXC that does not process transit data. Instead, incoming lightpaths destined to the current node are terminated and converted to the electronic domain, so that packets can be extracted, processed, and possibly retransmitted on outgoing lightpaths, if the current node is not the final destination of the data.

The cost introduced by this electronic processing operation at the intermediate nodes is significant. However, it enables a better use of the network resources and reduces the total network cost compared to P-to-P all-optical circuit-switched networks [6]. The main challenge with MH networks is to identify the optimal logical topology that minimizes the total network cost, while accommodating all the traffic requests. This link topology design, also referred to as the routing and wavelength assignment (RWA) problem, has been extensively studied in the literature [11–13]. It has been demonstrated that the identification of the optimal logical topology is computationally intractable for large size networks [11]. Therefore, several heuristic approaches were proposed in the literature [6].

10.2.2 The Super-Lightpath Approach

Another promising solution to achieve both optical transparency and sub-wavelength grooming is the super-lightpath concept [7]. This approach increases the grooming capacity of a regular P-to-P all-optical circuit-switched network, as it transforms the lightpath concept from a P-to-P pipe to a point-to-multipoint (P-to-MP) pipe. In other words, the source node of a super-lightpath does not limit its transmission to the end node of that lightpath; instead, it can transmit its traffic to all the intermediate nodes along the route. This allows the super-lightpath to carry multiple connections, resulting in better wavelength utilization.

The super-lightpath technique uses a simple Optical Time Division Multiplexing (OTDM) method, which permits splitting the bandwidth of a wavelength among several traffic flows. Accordingly, each bit in a given position of the fixed-size TDM frame, called bit slot, identifies a particular subchannel. Using a bit interleaver, the transmitter multiplexes sub-channels into the frame, and transmits the resulting stream into one lightpath. At the reception, each intermediate node splits the transit signal, synchronizes its receiver to a particular bit slot, and only receives data in that particular sub-channel.

The super-lightpath technique presents many advantages. First, it reduces the number of transmitters per node since the same transmitter will be used to send data to more than one receiver. Moreover, it improves the lightpath utilization. The main concern with this P-to-MP transmission method is related to the limited length of the super-lightpath. Specifically, a significant portion of the passing-through optical signal is tapped at each receiving intermediate node, and therefore, due to power limitations, the number of traversed nodes is limited.

10.2.3 The TWIN (Time-Domain Wavelength Interleaved Networking) Approach

Unlike the super-lightpath concept, which uses a P-to-MP approach to improve the traffic grooming capacity in a traditional P-to-P all-optical network, the TWIN technique adopts a MP-to-P approach [8]. Specifically, TWIN makes use of optical MP-to-P trees that are overlaid on top of the physical topology. In TWIN, a particular wavelength is assigned to each egress node to receive its data. Doing so, sources that have data to transmit to a particular destination, tune their transmitters to the particular wavelength assigned to that destination. As such, the optical signals from various sources to a particular destination can be merged at intermediate nodes. Thus, the TWIN approach requires special OXCs, which are able to merge incoming signals of the same wavelength to the same outgoing wavelength.

Despite the complex scheduling algorithms entailed by such an approach, the MP-to-P concept is in itself interesting. It avoids the limitations on the length of a super-lightpath introduced in the P-to-MP approach, since no splitting operations are performed.

Nevertheless, the MP-to-P concept as described in TWIN suffers from scalability issues. The assignment of multiple wavelengths to each egress node (according to the volume of its destined traffic) puts a serious stress on the number of wavelength channels required on each fiber link. Moreover, TWIN may lead to fiber link underutilization due to the lack of wavelength reuse, since a particular wavelength, wherever the link that it belongs to, can only be used to transmit to a specific egress node.

10.2.4 The Optical Light-Trails Approach

The light-trail (LT) is another optical circuit switching-based approach that aims at improving the grooming capacity of regular P-to-P all-optical networks. It minimizes active switching, maximizes wavelengths utilization, and offers protocol and bit rate transparency [9, 10]. So far, we have presented a P-to-P approach (MH), a P-to-MP approach (super-lightpath) and a MP-to-P approach (TWIN), all of which aim at achieving these goals. The LT solution is a multipoint-to-multipoint (MP-to-MP) approach where intermediate nodes can both receive and transmit data on the pass-through channel.

The basic operation in the LT approach is as follows. Each intermediate node i of the LT taps a sufficient amount of optical power from the incoming signal, using a splitter, in order to recover its corresponding packets sent by the upstream nodes. On the other side, with regard to transmission, the original transit signal is coupled with the local signal, by means of a coupler, before it continues its path to serve the remaining downstream nodes of the LT.

The main difficulty with this approach is the design of a MAC protocol that avoids collisions between transit and locally inserted packets. A simple MAC protocol based on in-band signalling was suggested in the original LT proposal [9]. Accordingly, each intermediate node i, wishing to transmit a packet, first sends a beacon signal to order downstream nodes to stop their activities on the shared medium. Then, after a guard band, it transmits its data packet. Note that, node i may receive a beacon signal from upstream nodes during its transmission of a beacon signal or a data packet. In this case, it preempts instantaneously its transmission and the truncated packet is lost.

The above concerns may have a negative impact on the performance of the LT approach. Indeed, the MAC scheme may result in low resource utilization due to the guard band, extra signaling packets and wasted truncated packets. Therefore, other works are now focusing on the development of more efficient MAC schemes adapted to the LT technology [14]. Also, additional mechanisms are required to avoid fairness issues among the sharing LT nodes [15]. Furthermore, since a significant portion of the signal is tapped at each intermediate node, the LT length may be limited. This limitation, however, can be overcome using a power compensator, such as a semiconductor optical amplifier (SOA). Finally, we note that packets received by an intermediate node are not removed from the LT, which prevents bandwidth reutilization by downstream nodes. This feature can be useful only when dealing with multicast applications.

10.3 The Distributed Aggregation Approach

As discussed in the previous section, methods based on multiple node reception, such as super-lightpath and LT, suffer from power limitations due to the required multiple splittings. Moreover, the multiple nodes reception feature in LT, is effective only when dealing with multicast applications due to the lack of bandwidth reutilization of the shared lightpath. In view of this, the MP-to-P strategy appears as the best choice to improve the grooming capacity of a lightpath. In this context, TWIN is a good candidate technique. However, TWIN suffers from inherent lack of scalability and wavelength reuse. In order to alleviate these shortcomings, we propose a new MP-to-P optical circuit switching-based solution, called the distributed aggregation (DA) scheme [2, 3].

The key idea underlying our proposed scheme is to allow sharing of a lightpath among several access nodes. Instead of limiting the access to the lightpath capacity at the ingress point, each node along the path can fill the lightpath on the fly according to its available capacity. This way, a lightpath can be shared by multiple connections with a common destination (i.e., Mp-to-P lightpaths). Wavelength routing is performed in a similar way as in all-optical networks, i.e., signals remain in the optical domain from end to end and are optically switched by intermediate nodes. Since the lightpath remains transparent at intermediate nodes, a MAC protocol is required to avoid collisions between transient optical packets and local ones injected into the lightpath [16]. Moreover, additional control mechanisms must be introduced to alleviate fairness problems, which are pronounced in the case of shared medium networks [17]. In what follows, we provide a detailed description of the proposed control mechanisms, their performance as well as the node architecture needed to support the DA feature.

10.3.1 Node Architecture

A typical node in a WDM network is shown in Fig. 10.1. It consists of an OXC part and an access station part. While the OXC performs wavelength routing and wavelength multiplexing/demultiplexing, the access station performs local traffic adding/dropping functionalities. Each OXC is connected to the access station, typically an MPLS/IP router, which can be the source or the destination of a traffic flow. Each access station is equipped with a certain number of transmitters and receivers (transceivers). Traffic originated at the access station is transmitted as an optical signal on one wavelength channel by virtue of a transmitter. Considering the DA, the access station can be either the origin of a lightpath or an intermediate node using an already established lightpath. In the latter case, the injected traffic by an intermediate node must have the same destination as that of the traversing lightpath. In this context, a MAC unit is required to avoid collisions between transient packets and local ones. In turn, the traffic destined to the access station is directed by the OXC to the access station, where it is converted from an optical signal to electronic data by means of a receiver.

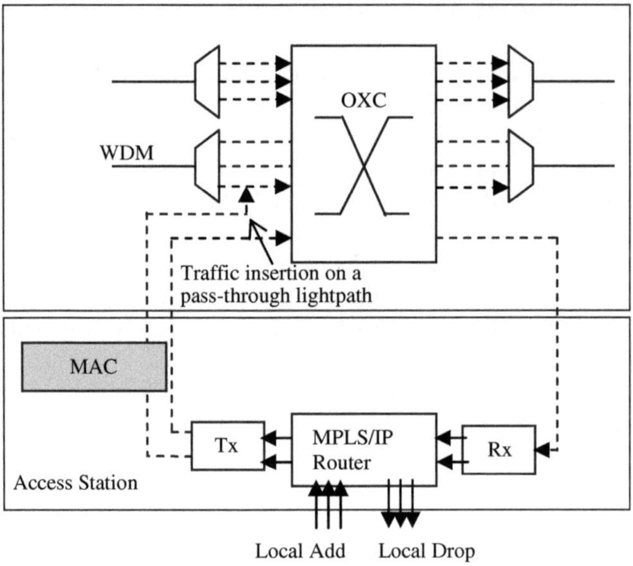

Fig. 10.1 Node architecture

Aggregating low-speed connections to high-capacity lightpaths is done by the MPLS/IP router according to the MAC unit decision. The advantages of this model are that: (1) it provides flexible bandwidth granularity for the traffic requests and (2) this MPLS/IP-over-WDM model has much less overhead than the SONET-over-WDM model, widely deployed in optical networks. Usually, the potential disadvantage of such a model is that the processing speed of the MPLS/IP router may not be fast enough compared to the vast amount of bandwidth provided by the optical fiber link. However, our scheme alleviates this issue since each MPLS/IP router processes only its local traffic. In other words, the transit traffic traveling through a WDM node remains at the optical layer, and it is not processed by the intermediate access nodes.

The merit of DA is that multiple connections with fractional demands can be multiplexed onto the same lightpath. As a result, the wasted bandwidth problem associated with pure wavelength routed networks is alleviated. In addition, due to the sharing of lightpaths, the number of admissible connections in the network is increased. Furthermore, the destination node terminates fewer lightpaths as connections from different nodes to the same destination are aggregated onto the same lightpath. In view of this, fewer physical components, such as wavelengths and transceivers, are used, resulting in savings on equipment. Moreover, in order to provide connections between all access node pairs using MP-to-P lightpaths, a total number of $O(N)$ lightpaths is required since only one lightpath per individual egress node could be sufficient. Thus, we alleviate the scalability issue encountered in traditional P-to-P all-optical wavelength routed networks (i.e., N-squared problem).

10.3.2 MAC Protocol

Let us consider N nodes placed in the unidirectional MP-to-P lightpath. Buffered packets at each access node are transmitted along the lightpath towards the node where the lightpath is terminated. Packets travel along the lightpath without any OEO conversion at intermediate nodes. Doing so, neither active optical devices nor electronic conversions are employed to handle the packet insertion on the shared MP-to-P lightpath. Instead, traffic control mechanisms are used at the electronic edge of the access nodes to avoid collisions with transit traffic.

In a fixed-slotted system with fixed-packet size, void (i.e., slot) filling can be carried out, by an intermediate node, immediately upon its detection, since the void duration is a multiple of the fixed-packet size duration. The detected void is therefore guaranteed to provide a minimum duration of one fixed-packet length. However in non slotted systems with variable packet length and arbitrary void duration, it is very likely for a collision to occur if a packet is immediately transmitted upon the detection of the beginning of a void. In our study, we adopt asynchronous transmission because we believe that it allows better use of resources compared with synchronous transmission. Asynchronous transmission fits better the traffic in high-speed networks, which is typically bursty.

To meet these requirements, we propose a new MAC protocol based on the void detection principle [16]. The MAC protocol detects a gap between two transient packets on the optical channel, and then it attempts to insert a local packet into the perceived gap. To do so, each access station must retain the transit traffic flow within the optical layer while monitoring the medium activity. Specifically, as shown in Fig. 10.2, each node first uses an optical splitter to separate the incoming signal into two parts: the main transit signal and its copy used for control purposes. With regard

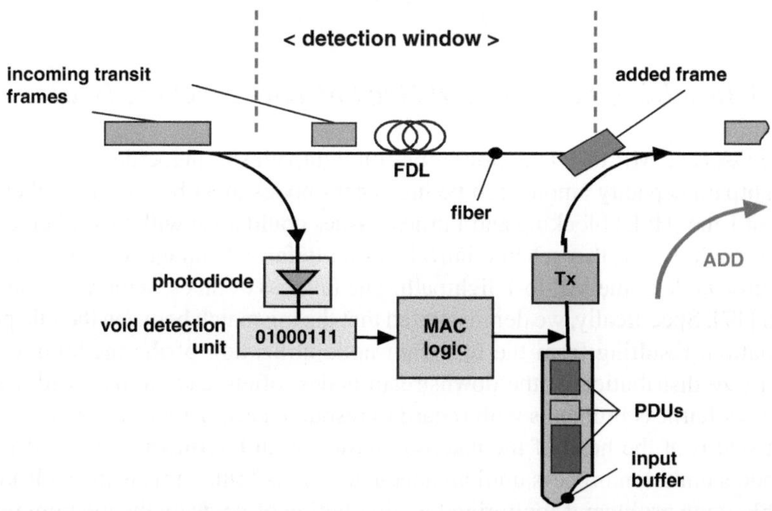

Fig. 10.2 Void detection-based MAC

to the control part, as in [18], low bit rate photodiodes (ph) –typically 155 MHz- are used to monitor the activity of the transit wavelengths. Once a free state of the medium is detected, the MAC unit measures the size of the progressing void. It is worth noting that signal splitting is done to monitor the medium activity (i.e., to know whether the medium is idle or busy) rather than to recognize the transit stream as with super-lightpath and LT schemes. This simply requires the tapping of a small part of the transit signal. Hence, the power penalty is relatively negligible. In [19], it is demonstrated that one can cascade up to 10 nodes without significant power penalty.

To be able to use a detected void, a Fiber Delay Line (FDL) is introduced on the transit path to delay the upstream flow by one maximum size frame duration augmented by the MAC processing time. The length of the FDL is therefore slightly larger than the Maximum Transmission Unit (MTU) size allowed within the network, in order to provide the MAC unit with sufficient time to listen and to measure the medium occupancy. The access station will begin injecting a packet to fill the void only if the null period is large enough (i.e., at least equal to the size of the packet to be inserted). Undelivered data will remain buffered in the electronic memory of the access station until a sufficient void space is detected. This way, collision free packet insertion on the transit lightpath from the add port is ensured.

We notice that FDL introduction at each intermediate node has a negligible impact on the end-to-end packet delay. Indeed, the extra delay introduced by each delay line does not exceed tens of μs. Considering a MP-to-P lightpath traversing several nodes, the total extra delay, introduced by all the FDLs along the route, is of the order of hundreds of μs, which is relatively negligible. Finally, it is worth noting that this access scheme relies only on passive components (couplers, FDL, ph) with relatively low cost. The cost introduced by the MAC unit is therefore negligible compared to the transceiver cost.

10.3.3 Resolving Fairness and Head of Line Blocking Issues

As the DA (i.e., MP-to-P insertion) relies on lightpath sharing, efficient partition of the lightpath capacity among competing access nodes must be ensured, otherwise, Head of Line (HoL) blocking and fairness issues could arise with this scheme.

It is obvious that this scheme introduces an unfair advantage to nodes closer to the source node of the MP-to-P lightpath. The fairness of this scheme was examined first in [17]. Specifically, we demonstrated that the mismatch between the idle period distribution, resulting from the upstream node utilization of the medium and the packet size distribution of the downstream nodes, often leads to bandwidth waste as well as fairness problems with regard to resource access. Once a packet of maximum size is at the head of the insertion buffer of an intermediate node, it blocks the node's emission process until an adequate void is found: this is the well-known HoL blocking problem. Monitoring the distribution of voids on the medium reveals a low probability of finding regular and sufficiently large gaps of free bandwidth.

Thus, sharing the bandwidth fairly but arbitrarily among nodes is not sufficient to ensure satisfactory results. The sharing process must thus be done smartly in order to preserve a maximum of useful bandwidth for the downstream nodes. In this context, we showed in [17] that the token bucket (TB) algorithm cannot resolve this issue. In the TB algorithm, the free bandwidth (stated in bit/s) allocated to each node is theoretically sufficient to handle the node's local traffic. However the main issue is that the free bandwidth is fragmented into unusable gaps. Hence, as a basic rule one should avoid a random division of the optical resource. To achieve this, we proposed the TCARD (Traffic Control Architecture using Remote Descriptors) mechanism [17].

In TCARD, each transmitting station is provided with anti-tokens that are used to prevent the station from transmitting a packet during a gap in the optical packet stream. These anti-tokens permit some of the gaps to go by unused, and therefore, they can be used by other downstream stations. The rate of generation of the anti-tokens at a station is set equal to the rate of the aggregate downstream transmission. Hence the key idea of TCARD is to force each node to preserve free bandwidth for its downstream neighbors in the form of gaps of size equal to the MTU size. This also avoids the HoL blocking problem, since downstream nodes can transmit large packets due to the reserved big-enough gaps.

To illustrate the TCARD mechanism, we present a simple three-node MP-to-P lightpath example. The nodes share a common channel that runs at 1 Gbit/s. We assume that the sustainable bit rate negotiated by each node and stipulated in its own service level specification is 0.3 Gbit/s. We consider traffic of variable packet size where the MTU is equal 1500 bytes. Considering the TCARD scheme, the first node must reserve 0.6 Gbit/s in average of available bandwidth for the downstream nodes, i.e., nodes 2 and 3. As explained before, the reserved bandwidth is representative of idle periods of 1500 bytes in order to comply with packets of maximum size. Thus the anti-tokens at node 1 are generated periodically at a rate equal to $(0.6 \cdot 10^9)/(1500 \cdot 8)$ anti-tokens/s. Note that a reserved void can be exploited by a downstream node either to transmit a packet of maximum size or to emit a burst of smaller frames. Furthermore, similarly to the first node, the second node reserves 0.3 Gbit/s of available bandwidth for the third node. The reserved bandwidth is also representative of voids of 1500 bytes.

10.3.4 Illustrative Example

To illustrate the DA mechanism, we consider the simple four-node bus network example shown in Fig. 10.3. Each fiber is supposed to have two wavelength channels. Three connection requests are to be served: (0,3), (1,3) and (2,3) with bandwidth requirements equal to 90%, 70% and 20% of the wavelength capacity, respectively.

In a P-to-P all-optical network case, only connections (0,3) and (1,3) will be served because of the resource limitations (the wavelength channels between (2,3) are already busy). The connection requested between node pair (2,3) will be rejected

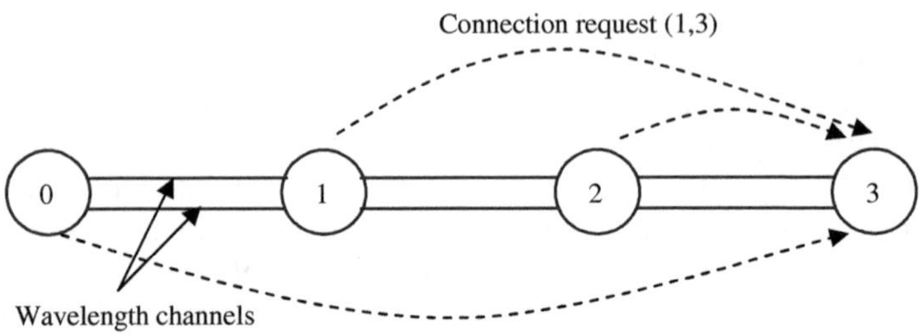

Fig. 10.3 A simple four-node demonstration network

even if the wavelengths between these two nodes are not fully used. To satisfy all the connection requests, a supplementary wavelength is required between pair nodes (2,3). In this case, a total number of 3 transmitters (Tx) and 3 receivers (Rx) is required in the network.

Thanks to its grooming capacity, an opaque network overcomes the above wavelength limitation. However, the network need in terms of transceivers increases significantly since 5 Tx and 5 Rx are required to satisfy all the connection requests within the network.

Likewise, the MH approach, which is a hybrid solution between the opaque and P-to-P all-optical circuit switched networks, allows all the connection requests satisfaction without requiring additional wavelength channels. In this case, three lightpaths are to be established: (0,3), (1,2) and (2,3). To achieve this, the network only needs 3 Tx and 3 Rx. It is easy to see that the MH approach overcomes the limitations of both opaque (i.e., high transceiver cost) and P-to-P all-optical (i.e., wavelength exhaustion) networks.

Finally, the DA scheme enables further equipment savings. It has the lowest transceiver cost since the network requires only 3 Tx and 2 Tx to carry all the connection requests. Specifically, two lightpths need to be established (0,3) and (1,3). This latter is shared by both (1,3) and (2,3) connections. Indeed, the second connection (2,3) will be carried by the spare capacity of the existing lightpath. Note that the lightpath $1 \rightarrow 2 \rightarrow 3$ is still routed in the optical domain at node 2, preserving the benefit of optical bypass. As such, we save 1 (5, respectively) terminal equipments compared to MH and P-to-P all-optical networks (opaque networks, respectively).

To further evaluate the gain introduced by the DA approach, the problem will be formulated using ILP in the next section. Then, a comparison with all other representative approaches will be presented based on their optimal solutions.

10.4 Impact on Network Blocking Probability: Resolving the Routing and Wavelength Assignment Problem Using ILP Formulation

In the previous section, we focused on the feasibility of the DA scheme by evaluating the performance of this multiple access method in terms of access delay and PLR. In this section, we rather evaluate the gain introduced by the DA scheme over classical approaches (P-to-P all-optical networks, opaque networks an MH networks) in terms of blocking probability. Specifically, we compute the average blocking probability of different sets of static traffic demands, under different strategies. The problem can be expressed in the form of an ILP problem within a mesh network as follows:

GIVEN

(1) A physical topology, consisting of nodes connected by physical links. In our model, each physical link represents two fibers that are used to communicate in opposite directions. So, the physical topology is completely defined by:

a- W : the number of wavelengths on each fiber link;
b- Q : the number of transmitters and receivers at each node.

(2) A $N \times N$ static traffic demand matrix, where N is the total number of network nodes.

FIND

The optimal virtual topology (i.e., set of lightpaths) maximizing the total network throughput (i.e., minimizing the total amount of blocked traffic)

Hence, according to our RWA optimization problem, lightpaths are established on the basis of maximizing the total network throughput. First, the problem will be treated in the light of our proposed solution considering that the DA is adopted within the network. Afterwards, it will be considered in the context of MH and P-to-P all-optical wavelength routed networks. In the latter case, we used the ILP formulation given in [6]. It is worth noting that in our model, the nodes do not have wavelength conversion capability. So, a lightpath must use the wavelength on each fiber along the route. Moreover, we do not allow the traffic from the same connection to be bifurcated over multiple lightpaths.

10.4.1 ILP Formulation of the RWA Problem in DA-Enabled Networks

In this section, we provide an ILP formulation of the RWA problem when the DA scheme is enabled. In this case, several connections from different sources to the same destination can be carried in the same lightpath. By extending the work in [6], we formulate the problem as an optimization problem.

With regard to the notations, we will use m and n to represent the source and destination nodes of a fiber link, i and j to denote the source and destination nodes of a lightpath, s and d to represent the source and destination nodes of a connection request. The rest of the notations used in our mathematical formulation are defined below:

Input parameters:

- N : total number of nodes in the network.
- W : number of wavelengths per fiber.
- P_{mn} : a binary variable that takes the value 1 if there is a physical optical fiber starting from node m and ending into node n.
- P_{mn}^{w} : number of wavelengths w on fiber Pmn ($P_{mn} = P_{mn}^{w}$).
- $Tx_i,\ Rx_i$: number of transmitters and receivers at node $i (i = 1, \ldots, N)$, respectively.
- Q_i : total number of transmitters and receivers at node i (i.e., $Q_i = Tx_i - Rx_i$).
- Λ : static traffic matrix of lightpath requests; the element of the matrix λ_{sd} denotes the capacity needed by the connection request from node s to node d, which can be a fraction of the lightpath capacity. In our study, we suppose that $\lambda_{sd} \in [0, 1]$, so at most one lightpath between every pair of nodes (s, d) is required to carry all the traffic requests.

Output variables:

— Variables of the virtual topology:

- V_{ij} : number of lightpaths from node i to node j in virtual topology.
- V_{ij}^{w} : number of lightpaths V_{ij} on wavelength w.
- V_{ij}^{s} : number of transit lightpaths between nodes i and j used by the intermediate node s for the transmission to the node j (with $s \neq i$).
- $V_{ij}^{s,w}$: number of transit lightpaths V_{ij}^{s} on wavelength w.

— Variables of the physical topology:

- $P_{mn}^{ij,w}$: a binary variable that takes the value 1 if one of the V_{ij}^{w} lightpaths is routed through the fiber link (m, n).

— Variables of the traffic forwarding:

- λ_{ij}^{sd} : the binary variable λ_{ij}^{sd} is 1 when the traffic flowing from node s to node d uses lightpath (i, j) as a virtual link, and 0 otherwise. Recall that the traffic from s to d is not bifurcated, i.e., all the traffic between s and d will flow through the same lightpath.
- $\lambda_{ij}^{sd,w}$: this binary variable takes the value 1 if the traffic flowing from node s to node d uses lightpath (i, j) on wavelength w as a virtual link.
- S_{sd} : the binary variable $S_{sd} = 1$ if the connection request from node s to node d has been successfully routed; otherwise, $S_{sd} = 0$.

The following formulation describes the DA-specific RWA problem:

- **Objective function**: Maximize the total successfully-routed traffic.

$$\text{Maximize} \sum_{s,d} \lambda_{sd} \cdot S_{sd} \tag{10.1}$$

- **Subject to**:
  — Virtual link (lightpath) constraints

$$\sum_j V_{ij} + \sum_{k,j,k \neq i} V_{kj}^i \leq Tx_i \qquad \forall i \tag{10.2}$$

Equation (10.2) limits the number of lightpaths originating from node i plus the number of transit lightpaths used by node i for transmission thanks to the DA feature, to the number of transmitters at that node.

$$\sum_i V_{ij} \leq Rx_j \qquad \forall j \tag{10.3}$$

Equation (10.3) limits the number of lightpaths terminated at node j to the number of receivers at that node.

$$Tx_i + Rx_i \leq Q_i \qquad \forall i \tag{10.4}$$

Equation (10.4) limits the number of transceivers at each node $i (i = 1, \ldots, N)$ to Q_i.

$$\sum_w V_{ij}^w = V_{ij} \qquad \forall i, j \tag{10.5}$$

Equation (10.5) shows that the lightpaths between (i, j) are composed of lightpaths on different wavelengths between nodes (i, j).

$$\sum_w V_{ij}^{s,w} = V_{ij}^s \qquad \forall i, j, s \text{ with } i \neq s \tag{10.6}$$

$$V_{ij}^{s,w} \leq V_{ij}^w \qquad \forall i, j, s, w \text{ with } i \neq s \tag{10.7}$$

$$\text{int} V_{ij}, V_{ij}^w, V_{ij}^s, V_{ij}^{s,w} \tag{10.8}$$

Equations (10.6), (10.7) ensure that an intermediate node s can only use an existing lightpath between node pair (i, j) for transmission to node j.

— Physical link constraints

$$\sum_m P_{mk}^{ij,w} = \sum_n P_{kn}^{ij,w} \qquad \forall i, j, k, w \text{ with } k \neq i, j \tag{10.9}$$

$$\sum_m P_{mi}^{ij,w} = 0 \qquad \forall i, j, w \tag{10.10}$$

$$\sum_n P_{jn}^{ij,w} = 0 \qquad \forall i, j, w \tag{10.11}$$

$$\sum_n P_{in}^{ij,w} = V_{ij}^w \qquad \forall i, j, w \tag{10.12}$$

$$\sum_m P_{mj}^{ij,w} = V_{ij}^w \qquad \forall i, j, w \tag{10.13}$$

Equations (10.9–10.13) are the multicommodity equations (flow conservation) that account for the routing of a lightpath from its origin to its termination. Note that (10.9–10.13) employ the wavelength-continuity constraint on the lightpath route. Accordingly, we ensure that for each lightpath exists a corresponding physical path that must depart from its source (10.12), reach its destination (10.13) and be continuous.

$$V_{ij}^{s,w} \leq \sum_m P_{ms}^{ij,w} \qquad \forall i, j, s, w \text{ with } i \neq s \tag{10.14}$$

Equation (10.14) ensures that the lightpath between node pair (i, j), on wavelength w used by node s for transmission to node j thanks to DA feature, passes through node s .

$$\sum_{i,j} P_{mn}^{ij,w} \leq P_{mn}^{w} \qquad \forall m, n, w \tag{10.15}$$

$$P_{mn}^{ij,w} \in \{0, 1\} \tag{10.16}$$

Equations (10.15) and (10.16) ensure that wavelength w on one fiber link (m, n) can only be present in at most one lightpath in the virtual topology.

— Traffic matrix constraints

Equations (10.17–10.24) are responsible for the routing of traffic requests on the virtual topology, and they take into account the fact that the aggregate traffic flowing through lightpaths cannot exceed the overall wavelength capacity.

$$\sum_i \lambda_{is}^{sd} = 0 \qquad \forall s, d \tag{10.17}$$

$$\sum_j \lambda_{dj}^{sd} = 0 \qquad \forall s, d \tag{10.18}$$

Equations (10.17) and (10.18) avoid the traffic coming in its source node or going out its destination node.

$$\lambda_{ij}^{sd} = 0 \qquad \text{if } j \neq d \ \forall s, d, i, j \tag{10.19}$$

This equation ensures that a connection can only traverse a single lightpath before reaching its final destination (all-optical networks constraint).

$$\sum_w \lambda_{ij}^{sd,w} = \lambda_{ij}^{sd} \qquad \forall s, d, i, j \tag{10.20}$$

$$\lambda_{ij}^{sd,w} \leq V_{ij}^{s,w} \qquad \text{if } i \neq s \ \forall s, i, j \tag{10.21}$$

Equation (10.21) states that a node s can use a path-through lightpath to transmit its traffic only if it has an available transmitter.

$$\sum_{s,d} \lambda_{sd} \cdot \lambda_{ij}^{sd,w} \leq V_{ij}^{w} \qquad \forall i, j, w \tag{10.22}$$

Equation (10.22) ensures that the aggregate traffic flowing through a lightpath cannot exceed its overall capacity.

$$\sum_{i} \lambda_{id}^{sd} = S_{sd} \qquad \forall s, d \tag{10.23}$$

$$S_{sd} \in \{0, 1\} \tag{10.24}$$

Equations (10.23) and (10.24) stipulate that a connection is successfully served to its destination if it is carried by one of the lightpaths that terminates at that destination.

10.4.2 Illustrative Results from the ILP Formulation

This section instantiates the traffic routing and grooming problem using the physical topology of the network depicted in Fig. 10.4. In this example, we assume that a connection needs to be established between each pair of nodes in the network. In terms of capacity, the traffic demand of each connection is represented by a random fractional number uniformly distributed in the interval [0,1].

Table 10.1 shows the results regarding the network throughput and the associated number of established lightpaths. These results are obtained using a commercial ILP solver, "CPLEX", taking into consideration different network resource parameters. The reported results are averaged over 100 traffic demand matrixes. In Table 10.1, Q denotes the number of transceivers at each node and W denotes the number of wavelengths per fiber. In the P-to-P all-optical and MH networks, we used the ILP formulation given in [6]. When the DA is enabled, we run our ILP presented above.

According to the results presented in Table 10.1, it is clear that when the number of tunable transceivers at each node is increased from 2 to 4, the network throughput increases significantly. This throughput increase is observed under all strategies. However, it is important to point out that this increase is more significant

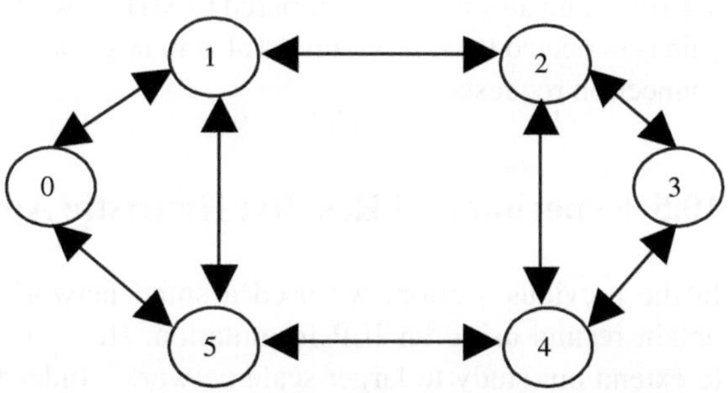

Fig. 10.4 A six-node network

Table 10.1 Network throughput and associated number of established lightpaths

	P-to-P all-optical networks		MH networks		DA-enabled networks	
	Throughput	Lightpaths	Throughput	Lightpaths	Throughput	Lightpaths
$Q = 2, W = 2$	40%	12	40%	12	40%	6
$Q = 3, W = 2$	60%	18	60%	18	60%	13
$Q = 4, W = 2$	66%	20	80%	24	80%	11
$Q = 5, W = 2$	66%	20	80%	24	94%	13
$Q = 3, W = 3$	60%	18	60%	18	60%	8
$Q = 4, W = 3$	74%	24	90%	24	90%	15
$Q = 5, W = 3$	74%	24	94%	28	100%	21

when MH or DA-enabled networks are considered. But when the number of tunable transceivers at each node increases from 4 to 5, the network throughput improves only in the DA case. In fact, with DA, the capacity left in the already established lightpaths is used to carry new connection requests. The resource utilization is thus improved. On the other hand, in the classical cases (i.e., P-to-P all-optical and MH), there are not enough wavelengths to setup more P-to-P lightpaths in order to carry the connection requests which were blocked.

When the number of transceivers approaches that of wavelengths, all approaches present the same behavior. As a result, the same throughput results are obtained. This is shown in Table 10.1 for the case where $Q = 2$, $W = 2$ and $Q = 3$, $W = 3$. These results are expected since in this case the number of transceivers is not enough to setup more lightpaths or to share established ones in order to carry more connection requests. Even though the throughput is the same, still the number of lightpaths that should be managed in the network decreases significantly when using the DA approach. The number of MP-to-P lightpaths constitutes a percentage of 63% from that of P-to-P lightpaths. Building on these results, it is clear that the DA approach enables the establishment of a given set of routes in a scalable fashion when compared to classical approaches.

On the other hand, when the number of transceivers becomes important compared to that of wavelengths, more lightpaths are shared in the DA case to carry the connection requests. Hence, the utilization percentage of the lightpaths increases and the number of rejected connections decreases. This is expressed in Table 10.1 by the better throughput values obtained in the DA approach compared to classical approaches. In this example, the gain is over 25% compared to P-to-P all-optical networks and around 10% compared to MH networks, which is very significant. The gain is expected to be pronounced also in large networks with much more nodes and connection requests.

10.5 Experimental Results: Heuristic Approach

In the previous section, we used a small network topology as an illustration to obtain results using an ILP formulation. Here, we will use a heuristic approach to extend our study to larger scale networks. Indeed, the DA-aware RWA problem is NP-complete since it represents a generalization of the well-known NP-complete standard RWA problem [11], in the sense that it includes the RWA standard problem as a particular case. More specifically, if we assume that each connection request requires the full capacity of a lightpath, our DA-aware RWA problem becomes the standard RWA optimization problem.

To extend our study to large networks, we developed a new discrete-event simulation tool. Accordingly, we compute the blocking probability of dynamically arriving connection requests, under different strategies (P-to-P all-optical networks, opaque networks, MH networks, and DA-enabled networks). With the elaborated tool, we also use realistic dynamic traffic demands instead of the static traffic patterns.

Later, as a second criterion of comparison, we will quantify through network dimensioning analysis the network costs entailed by the various approaches. These costs include that of the transceivers required at the access station level, as well as the number of OXC ports. A new heuristic algorithm is developed for that purpose.

10.5.1 A- Blocking Probability Comparison

In this section, we evaluate the blocking probability under different strategies. We simulate the following schemes: (1) P-to-P all-optical networks; (2) opaque networks; (3) MH networks; (4) DA-enabled networks (i.e., MP-to-P approach); and finally (5) a hybrid variant of networks combining the MH and MP-to-P approaches.

The following assumptions were made in our simulations: (1) the US backbone shown in Fig. 10.5 is used; (2) Each link in the network represents two fibers that are used to communicate in opposite directions. Each fiber carries 32 wavelengths; (3) Each node is equipped with 20 transceivers and 40 OXC interfaces; (4) The shortest path adaptive routing is used; (4) The first fit (FF) wavelength assignment approach is adopted; (5) Connection requests arrive at each ingress node following a Poisson distribution, and the holding time of each request is exponentially distributed. The total traffic load offered to the network by each node is $\rho = \lambda/\mu$, where λ and μ are the arrival and departure rates at each ingress node, respectively; (6) The destination node of each arriving connection is randomly chosen among the *N-1* remaining edge

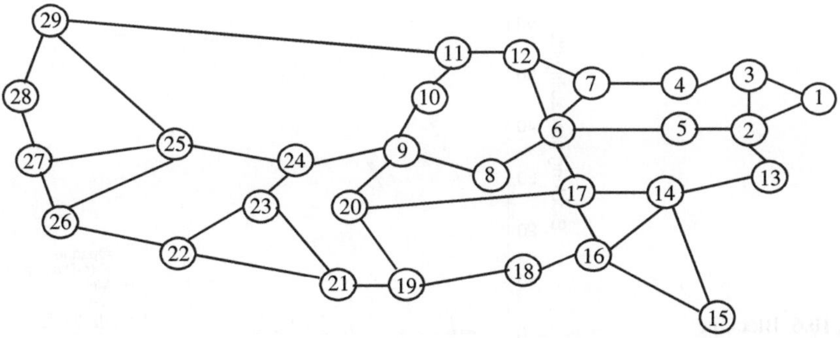

Fig. 10.5 The US optical backbone

nodes of the network; (6) The bandwidth requirement of each connection request λ_{sd} is randomly chosen in the interval [0,1], so at most one lightpath is needed to carry any traffic request. We note that, in our simulations, we do not allow the traffic from the same connection to be bifurcated over multiple lightpaths. Finally, each value of the blocking probability has been computed over multiple simulations to achieve very narrow 97.5% confidence intervals.

In the optical context, each arriving connection tries first to use the existing virtual topology (i.e., already established lightpaths). If the available bandwidth of existing lightpaths is not sufficient, the connection will try to establish new lightpaths subject to transceiver, OXC port and wavelength constraints. Specifically, when the DA (i.e., MP-to-P) case is considered, the ingress node s of an arriving connection request with destination d, looks first for a pass-through lightpath traveling towards the same egress node d which has sufficient available bandwidth. Otherwise, node s tries to establish a new lightpath subject to resource availability. If there are not enough resources to satisfy the connection request, it is simply blocked. In the same way, when MH approach is adopted, the source node s first tries to find an available route through existing lightpaths. In this case, the connection may span multiple lightpaths before reaching its destination. If such route is not available, the connection tries to establish the missing lightpaths (end-to-end lightpath, or missing segments along the route) to reach its destination.

In our simulations, lightpaths are routed using an adaptive routing approach since it is the most flexible approach. Doing so, the shortest path between the source and destination nodes is dynamically calculated, according to the current network state. Once the route has been chosen for a lightpath, we use the FF approach to assign wavelengths to each lightpath such that any two lightpaths passing through the same physical link are assigned different wavelengths due to the wavelength-continuity constraint.

Figure 10.6 plots different blocking probabilities as a function of the network load ρ. We observe that the opaque strategy always leads to the maximum blocking

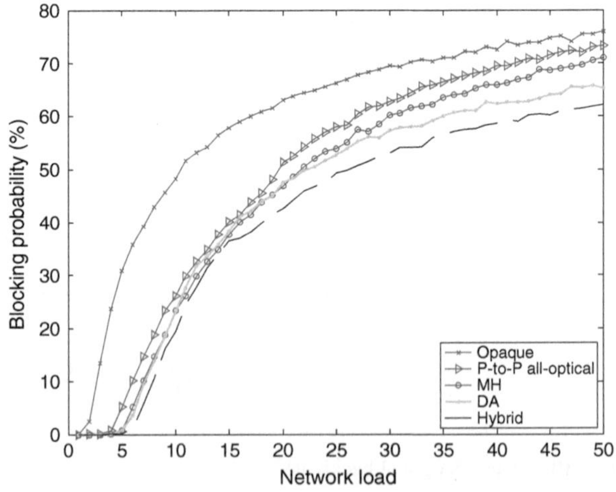

Fig. 10.6 Blocking probability evolution with the network load

probability. This is mainly due to the lack of available transceivers. Indeed, the total network capacity (in terms of transceiver equipments) is quickly exhausted, since the nodes do not permit transit connections to pass through optically. The P-to-P all-optical circuit-switched strategy alleviates slightly this problem due to the optical transparency. Even though, the blocking probability remains relatively high due to the great number of P-to-P lightpaths that are required in this case. These lightpaths require a large number of OXC interfaces and wavelengths. The MH and MP-to-P schemes reduce significantly the blocking probability since they alleviate the scalability issue of the P-to-P all-optical networks by increasing their grooming capacity.

In addition, the MP-to-P scheme outperforms the MH scheme, since it requires less transceivers and OXC interfaces. Indeed, the MP-to-P scheme improves the P-to-P all-optical circuit-switched network grooming capacity while conserving its entire transparency as opposed to the MH approach, where electronic grooming is needed at some network nodes. This active electronic processing operation enables an MH network to save in components over P-to-P all-optical and opaque networks, but requires additional equipments, such as OXC interfaces and transceivers, when compared to the passive MP-to-P insertion. In addition, the MP-to-P scheme outperforms the MH scheme, since it requires less transceivers and OXC interfaces. Finally, we notice that the hybrid strategy, combining the MH and DA schemes, always achieves the best results.

Figure 10.7 plots the blocking probability as a function of the bandwidth requirement λ_{sd} of each connection request. In this case, we consider a uniform traffic matrix, i.e., $\lambda_{sd} = \tau \ \forall \ s$ and d, where τ ranges from 0 to 1, and $\rho = 10$. This figure illustrates the general trade-off among the different strategies. According to the value of τ, we get different optimal solutions. At one extreme, when each node transmits close to the wavelength capacity to every other node, the P-to-P all-optical circuit-switched approach is the best solution as the network is already well utilized

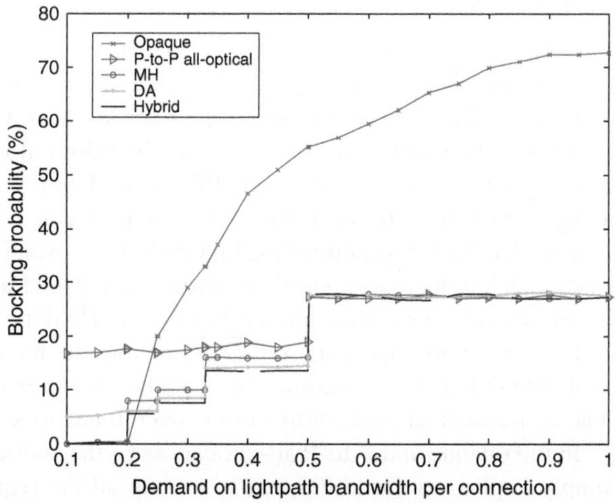

Fig. 10.7 Blocking probability evolution with the bandwidth requirement per connection

without grooming. At the other extreme, when the total demand from each node is a small fraction of the wavelength capacity, opaque strategy stands out as the best solution due to its grooming capability. In most cases, when the demand is moderate to normal, MH and MP-to-P schemes generally present the best solutions, with an advantage to the MP-to-P scheme. Finally, we underline that the hybrid solution enables achieving this trade-off, whatever the value of τ. It always leads to the minimal blocking probability. This solution represents therefore a sensible choice for next-generation networks. We note that MH, MP-to-P and P-to-P all-optical strategies achieve almost the same results when $\tau > 1/2$. This is due to the fact that we do not allow traffic belonging to the same connection request to be bifurcated over multiple lightpaths. In doing so, grooming multiple connections on the same lightpath is no more possible when $\tau > 1/2$.

10.5.2 Network Cost Comparison

In this section, the comparison between the different strategies is tackled from a different perspective. We aim at evaluating the cost savings accomplished by the DA scheme over existing solutions. To achieve this, we dimension the optical US backbone (Fig. 10.5), under different strategies.

10.5.2.1 Procedures and Algorithms for Network Dimensioning

Network planning has been conducted following the logical process shown below. The inputs of the analysis are:

(1) The network topology.
(2) The traffic matrix.
(3) The adopted routing scheme, which is the shortest path algorithm in our case.
(4) The adopted wavelength assignment approach, which is the first fit (FF) in our work.

Network dimensioning is achieved by evaluating the OXC and IP router dimensions by means of heuristic algorithms, which are used to map the different lightpaths needed to forward all the traffic requests within the network. A lightpath is established between nodes by setting up the OXCs along the route between nodes. Each lightpath needs a dedicated OXC port when traversing an intermediate node along its root. In addition, a transmitter is required at the ingress node and a receiver is needed at the egress node of the lightpath. Moreover, in the distributed aggregation case, each intermediate node along the path using the traversing lightpath to transmit its traffic needs also a transmitter. Let $\overline{Tx}$ and $\overline{Rx}$ denote the number of transmitter and receiver ports per node. Let $\overline{OXC}$ denote the number of OXC ports per node as shown in Fig. 10.8. We omit, here, the number of wavelengths since we consider that the transceiver cost dominates the overall network cost.

In the opaque and P-to-P all-optical cases, the shortest path routing algorithm is simply applied to the traffic matrix to map all the required lightpaths. Hence, we

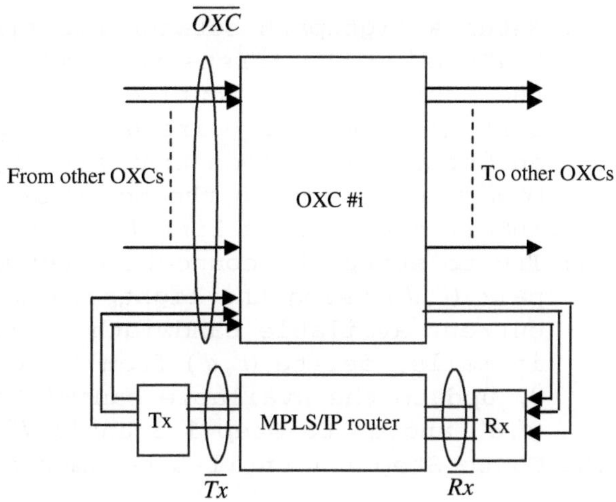

Fig. 10.8 Generic dimensioning parameters of a node

deal with exact (optimal) dimensioning results. In contrast, in MH and MP-to-P (i.e., DA) cases, we need heuristic algorithms. As such, the obtained results can be considered as an upper bound of the optimal network cost. Specifically, when MH strategy is considered, we apply the MST (Maximizing single hop traffic) heuristic algorithm [6]. Note that we also run simulations using other heuristics and found the results to be qualitatively similar. Finally, in the DA case, we propose a new heuristic algorithm, called the MTA (Maximizing Traffic Aggregation) algorithm, in order to plan the MP-to-P lightpaths. Then the number of OXC ports, transmitters and receivers are determined.

The basic operation of the MTA algorithm is described as follows. Let λ_{sd} denote the aggregate traffic between node pair s and d, which has not been yet carried. As explained before, λ_{sd} can be a fraction of the lightpath capacity. In our study, we suppose that $\lambda_{sd} \in [0, 1]$, so at most one lightpath between every pair of nodes (s, d) is required to carry all the traffic requests. Let $H(s, d)$ denote the hop distance in the physical topology between nodes pair (s, d). The MTA algorithm attempts to establish lightpaths between source-destination pairs where there is any traffic and with the highest $H(s, d)$ values. The connection request between s and d will be supported by the newly established lightpath. Afterwards, the algorithm will try to satisfy connection requests, as much as possible, originating from intermediate nodes and traveling to the same destination d, based on the currently available capacity of the lightpath (s, d). This heuristic tries therefore to establish lightpaths between the farthest node pair in an attempt to allow a virtual topology to collect the maximum possible traffic at the intermediate nodes. The pseudo-code for this heuristics is presented hereafter:

`Step 1: Construct virtual topology:`

`1.1: Sort all the node pairs` (s,d) `(where` $\lambda_{sd} \neq 0$`) according to the hop distance` $H(s,d)$ `and insert them into a list` L `in descending order.`

```
1.2: Setup a lightpath between the first pair of nodes
     (s', d') using the first-fit wavelength assignment and
     the shortest-path routing, let λ_{s'd'} = 0.
1.3: Sort all the node pairs (i, d') (where λ_{id'} ≠ 0 and i is
     an intermediate node traversed by the lightpath
     (s', d')) according to the hop distance H(i, d') and
     insert them into a list L' in descending order.
1.4: Try to setup the connection between the first node
     pair (i', d') using the lightpath (s', d'), subject to the
     current available bandwidth on lightpath (s', d'). If
     it fails, delete (i', d') from L'; otherwise, let λ_{i'd'} =
     0, update the available bandwidth of the lightpath
     (s', d') and go to Step 1.3 until L' becomes empty.
1.5: Go to Step 1.1 until L becomes empty.

Step 2: Evaluate the required number of transceivers
        and OXC ports to route all the connection
        requests based on the obtained virtual network
        topology.
```

To present the heuristic more formally, as depicted in Fig. 10.9, we define the following terms:

- Let $F(V, E)$ be a graph corresponding to the physical topology, where V is the set of vertices (i.e., network nodes) and E is the set of edges (i.e., fiber links).
- Let $\Pi(F, s, d)$ be a function that returns the shortest path from s to d.
- Let L, as defined above, be the set of all connection requests, i.e., $L = \{(s, d) \in V^2,\ \lambda_{\mathrm{sd}} \neq 0\}$. The list L is ordered in descending order according to the hop distance $\Pi(F, s, d)$.
- Let $\Phi(k, \Pi(F, s, d))$ be a function that returns the kth link on the shortest path between s and d.
- Let $S(l)$ and $D(l)$ be the functions that return the originating node and terminating node of the link l, respectively.
- Let $A(i, j) \in [0, 1]$ be the residual bandwidth on lightpath (i, j).
- We denote by $Tx(i)$, $Rx(i)$ and $OXC(i)$ the number of transmitters, receivers and OXC interfaces required at node $i(i = 1, \ldots, N)$, respectively. We outline that $Tx(i)$, $Rx(i)$ and $OXC(i)$ are the output dimensioning results of the heuristic.

Steps 1–7 create the end to end MP-to-P lightpath and update its available bandwidth. Then, the intermediate node connections are aggregated into the pass-through lightpath (Steps 8). It is worth noting that the asymptotic complexity of the MTA algorithm is $O\left(|V|^2 \log(|V|) + |V|\,|E|\right)$, as it requires only the knowledge of all the shortest paths on the physical topology, which can be obtained using the Dijkstra algorithm.

Algorithm: MTA
Input: Static traffic matrix and the physical topology $F(V, E)$
Output: Number of transceivers and OXC interfaces required to accommodate the input traffic matrix
BEGIN
1. Extract the first couple (s', d') from L, subject to:
   $|\Pi(F, s', d')| \geq |\Pi(F, s, d)| \; \forall \, (s, d) \in L$
2. $A(s', d') = 1$
3. $Tx(s') = -1$
4. $Rx(d') = -1$
5. For k = 1 to $|\Pi(F, s', d')|$
   $l = \Phi(k, \Pi(F, s', d'))$
   $OutputOXC(S(l)) = -1$
   $InputOXC(D(1)) = -1$
6. $L = L \backslash (s', d')$
7. $A(s', d') = A(s', d') - \lambda_{s'd'}$
8. For k = 2 to $|\Pi(F, s', d')|$
   $l = \Phi(k, \Pi(F, s', d'))$
   $l' = S(l)$
   if ($\lambda_{i'd'} \neq 0$ and $\lambda i'd' \leq A(s', d')$)
   $Tx(i') = -1$
   $A(s', d') = A(s', d') - \lambda_{i'd'}$
   $L = L \backslash (i', d')$
9. if $L \neq \emptyset$ go to 1
END

Fig. 10.9 MTA planning heuristic

10.5.2.2 Dimensioning Results and Comparison

Table 10.2 reports the dimensioning results of the network when using the studied strategies. The reported results are averaged over multiple randomly generated traffic matrices, so that we ensure very narrow 97.5% confidence intervals. A detailed representation of the results reported in Table 10.2 is given in Figs. 10.10 and 10.11. The figures depict the dimensioning results corresponding to each node of the network.

Table 10.2 shows that the opaque network has the highest transceiver cost. This result is obvious since opaque nodes do not have optical pass-through. The P-to-P

Table 10.2 Dimensioning results

	# Tx	# Rx	#lightpaths	#OXC ports	load/lightpath	hops/lightpath
Opaque	1497	1497	1497	—	97,06%	1
P-to-P all-optical	812	812	812	3722	49,54%	3,58
MH	734	734	734	3082	74,20%	3,20
DA	812	540	540	2626	75,04%	3,86

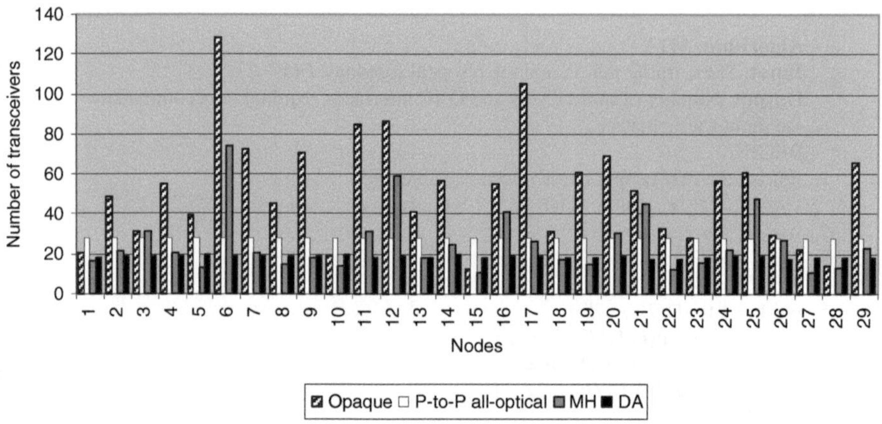

Fig. 10.10 Transceivers needed per node under different strategies

strategy reduces considerably the transceiver requirements due to the network transparency. However, the P-to-P all-optical network still suffers from its inherent issue regarding the transceiver and wavelength underutilization. The MH and MP-to-P approaches alleviate this issue and enable thus further transceiver cost reduction. Indeed, MH and the DA schemes improve the P-to-P all-optical network grooming capacity, while conserving the transparency propriety as opposed to opaque networks where electronic grooming is needed at each intermediate node. Specifically, the transparency of P-to-P is totally conserved with the DA scheme and partially conserved in MH networks.

The DA scheme allows the aggregation in the same lightpath of multiple connections traveling to a common destination. Consequently, the number of MP-to-P lightpaths (or receivers) required in the network to handle all the traffic requests is reduced compared to P-to-P all-optical networks. Note that, the number of MP-to-P lightpaths that should be managed by the network is equal to the number of

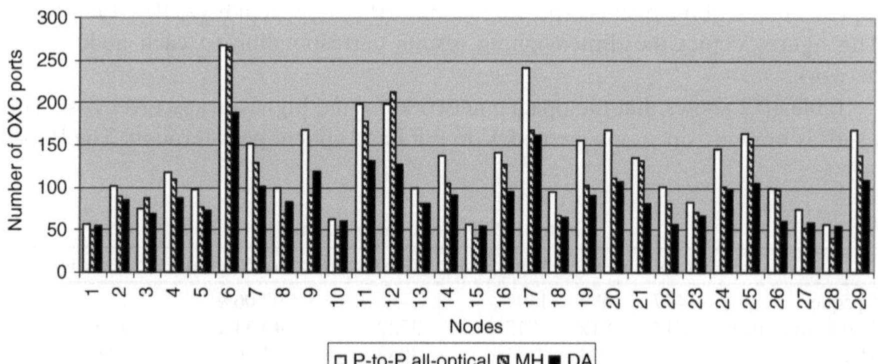

Fig. 10.11 OXC ports needed per node under different strategies

receivers. The gain obtained in this case is above 33%. Moreover, as the number of lightpaths is reduced when the DA is allowed, the number of OXC ports is also reduced. The gain recorded is beyond 29%. This latter gain is less important than the one obtained when dealing with receivers since the number of OXC ports depends not only on the number of established lightpaths but also on the number of hops per lightpath. Indeed, the average number of hops per lightpath is 3,58 in the P-to-P case, whereas it is 3,86 in the MP-to-P case. These results show how the distributed aggregation scheme alleviates the scalability issues encountered with the P-to-P all-optical networks.

Compared to MH networks, the DA scheme leads also to significant cost savings: around 15% of OXC ports and 8% of transceivers are saved. The DA scheme increases the grooming capacity of P-to-P all-optical networks, while avoiding the extra electronic processing entailed by the MH approach. This active electronic operation reduces the cost of an MH network over P-to-P all-optical networks, but it introduces additional costs when compared to DA-enabled all-optical networks.

Finally, it is useful to compare the lightpath load entailed by the different strategies (see Table 10.2). As expected, opaque, MH and MP-to-P outperform the P-to-P case. Moreover, MH and MP-to-P networks perform slightly worse than opaque networks. This result is obvious since opaque networks have the maximal grooming capability. In the P-to-P case, the average load of a lightpath is 49,54%. This result emphasizes the already mentioned problem of resource under-utilization. The efficiency of such a strategy is possible only when there is enough traffic between pair nodes to fill in the entire capacity of wavelengths. DA scheme alleviates this issue, while the average load of a lightpath reaches 75% in this case.

10.6 Conclusion

In this chapter, we have presented, Distributed Aggregation, a novel solution for sub-wavelength grooming in all-optical networks. The proposed solution, which is an alternative to optical packet switching technology, aims at reconciling the two opposite requirements of packet switching and optical transparency. This is achieved by allowing multiple connections traveling from different nodes to a common destination to be aggregated into the same lightpath. A comparison between our scheme and existing solutions was given. Results obtained from the ILP and heuristic algorithms showed that the DA scheme increases the total throughput in the network. An increase of approximately 25% was recorded. Besides, we compared the results of the US optical backbone dimensioning using all compared strategies. The analysis revealed that our proposed approach reduces significantly the network cost. Specifically, compared to classical all-optical networking approaches, around 30% of the receiver and OXC ports are saved when the distributed aggregation is used. This technique is proven to be particularly effective when the bandwidth requirements of connections between node pairs are fractions of the lightpath capacity. As a result, the distributed aggregation scheme reduces the wasted bandwidth problem and alleviates the scalability issue encountered in all-optical wavelength networks while preserving the benefits of optical by pass.

References

1. M. J. O'Mahony, D. Simeonidou, D. K. Hunter, and A. Tzanakak, "The application of optical packet switching in future communication networks", IEEE Commun. Mag., Vol. 39, Issue 3, pp. 128–135, March 2001.
2. N. Bouabdallah, E. Dotaro, N. LeSauze, L. Ciavaglia, and G. Pujolle, "Distributed aggregation in all-optical wavelength routed networks", Proc. of IEEE ICC '2004, Paris, France, June 2004.
3. N. Bouabdallah, "Sub-wavelength solutions for next-generation optical networks", IEEE Commun. Mag., Vol. 45, Issue 8, pp. 36–43, August 2007.
4. H. Zhang, J. P. Jue, and B. Mukherjee, "Capacity allocation and contention resolution in a photonic slot routing all-optical WDM mesh network", IEEE/OSA J. Light. Tech., Vol. 18, Issue 12, December 2000.
5. Y. Chen, C. Qaio, and X. Yu, "Optical burst switching: A New area in optical networking research", IEEE Network, Vol. 18, pp. 16–23, May 2004.
6. K. Zhu, and B. Mukherjee, "Traffic grooming in an optical WDM mesh network", IEEE J. Select. Areas Commun., Vol. 20, pp. 122–133, January 2002.
7. M. Mellia, E. Leonardi, M. Feletig, R. Gaudino, and F. Neri, "Exploiting OTDM technology in WDM networks", in Proc. IEEE INFOCOM' 2002, pp. 1822–183, New York, USA, June 2002.
8. I. Widjaja, I. Saniee, R. Giles, and D. Mitra, "Light core and intelligent edge for a flexible, thin-layered, and cost-effective optical transport network", IEEE Opt. Commun., Vol. 41, Issue 5, pp. S30–S36, May 2003.
9. A. Gumaste and I. Chlamtac, "Light-trails: A novel conceptual framework for conducting optical communications", Wksp. High Perf. Switching and Routing, pp. 251–56, June 2003.
10. A. Gumaste, "Light-trails and light-frame architectures for optical networks", Ph.D. thesis, Fall 2003, UT-Dallas; at: www.cotrion.com/light-trails.
11. B. Mukherjee, Optical Communication Networks. New York: Mc-Graw-Hill, 1997.
12. I. Chlamtac, A. Faragó, and T. Zhang, "Lightpath (wavelength) routing in large WDM networks", IEEE J. Select. Areas Commun., Vol. 14, pp. 909–913, June 1996.
13. D. Banerjee and B. Mukherjee, "Wavelength-routed optical networks: Linear formulation, resource budgeting tradeoffs, and a reconfiguration study", IEEE/ACM Trans. Networking, Vol. 8, pp. 598–607, October 2000.
14. S. Balasubramanian, A. Kamal, and A. K. Somani, "Medium access control protocols for light-trail and light-bus networks", Proc. 8th IFIP Working Conf. Opt. Net. Design and Modeling, February 2004.
15. N. A. VanderHorn, M. Mina, and A. K. Somani "Light-trails: A passive optical networking solution for wavelength sharing in the metro", Wksp. High Capacity Opt. Net. and Enabling Technologies, December 2004.
16. N. Bouabdallah, L. Ciavaglia, E. Dotaro, and N. Le Sauze, "Matching fairness and performance by preventive traffic control in optical multiple access networks", In Proc. OptiComm' 2003, Dallas, pp. 424–429, October 2003.
17. N. Bouandallah, A. L. Beylot, E. Dotaro, and G.Pujolle, "Resolving the fairness issues in bus-based optical access networks", IEEE J. Select. Areas Commun., Vol. 23, Issue 8, pp. 1444–1457, August 2005.
18. R. Gaudino et al., "RINGO: A WDM ring optical packet network demonstrator", In Proc. of ECOC' 2001, Amsterdam, Netherlands, Vol. 4, pp. 620–621, September 2001.
19. N. Le Sauze et al., "A novel, low cost optical packet metropolitan ring architecture", in Proc. Of ECOC' 2001, Amsterdam, Netherlands, September 2001.

Chapter 11
Guaranteed Quality of Recovery in WDM Mesh Networks

I-Shyan Hwang, I-Feng Huang and Hung-Jing Shie

Abstract This study proposes a mechanism of guaranteed quality of recovery (GQoR) for Wavelength Division Multiplexing (WDM) mesh networks. Four GQoR levels are used to support customized services, and each of them is mapped to the adaptive recovery methodology. Once a failure occurs, the control system activates the recovery mechanism in compliance with the GQoR level. If the protection procedure fails as well, the proposed algorithm will then execute the restoration mechanism. Consequently, the recovery success rate is increased. This paper examines the shared segment recovery methods to establish backup path; therefore, it is well suited for large-scale networks and also increases the bandwidth utilization of the networks. Furthermore, a node deals only with its own routing information by employing the distributed control, so the fault recovery procedure can be speeded up. Simulation results reveal that the proposed method has greater performance of lower blocking probability and mean hop number than other methods previously reported in the literature.

Keywords WDM · Guaranteed quality of recovery · Shared segment recovery · Survivability

11.1 Introduction

Wavelength Division Multiplexing (WDM) [1, 2] technology divides the tremendous bandwidth in a single fibre into many independent channels. All channels can transmit information across the fibre in parallel. Factors such as construction work, rodents, fires or human error may cut the fibre, which may lead to fibre failure and traffic loss. Managing faults in optical networks, including fault diagnosis and recovery, has thus become very important. In fault diagnosis, hardware components detect network anomaly, and the failure is pinpointed from the alarms received by the management system. Then, in fault recovery, the failed path is detoured to the

I-S. Hwang (✉)
Department of Computer Science and Engineering, Yuan-Ze University, Chung-Li, Taiwan

M. Ma. (ed.), *Current Research Progress of Optical Networks*,
DOI 10.1007/978-1-4020-9889-5_11,

backup path. The upstream node from the failure point is notified of the fault and the fault recovery mechanism is initiated subsequently. Multiple fault recovery paths may be available in the mesh networks; therefore, the recovery algorithm must determine the adaptive paths to detour. The *fault recovery* scheme can be divided into two types – *fault protection* that pre-calculates the backup paths before failure occurs, and *fault restoration* that calculates the backup paths dynamically after the failure has occurred. The merits of fault protection are that the backup paths are calculated in advance to save time needed to search through routes. However, this approach requires much spare capacity of bandwidth to protect networks quickly, and the backup paths reserved for fault protection may not be optimal routes. Typically, a fault restoration mechanism must be triggered to make the adaptive restoration paths. Although the restoration paths need not be pre-calculated, computing the adaptive restoration path will take longer time than fault protection after failures occur. Depending on where a detour originates, the fault recovery technique can be classified into *link-based*, *path-based* or *segment-based* (or called *subpath-based*) recovery methods [3]. The link-based method employs local detouring, while the path-based method employs end-to-end detouring. The link-based method can make faster responding than path-based method. However, link-based method has lower recovery success rate than path-based method. The segment-based method, which divides a path into several segments, and detours reroute traffic on the selected segment. This method has the benefits of fast recovery and improving recovery success rate. For various fault recovery requests, the recovery technique can be either dedicated or shared in *1 + 1*, *1:1*, *1:N* and *M:N* recovery policies [4]. For 1 + 1 policy, as dedicated facility recovery, traffic is passing through both the working and backup paths. Upon failure notification, the traffic on the backup path becomes the active traffic. Therefore, the resources on both working and backup paths are fully reserved. It is the fastest protection switched recovery mechanism, but also the most expensive in terms of resources. For 1:1 policy, it is similar to 1 + 1 policy, but the traffic is passing through the working path only. For 1:N policy, as shared facility recovery, N working paths are protected using a backup path. For M:N policy, M backup entities are shared among N working resources. As a result, recovery channels are shared among different failure scenarios, and therefore shared facility recovery is more capacity-efficient when compared with dedicated facility recovery. Shared Risk Link Group (SRLG) [5] is a link-state that defines the availability of protection resources to a working path. It stipulates that any two or more working paths sharing the same risk of failure cannot make use of the same protection resource. The basic operation for deriving the SRLG for a link or a node is to identify the network resources that cannot be taken for the protection purpose by newly arrived working paths traversing the link or node. The purpose of the SRLG constraint is to guarantee 100% restorability for failure of any single link or node in the network.

Quality of Protection (QoP) is a mechanism to classify the protection service into several levels depended on customer's request in communication networks. Some pioneers explore QoP mechanism and classify into either three [6, 7] or four [8] service levels. The reliability of service [6] addresses three levels of fault protection for ATM networks. Two of the virtual paths could have backup paths, one with *dedicated redundant capacity* and the other with *shared spare capacity*. The third virtual path could be *unprotected*, but in the event of failure, restoration could be performed dynamically. The recent studies [7, 8] present different service levels of fault protection for WDM networks. The classification of QoP service of [7] is similar to that of [6]; moreover, the SRLG constraint is considered for fault protection design in the literature. In the research of [8], the service class is divided into four levels. The first three levels are the same as that of [6], but the fourth level utilizes protection bandwidth under normal circumstances and is preempted when other lightpaths need to be protected.

Since networks become larger and more complex, the QoP mechanism is insufficient for present applications. Besides, the segment-based recovery method has better performance than that of path-based or link-based recovery method, and the shared facility recovery method has higher bandwidth utilization. Furthermore, if a fault has one more chance to detour, the recovery success rate will increase. The other idea is to create or to reserve a new backup path to certify networking recoverability after the original backup path is used. The proposed guaranteed quality of recovery (GQoR) aims to support different services for fault recovery in WDM mesh networks and to guarantee both recovery time and backup capacity in the certain level to satisfy customer's request. Therefore, not only the dedicated protection, but the segment method, the shared facility recovery, the restoration mechanism and the SRLG constraint are also considered. The first level of GQoR is the $1 + 1$ dedicated protection. The second level of GQoR is the shared segment protection. The third level of GQoR is the shared segment restoration. The fourth level of GQoR is the reroute or preemption. When a failure occurs, the upstream node from the failure point activates the recovery mechanism in compliance with the GQoR level. If the level 1 and level 2 protection procedures fails, the proposed GQoR algorithm will then execute the level 3 segment restoration mechanism. Consequently, there are two opportunities to detour when a failure occurs, and the recovery success rate will be significantly increased. Moreover, the distributed control is employed for the proposed algorithm, so the fault recovery procedure can be speeded up.

The rest of this chapter is organized as follows. Section 11.2 describes the assumptions and definitions of this paper. Section 11.3 addresses the proposed GQoR algorithm and fault recovery method that deals with link failure [9, 10], node failure and channel failure [11]. Section 11.4 shows and discusses the simulation results in terms of the blocking probability and the mean hop number comparison for the proposed GQoR mechanism vs. QoP mechanism [8]. Section 11.5 draws conclusions and offers suggestions for the direction of future research.

11.2 Assumptions and Definitions

In this study, the nodes are assumed having capability of wavelength conversion in the networks. Furthermore, the parameter q of the GQoR will be delivered to every node along the working path when a new route is creating. If a route is completely established, all nodes along the working and backup paths will obtain the path information, and then the path information will be stored in the database called Recovery Table in each node. Moreover, the GQoR mechanism will be further explained in this paper, since only the concepts are addressed in the authors' previous works such as implementation of distributed control for overlapped and non-overlapped segment protection algorithms (OSP and NOSP) [12] and Dynamic Multiple Ring Algorithm (DMRA) [13].

11.2.1 Classification of GQoR Mechanism

The proposed GQoR mechanism which is divided into four levels is shown in Table 11.1, and the definition of GQoR levels is addressed in details as follows.

A. *Global Protection*: The level 1 recovery has the highest priority, and the dedicated 1 + 1 protection is applied to achieve the protection requirement. Once the working path is completely created during the request, the network will establish a disjoint path called a dedicated backup path to protect the working path. Furthermore, the SRLG constraint is considered for this level. After these two paths have been built, the data will be delivered through them simultaneously. If the failure occurs somewhere in the working path, the traffic on the backup path will become active traffic.

B. *Segment Protection*: The segment protection is considered to be the second priority and the created backup path may be shared with other ones. The implementation of segment protection using distributed control is introduced in [12]. The shared facility method and the SRLG constraint are considered in this level. Two different types of segment protection are investigated [14, 16] based on the capability of protection.

- *Overlapped Segment Protection* (OSP): For OSP method, two adjacent backup paths overlap to protect the same working link, as shown in Fig. 11.1(a). This method has high protection ability, but sometimes the objective of overlapping just a link between two adjacent backup segments cannot be achieved1 [4, 15].
- *Non-overlapped Segment Protection* (NOSP): For NOSP method, two adjacent backup paths do not overlap to protect the same working link, as shown in Fig. 11.1(b). The benefit of the NOSP method is simple and economic, but it is less protection ability if a beginning node of any backup path fails in the working path [16].

Table 11.1 The classification of GQoR

Level	Recovery mode	Description
1	Global Protection	1 + 1 dedicated protection
2	Segment Protection	shared segment protection
3	Segment Restoration	shared segment restoration
4	Reroute or Preemption	It will normally do end to end reroute if a fault occurs, but the reroute path may be preempted by level 1 $\sim$ 3 if resource is insufficient.

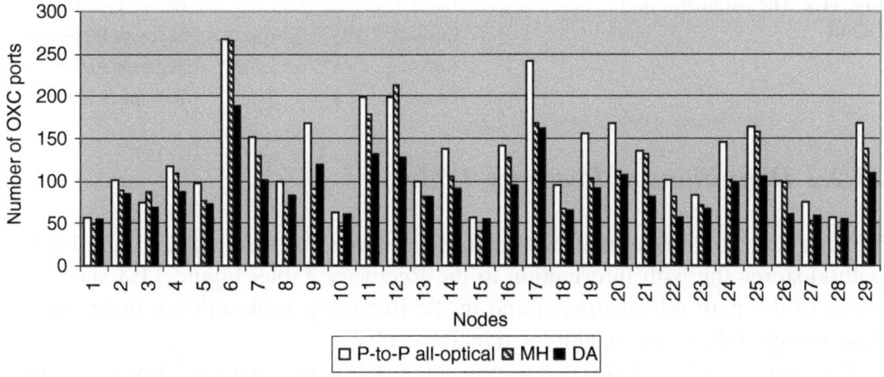

Fig. 11.1 Overlapped and non-overlapped segment protection paths

C. *Segment Restoration*: The level 3 recovery method does not apply to the pre-calculated protection path, instead the restoration mechanism of DMRA [13] to recover from the failure. For brief address of DMRA, the nodes can use distributed control to find neighbouring nodes and establish relationships between nodes to construct several logical rings. Each logical ring may share a single path or node in the network and cover all links. Nodes can locate the fault and then restoration paths will be chosen from the logical rings according to the cost function. The selected restoration paths are appropriate transmission routes around the faulty point when failure occurs. Therefore, farther nodes and links are not impacted. All candidate restoration paths share the loads induced by the fault, so to utilize the network resources effectively and to increase the connectivity rate. However, the restoration path is calculated after the fault occurs, so the restoration time in this level is greater than that of the previous two levels.

D. *Reroute or Preemption*: The level 4 recovery method does not utilize any protection or restoration method. Once a failure occurs, the rerouting mechanism is activated. Nevertheless, if the network capacity is insufficient to cause blocking in the level 1 to level 3 recovery mechanisms, the level 4 routes will be torn down to release the resources for any other high-level recovery mechanism.

When a node in the network receives the request to establish a new route, the node creates an appropriate working path. At the same time, the node also establishes a dedicated backup path for level 1 Global Protection, and reserves segment backup paths for level 2 Segment Protection. Later, the path information, which includes GQoR parameter q, is delivered to all nodes in the working and backup paths. Each node will write the path information into Recovery Table. Figure 11.2 shows the q values of GQoR levels. When q is equal to 1, the recovery method belongs to level 1 and the dedicated protection will be supported. When q is equal to 2.1, the recovery method belongs to level 2 and the OSP algorithm is utilized. When q is equal to 2.2, the recovery method also belongs to level 2, but the NOSP algorithm is applied. When q is equal to 3, the recovery method belongs to level 3 and DMRA mechanism is used. When q is equal to 4, the level 4 recovery method is employed, and the end to end rerouting is prepared for the failure.

Fig. 11.2 The parameter q in GQoR

$$\begin{cases} \text{Level 1} & : \quad q = 1 & \text{Global Protection} \\ \text{Level 2} & : \quad q = 2.1 \text{ or } 2.2 & \text{Segment Protection} \\ \text{Level 3} & : \quad q = 3 & \text{Segment Restoration} \\ \text{Level 4} & : \quad q = 4 & \text{Reroute or Preemption} \end{cases}$$

11.2.2 Definitions of Recovery Table

When a new route is established, each node along the working path and backup path(s) stores the path information to the Recovery Table. Figure 11.3 shows the terms of the path information stored in the Recovery Table in each node and the description of the terms is addressed in Table 11.2.

For example, Fig. 11.4(a) is a simple network topology and each link is assumed having three channels, λ_1, λ_2, and λ_3. The working path is a-c-e, and the backup paths are a-b-c and c-d-e by using NOSP method. Figure 11.4(b) shows the path information of Recovery Table in each related node. In the first row of node (a), the W/B field is set to W to represent a working path. The set of nodes of working path will be recorded in the path field as a-c-e and the assigned wavelengths are recorded in field w as λ_1 by the system RWA mechanism. The field q records the GQoR recovery method mapped q value as 2.2 by using NOSP method. The set of nodes of backup path that pertains to the working path is written to the Bpath field as a-b-c and the assigned wavelength is written to Wb field as λ_2 by the NOSP mechanism. The B_B node field stores the beginning nodes of each backup path. Furthermore, the backup path information will be filled in the second row of node (a), and the other related nodes (b), (c) and (d) will do the same process as well. When a node receives a recovery request, it simply checks the path information in the Recovery Table, and then begins the recovery mechanism. If the link a-c is cut off, node (a) will obtain backup path a-b-c and wavelength λ_2 retrieved from Recovery Table. If link c-e is cut off, node (c) will get backup path c-d-e and wavelength λ_3 to recover the fault.

Fig. 11.3 Recovery table

W/B path	path	w	q	Bpath	Wb	B_B node
...	...	...	...	...	...	...

Table 11.2 Description of terminologies of recovery table

Terminology	Description
W/B path	determining whether the path is a working path or a backup path. "W" represents a working path, and "B" depicts a backup path.
path	set of nodes along working or backup path
w	assigned wavelengths for the path
q	recovery level of the working path
Bpath	set of nodes along the backup path which pertains to a working path
Wb	wavelength of the backup path
B_B node	beginning nodes of each backup path

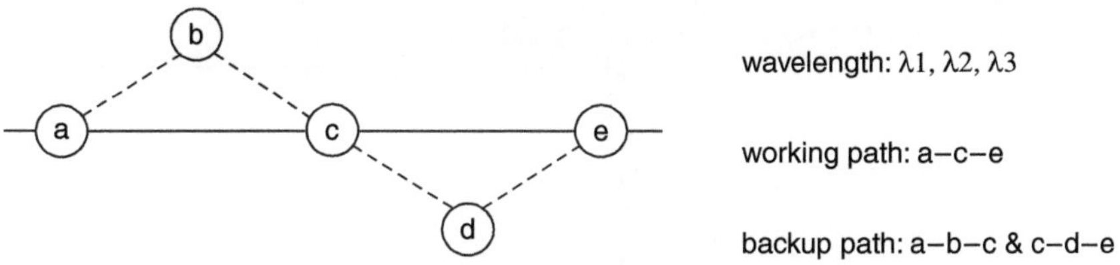

(a)

Node (a)

W/B path	path	w	q	Bpath	Wb	B_B node
W	a–c–c	λ1	2.2	a–b–c	λ2	a,c
B	a–b–c	λ2	none	none	none	none

Node (b)

W/B path	path	w	q	Bpath	Wb	B_B node
B	a–b–c	λ2	none	none	none	none

Node (c)

W/B path	path	w	q	Bpath	Wb	B_B node
W	a–c–c	λ1	2.2	c–d–e	λ3	a,c
B	c–d–e	λ3	none	none	none	none

Node (d)

W/B path	path	w	q	Bpath	Wb	B_B node
B	c–d–e	λ3	none	none	none	none

(b)

Fig. 11.4 Example of recovery table

11.3 GQoR Mechanism and Fault Recovery

The GQoR main algorithm and its subroutines are described in details in this section. The fault recovery in the events of link failure, node failure and channel failure are also discussed in [16].

11.3.1 Main GQoR Recovery Mechanism

The distributed control is designed for the proposed GQoR mechanism. When a fault is detected, the upstream node from the failure point will be notified, and then the node generates a beginning-token which gives a right to begin the recovery mechanism. After the GQoR mechanism begins, the recovery methods, q value will be retrieved from the Recovery Table, and then execute the mapped recovery subroutine. If the GQoR mechanism succeeds in recovery, the beginning-token will be discarded and the transmission will continue. If the recovery method is either the

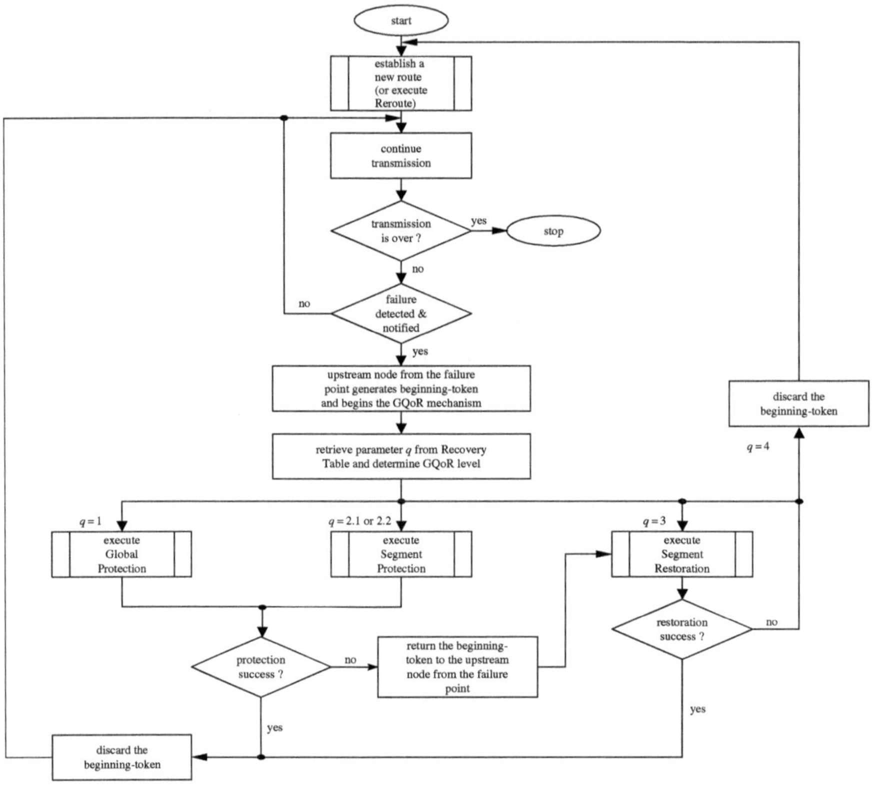

Fig. 11.5 Flowchart of main GQoR recovery mechanism

Global Protection or the Segment Protection, there is one more chance to recover by executing Segment Restoration method when the protection process fails. If the recovery method is the Reroute or the Segment Restoration method fails, a new route will substitute the old one. Figure 11.5 shows the flowchart of the main GQoR recovery mechanism. The detailed descriptions of each GQoR level will be depicted as follows.

11.3.2 GQoR Recovery Subroutines

Figure 11.6 shows the flowchart of subroutine – execute Global Protection. When this subroutine executes, the node, which owns the beginning-token, will check if it is the source node. If it is not the source node, the beginning-token will be delivered to the source node. Consequently, the source node can activate the backup path. Later, the source node will begin to create a new backup path. However, if the resources are not available even though level 4 resources have taken account of, the recovery level will be degraded to level 3.

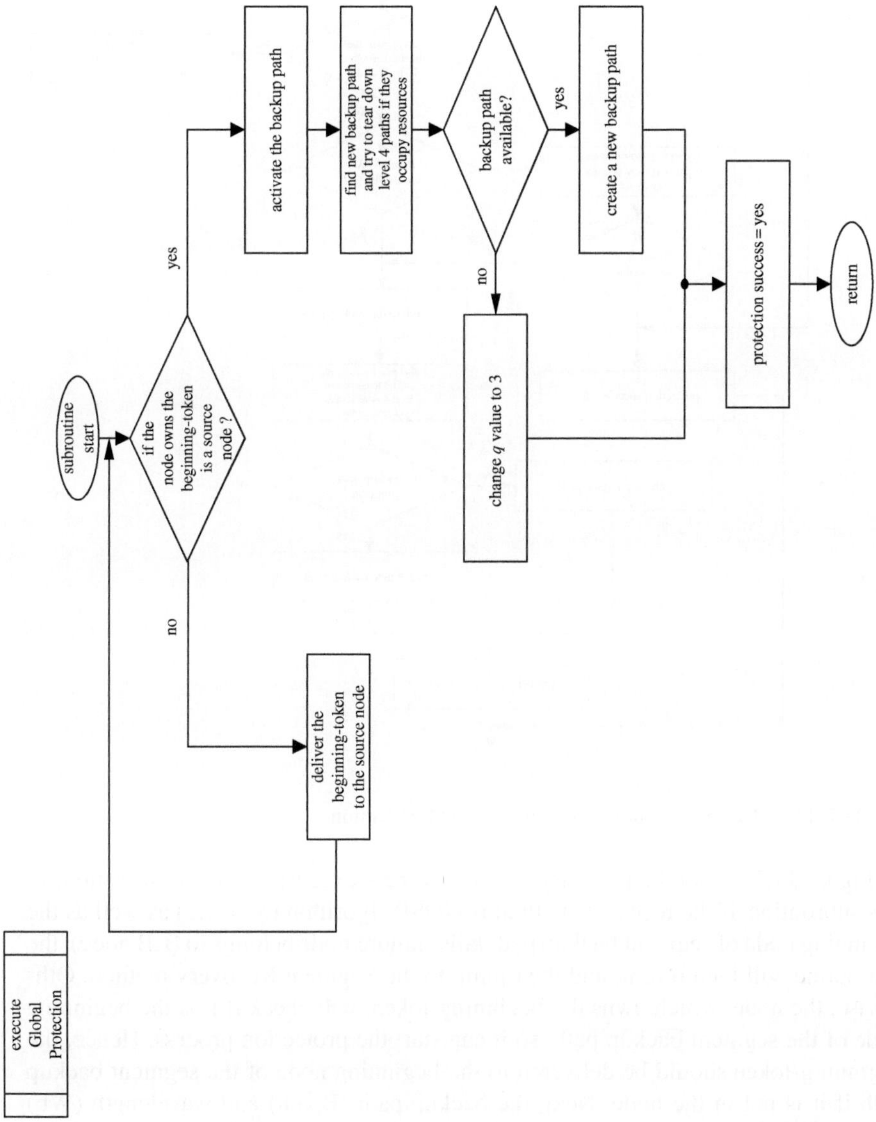

Fig. 11.6 Flowchart of subroutine – execute global protection

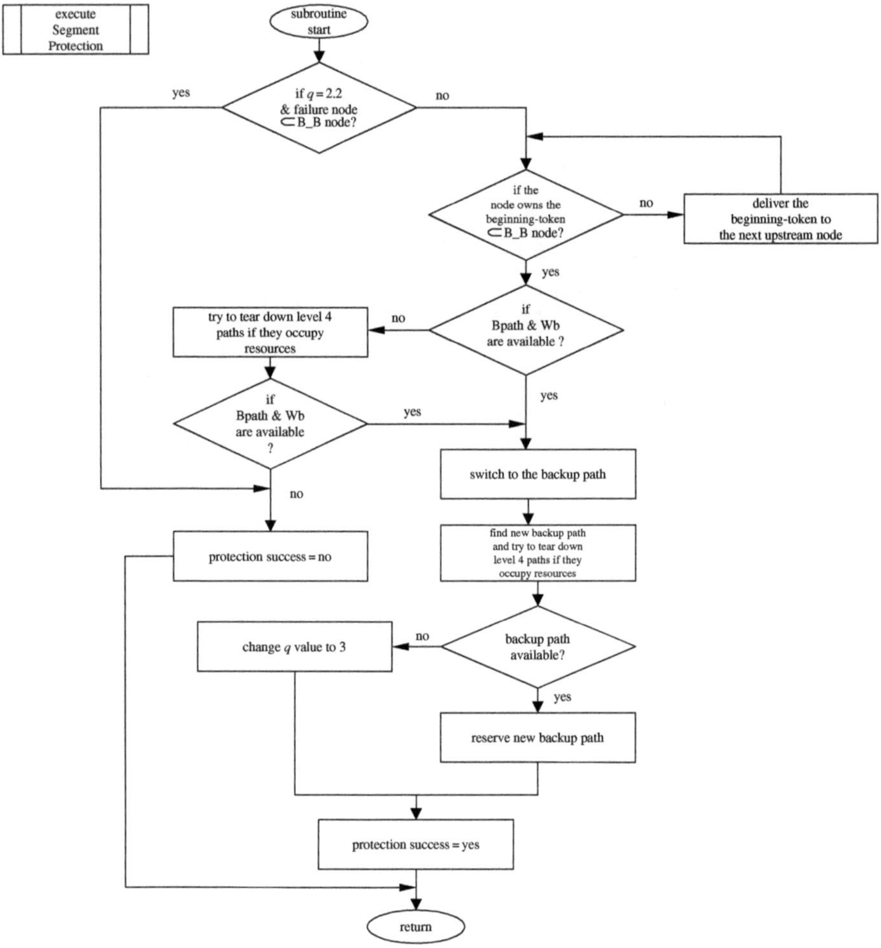

Fig. 11.7 Flowchart of subroutine – execute segment protection

Figure 11.7 shows the flowchart of subroutine – execute Segment Protection. In this subroutine, if the recovery method is NOSP algorithm ($q = 2.2$) as well as the beginning node of segment backup path fails (failure node belongs to B_B node), the subroutine will then return, and then jump to the Segment Recovery method. Otherwise, the node, which owns the beginning-token, will check if it is the beginning node of the segment backup path, so it can start the protection process. Hence, the beginning-token should be delivered to the beginning node of the segment backup path if it is not in the node. Next, the backup path (Bpath) and wavelength (Wb) need to check for availability. If they are not available, the subroutine will try to drop some level 4 paths when they occupy resources, and then check the segment backup path and wavelength(s) again before activating the segment backup path. If it is available, then the node will switch traffic to the backup path. Later, a new segment backup path will be found and be reserved. However, if the resources are

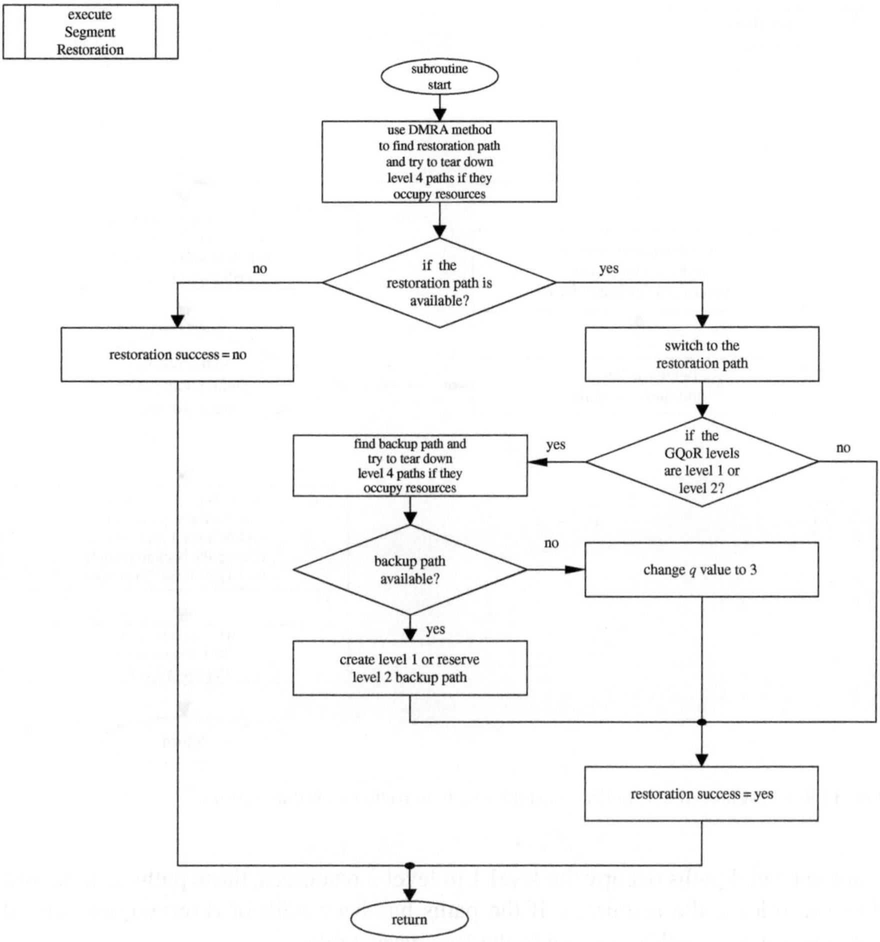

Fig. 11.8 Flowchart of subroutine – execute segment restoration

not available even though level 4 resources has been considered, the recovery level will be degraded to level 3.

Figure 11.8 shows the flowchart of subroutine – execute Segment Restoration. In this subroutine, the DMRA mechanism [13] is used to find the adaptive segment restoration path. If some level 4 paths occupy the resources, the subroutine will try to tear down these paths and find the restoration path again. After the restoration path is found, it will be activated to the working path. Later, a new backup path will be created for level 1 Global Protection or reserved for level 2 Segment Protection.

Figure 11.9 shows the flowchart of subroutine – establish a new route or execute Reroute. In this subroutine, the optimal working path will be established and backup path(s) will be built or be reserved depending on the recovery level. If the paths are not available, the connection will defers for τ mini-seconds, which is randomly generated from 0 to 100 ms in our simulation to wait for available resources. Moreover,

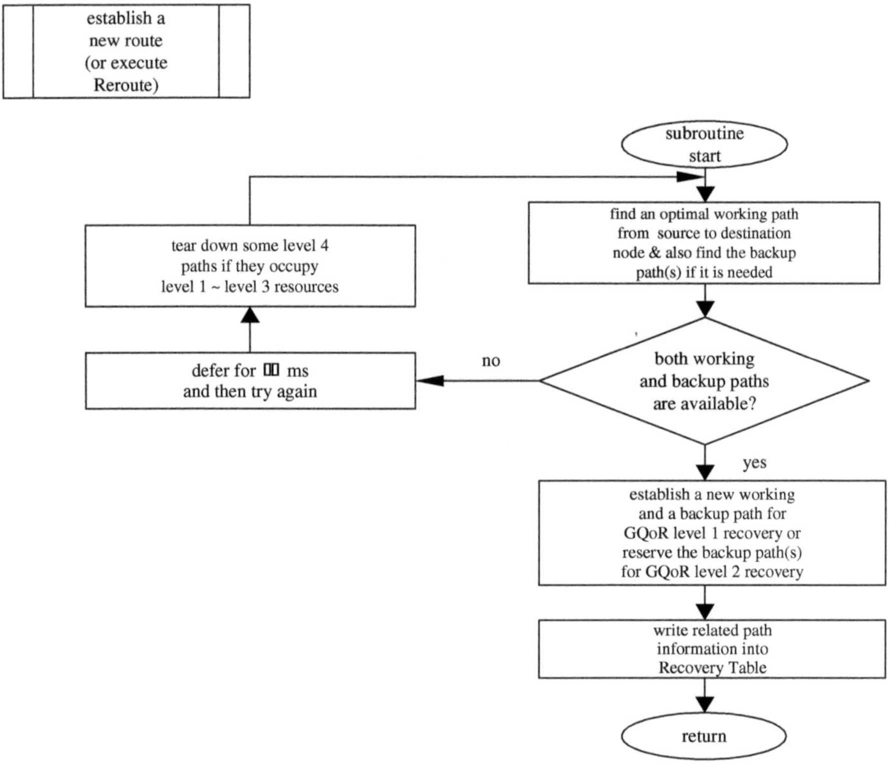

Fig. 11.9 Flowchart of subroutine – establish a new route or execute reroute

if some level 4 paths occupy the level 1 to level 3 resources, these paths will be torn down to release the resources. If the paths has been built or reserved, the related path information will be written to the Recovery Table.

11.3.3 Fault Recovery in Link, Node, and Channel Fault

For the case of link failure, the upstream node from the failure point is notified the fault and the GQoR mechanism begins. In this event, the network topology is still in its entirety, so the extra consideration is not necessary for GQoR mechanism.

When a fault occurs in the node, the network topology is destroyed and many links will be broken simultaneously. The level 1 Global Protection works well to recover the fault, because its backup path is a disjoint and dedicated path. For the level 2 Segment Protection, if the fault occurs in the beginning node of any segment backup path when the NOSP algorithm is used, the segment backup path is destroyed and the fault can not be recovered in this level. Therefore, the GQoR mechanism jumps to the level 3 Segment Restoration mechanism to avoid this problem. In level 3, the DMRA [13] algorithm can immediately build the new network topology and find an adaptive restoration path. For level 4 Reroute, a new route and the backup paths will be created if the resources are enough.

If a fault occurs in a channel, the upstream node from the failure point will select another channel to detour to the original link, since the network framework is not destroyed. If no channel can be used at all, the situation is identical to a link fault, and the recovery procedure is the same as that of link fault recovery.

11.4 Simulation Results

The performance of the proposed algorithm herein is studied by simulating the mesh-based NSFNet, USANET, Mesh 6×6 (6 nodes and 15 links), and Mesh 9×9 (9 nodes and 36 links) under incremental traffic. In the experiments, each link has 12 wavelengths, and each wavelength provides 10 Gbps. The 11th and 12th wavelengths are reserved for bi-directional control channels. Simulation programs are developed using the OPNET, and the simulation scenarios present metrics of blocking probability and mean hop number. The definition of blocking probability is the total unsuccessful recovery number divided by the total recovery requests. The lower the blocking probability is which means the recovery successful rate is higher, and better the performance of algorithm will be. The mean hop number is calculated from the upstream node of the failure point to the beginning node of the backup path and adds hop numbers in the backup path. Therefore, the mean hop number is a metric to represent the difference in recovery time and expense. The mean hop number is dependent on the number of segments in a path and the length of the backup path, and it will be small if there are many segment numbers and short backup paths. The traffic load is generated uniformly from average 10% of entire network until it reaches 80% of the load, and it is increased 10% each time. Furthermore, for each incremental traffic load, each level of GQoR request is generated randomly in proportion to 20% for level 1, 20% for level 2 – OSP algorithm, 10% for level 2 – NOSP algorithm, 30% for level 3 and 20% for level 4.

The comparison between the proposed GQoR mechanism and the four layers QoP mechanism in [8] are shown as follows. The simulation scenarios include three types of network failure, link fault, node fault and channel fault in different network topologies. In each incremental of the traffic load, a single fault will be set randomly throughout the network and then executes recovery algorithms to record results. After evaluation of ten times in the same scenario, the blocking probability and mean hop number are calculated and stored in the database.

Figures 11.10, 11.11 and 11.12 shows the performance of blocking probability comparison for the proposed GQoR mechanism vs. QoP mechanism under the events of link failure, node failure and channel failure. As shown these three figures, the proposed GQoR mechanism produced a lower blocking probability than the four layers QoP algorithms, especially in the traffic load between 40% and 70% with a difference from 0.05 to 0.2. This situation can be explained that the failure and channel failure. As shown these three figures, the proposed GQoR mechanism produced a lower blocking probability than the four layers QoP algorithms, especially

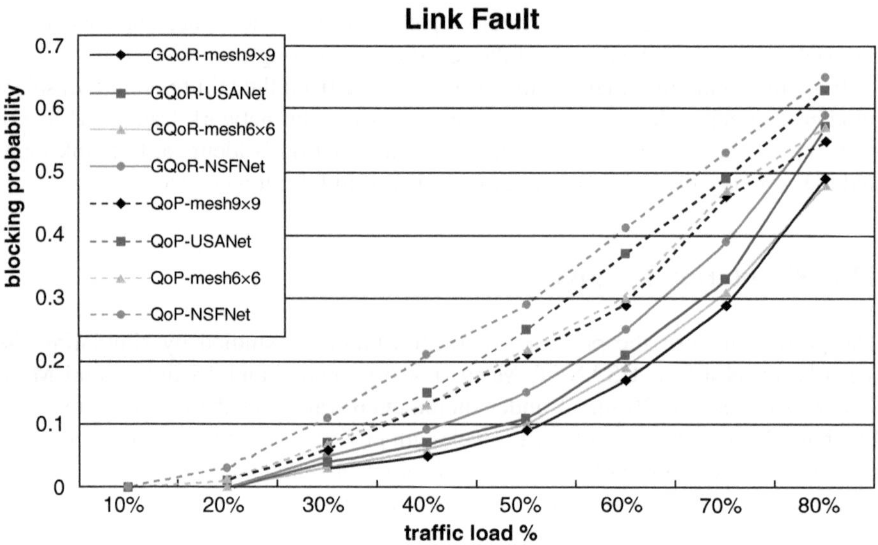

Fig. 11.10 Blocking probability comparison for the proposed GQoR mechanism vs. QoP mechanism in event of link fault

in the traffic load between 40% and 70% with a difference from 0.05 to 0.2. This situation can be explained that the OSP and NOSP algorithms perform better blocking probability, and the restoration mechanism will follow if protection methods fail, so the proposed GQoR mechanism has lower blocking probability than that of QoP. In the channel failure, the performance of blocking probability for proposed GQoR

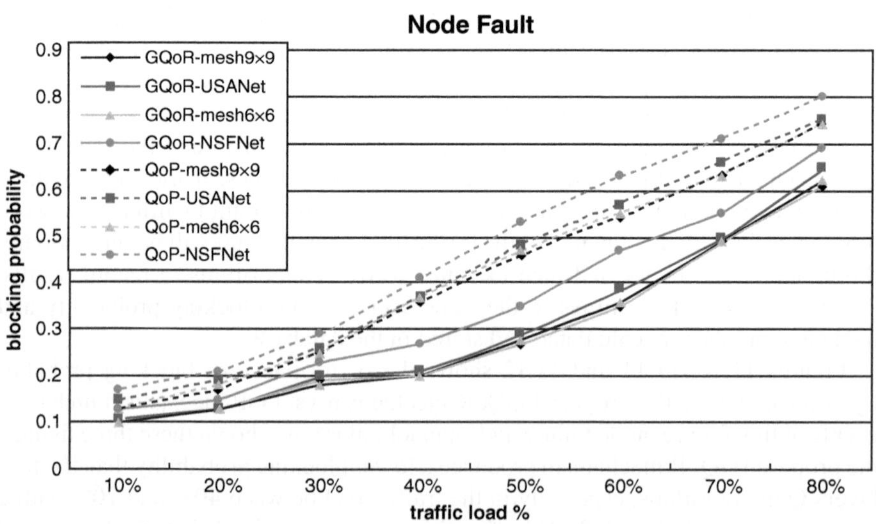

Fig. 11.11 Blocking probability comparison for the proposed GQoR mechanism vs. QoP mechanism in the event of node fault

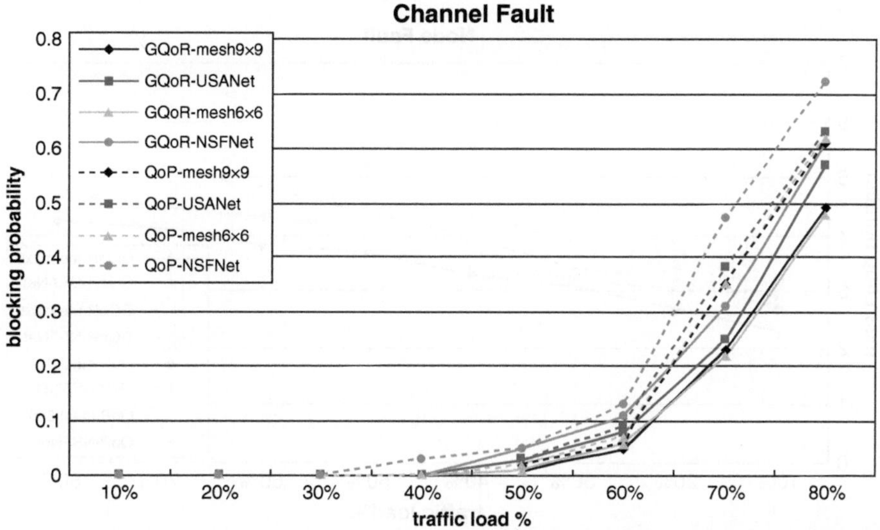

Fig. 11.12 Blocking probability comparison for the proposed GQoR mechanism vs. QoP mechanism in the event of channel fault

mechanism is better than that of QoP as well. However, the recovery mechanism may be utilized if the traffic load is large, so the change is more obvious when traffic load is greater than 60%.

Figures 11.13, 11.14 and 11.15 show the performance of mean hop number comparison for the proposed GQoR mechanism vs. QoP mechanism under the events

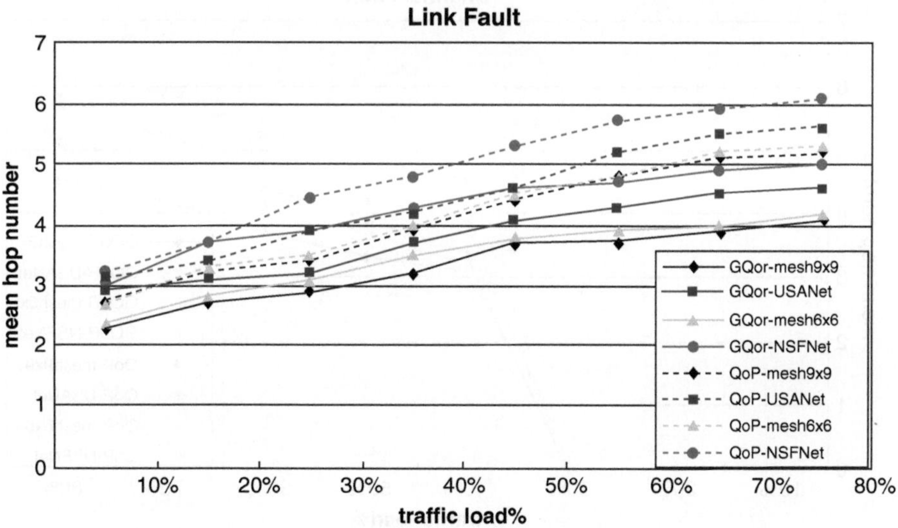

Fig. 11.13 Mean hop number comparison for the proposed GQoR mechanism vs. QoP mechanism in the event of link fault

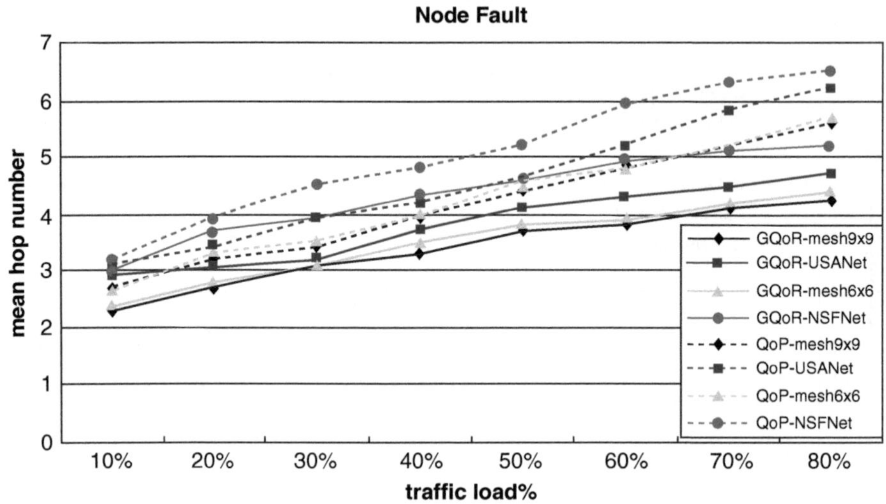

Fig. 11.14 Mean hop number comparison for the proposed GQoR mechanism vs. QoP mechanism in the event of node fault

of link failure, node failure and channel failure. The results show that the proposed GQoR mechanism has better performance in the mean hop number than that of QoP. For the failure in the protection procedure, the restoration mechanism will activate; therefore, the mean hop number may be increased. However, the proportion of running a restoration mechanism in GQoR level 3 is not high, so the mean hop number

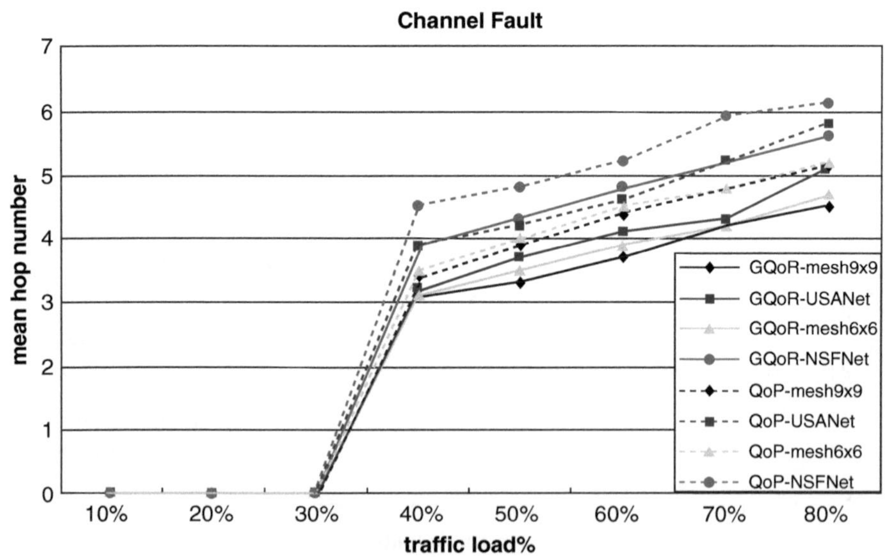

Fig. 11.15 Mean hop number comparison for the proposed GQoR mechanism vs. QoP mechanism in the event of channel fault

is still low overall. There are about 0.5 hop differences in the cases of link and node failure for the same topology as shown in Figs. 11.13 and 11.14. In channel failure, because resources are sufficient and the failure can be recovered by wavelength converting, the mean hop number is similar in these two cases when traffic load is less than 40%. Furthermore, some paths need to be recovered when traffic load is greater than 40%, so the results are more apparent and the difference in these two cases is about 0.5 hops in the same topology.

11.5 Conclusion and Future Work

In this study, a guaranteed quality of recovery (GQoR) mechanism is proposed. Four classes of GQoR level are applied according to the customer's request, and each of them is mapped to the adaptive recovery methodology. Once a fault occurs, the control system can select the recovery method which corresponds to the GQoR level. If the protection procedure fails, the proposed algorithm will execute the restoration mechanism to recover again. Consequently, there are two opportunities to recover when a failure occurs, and the recovery success rate is increased. The other contribution for the proposed mechanism is to create or to reserve a new backup path to certify networking recoverability when the original backup path is used. In this study, the shared segment recovery and distributed control techniques are applied to the proposed mechanism, so the performance of the recovery time and the bandwidth utilization can be improved. For these reasons, the data loss rate and the system building cost are reduced. The simulation results reveal that the proposed mechanism has greater performance of blocking probability and mean hop number than those of the other QoP methods. These results can be explained that the segment protection algorithm performs better than path protection algorithm, and the restoration mechanism follows if the protection procedure fails. This research proposes a fault recovery service model for WDM mesh networks and the proposed method can be practically implemented to embed in the network management system. Moreover, the potential for further research is significant on the mathematic model analysis and may involve cooperating with and intelligent network management.

References

1. C.A. Brackett, Dense Wavelength Division Multiplexing Networks: Principles and Applications, *IEEE Journal on Selected Areas in Communications*, **8**(6), 948–964 (1990).
2. J.R. Kiniry, Wavelength Division Multiplexing: Ultra High-Speed Fiber Optics, *IEEE Internet Computing*, **2**(2), 13–15 (1998).
3. J. Wang, L. Sahasrabuddhe, and B. Mukherjee, Path vs. Subpath vs. Link Restoration for Fault Management in IP-over-WDM Networks: Performance Comparisons using GMPLS Control Signaling, *IEEE Communication Magazine*, **40**(11), 2–9 (2002).
4. S. Lee, D. Griffith, and N.O. Song, A New Analytical Model of Shared Backup Path Provisioning in GMPLS Networks, *Photonic Network Communications*, **4**(3/4), 271–283 (2002).
5. D. Papadimitriou et al. Inference of Shared Risk Link groups, Internet Draft RFC 5378 (2008).

6. P. Veitch, I. Hawker, and G. Smith, Administration of Restorable Virtual Path Mesh Networks, *IEEE Communications Magazine*, **34**(12), 96–102 (1996).
7. R. Ramamurthy et al. Capacity Performance of Dynamic Provisioning in Optical Networks, *Journal of Lightwave Technology*, **19**(1), 40–48 (2001).
8. O. Gerstel and R. Ramaswami, Optical Layer Survivability – An Implementation Perspective, *IEEE Journal on Selected Areas in Communications*, **18**(10), 1885–1899 (2000).
9. O. Crochat, and J.L. Boudec, Design protection for WDM optical networks, *IEEE Journal on Selected Areas in Communications*, **16**(7), 1158–1165 (1998).
10. Y. Miyao, and H. Saito, Optimal Design and Evaluation of Survivable WDM transport networks, *IEEE Journal on Selected Areas in Communications*, **16**(7), 1190–1198 (1998).
11. P. Gadiraju and H.T. Mouftah, Channel Protection in WDM Mesh Networks, *IEEE Workshop on High Performance Switching and Routing*, 26–30 (2001).
12. H.J. Shie, Quality of Protection (QoP) Guarantee in WDM Mesh Network, *M.S. Thesis of Department of Computer Science and Engineering*, Yuan-Ze University, (2004).
13. I.S. Hwang, I.F. Huang, and C.C. Chien, A Novel Dynamic Fault Restoration Mechanism using Multiple Rings Approach in WDM Mesh Network, *Photonic Network Communications*, **10**(1), 87–105 (2005).
14. C.V. Saradhi, and C.S.R. Murthy, Segmented Protection Paths in WDM Mesh Networks, *Workshop on High Performance Switching and Routing*, 311–316 (2003).
15. R. He, H. Wen, G. Wang, and L. Li, Dynamic Sub-Path Protection Algorithm for Multi-Granularity Traffic in WDM Mesh Networks, *International Conference on Communication Technology*, 1, 697–701 (2003).
16. D. Xu, Y. Xiong, and C. Qiao, Novel Algorithms for Shared Segment Protection, *IEEE Journal on Selected Areas in Communications*, **21**(8), 1320–1331 (2003).

Chapter 12
TCP-Oriented Restoration Objectives for SONET/SDH Networks

Qiang Ye and Mike H. MacGregor

Abstract The de facto requirement in SONET/SDH is to restore failures in 50 milliseconds or less – this was derived from the requirements of conventional telephone traffic. Unfortunately this same standard has been forced onto the SONET/SDH transport systems supporting the Internet. In today's Internet, the majority of the bandwidth is consumed by P2P file transfer using TCP as the transport layer protocol. Network operators have consistently reported that up to 80% of the total traffic in their networks is P2P traffic. This percentage is expected to increase significantly in the near future because of subscriber adoption and increasing file sizes. Thus, the proper restoration objective for SONET/SDH networks carrying Internet traffic should be based on the requirements of TCP-based P2P file transfer. In this study we consider the reaction of TCP to a failure in a continental-scale network. Our goal is to determine whether there are particular values for failure duration at which file transfer times increase markedly. Such values would indicate significant objectives for the restoration of SONET/SDH networks. We studied the resilience behavior of SACK, NewReno, and Reno TCP in the case of a single TCP session and multiple TCP flows. Our experimental results show that the 50 millisecond target is overly aggressive. Considering the current migration of client access from low-rate ADSL to high-rate Fast Ethernet or ADSL2+ and receive windows from 16 KB to 64 KB or even larger we recommend 1 second as the restoration target for Internet backbone links.

Keywords Restoration objectives · TCP · SONET/SDH · Internet backbone

12.1 Introduction

The internet is a revolutionary technology that has changed our life dramatically. After being around for several decades, it has become the information infrastructure supporting various critical aspects of our daily life, such as banking and finance,

Q. Ye (✉)
Department of Computer Science and Information Technology, UPEI, Charlottetown, PE, Canada C1A 4P3

M. Ma. (ed.), *Current Research Progress of Optical Networks*,
DOI 10.1007/978-1-4020-9889-5_12,

government services, etc. As we move more and more critical applications onto the Internet, a high-quality service is expected by Internet users. However, Internet malfunctions that result in packet losses are not uncommon despite the popularity of the Internet. Physical failures are relatively frequent in the Internet: a top-tier carrier will, on average, experience one fiber failure every three hours [1]. Of course, simple congestion events in the Internet are even more common than failures. Thus, Internet restoration, the ability to recover quickly from Internet malfunctions, has been a very important issue.

SONET/SDH has been the dominant technology used to build Internet backbones. The restoration capability of SONET/SDH determines how efficiently the Internet can recover from malfunctions. The default restoration objective in SONET/SDH is for restoration to occur in 50 milliseconds or less [2, 3]. However, this traditional 50 msec objective was originally adopted as the result of considering the impact of outage duration on voice calls in traditional telephone networks. Outages of greater than 50 msec will likely result in many calls being dropped, due to various voice switch design parameters. Once these calls have been dropped, there is the potential for an inrush of reattempts which has the potential to overload and crash the switching network. Although the 50 msec requirement was important to traditional telephone networks, the same considerations do not necessarily apply to Internet traffic. Despite this, the same 50 msec objective has been assumed in the development of data networks. Now that the volume of data traffic has surpassed voice we would like to know whether this target is too liberal or too exacting, in the context of providing service for Internet traffic over SONET/SDH networks.

In the Internet, there are many different applications. Some of them, such as HTTP and email, use TCP as the transport layer protocol. For others, such as online audio and video, UDP is the default transport protocol. In recent years, with the extreme success of Peer-to-Peer (P2P) applications, the traffic mix in the Internet has changed much [4–6]. A few months after Napster offered the platform for file sharing in 1999, more than 20% of the traffic on IP networks in the US became Napster P2P traffic [4]. Cisco estimated that nowadays 70% or more of broadband bandwidth is consumed by P2P downloads of music, video, games, etc [5]. Network operators have consistently reported that a very large portion of the total traffic in their networks is P2P traffic, sometimes P2P traffic even reaches 80% at non-peak times [6]. Bandwidth consumption by P2P will likely arrive at a higher percentage in the near future because of subscriber adoption and increasing file sizes. Note that almost all P2P traffic results from large file transfer using TCP as the transport layer protocol. Surely when we propose restoration objectives for Internet backbone links, we should take into account both TCP-based and UDP-based applications. However, considering the fact that the majority of current bandwidth is consumed by P2P file transfer using TCP, we should give special attention to the restoration objective required by TCP-based file transfer.

This paper focuses on the restoration objectives for backbone links from the perspective of TCP-based file transfer. That is, the goal of this study is to find out the restoration requirements of file transfer applications. Since all these applications use TCP as the transport layer protocol, essentially our goal is to study how TCP resilience mechanisms react to outages in the absence of any other compensating mechanisms such as rerouting. These results are fundamental to designing any restoration mechanisms for SONET/SDH networks carrying Internet traffic.

In this paper we first consider the reaction of a single TCP session to network link failures. Interactions between multiple TCP flows in the case of network failure are presented afterwards. Our experimental results show that the traditional 50 msec objective is overly aggressive. For different client access rates, varied values can be chosen as the restoration objectives. For low-rate access, such as Dial-Up and DS0, 200 msec, instead of 50 msec, should be set as the restoration objective. For medium-rate access, such as ADSL and DS1, 100 msec is a more appropriate objective. For high-rate access, such as Fast Ethernet and ADSL2+, 1 second should be used as the restoration objective. Considering the current migration of client access from low-rate ADSL to high-rate Fast Ethernet or ADSL2+ and receive windows from 16 KB to 64KB or even larger we recommend 1 second as the restoration target for Internet backbone links.

The rest of the paper is organized as follows. Section 12.2 gives the background of TCP resilience mechanisms and In Section 12.3 we discuss the behavior of TCP in the case of network failures. Section 12.4 contains our detailed recommendations on TCP-oriented restoration objectives for SDH/SONET networks. The paper closes with our conclusions and recommendations in Section 12.5.

12.2 Resilience Mechanisms in TCP

TCP does not have any resilience mechanisms that are designed specifically to deal with network failures. From the viewpoint of TCP, there is no difference between network failure and network congestion. As a result, when part of the network fails and some segments are dropped, TCP will assume that there is congestion somewhere in the network, and the TCP congestion control mechanisms will start dealing with the segment loss.

TCP congestion control mechanisms have improved over time. The main versions of TCP are Tahoe TCP, Reno TCP, NewReno TCP and SACK TCP. Tahoe TCP is the oldest version and only a few old systems use it. Reno TCP, NewReno TCP and SACK TCP are widely implemented [7]. This paper focuses on SACK, NewReno and Reno TCP because they are the newer versions and are more widely deployed. Details about TCP congestion control can be found in [8–12]. In our experiments, the TCP implementation conforms to the one illustrated in [11].

With the mechanism of cumulative acknowledgements used in Reno TCP, a Reno TCP sender can only learn about a single lost segment per round trip time (RTT). Thus Reno TCP may experience poor performance in the case of multiple segment losses. NewReno TCP includes a small but effective change to Reno TCP that eliminates Reno's wait for timeout when multiple segments are lost in a transmission window. When multiple segments are lost in a transmission window, NewReno TCP can recover without a timeout, retransmitting one lost segment per one round trip time.

SACK TCP is also an enhanced version of Reno TCP. Selective Acknowledgement (SACK), together with selective repeat retransmission, can help improve TCP performance when multiple segments are dropped within one window of data. Selective Acknowledgement is achieved by adding a list of the contiguous blocks that have been received by the receiver. When a valid segment that is in the receive window but not at the left edge (i.e. not the next expected segment) arrives at a TCP receiver, the receiver sends back a selective acknowledgement to inform the sender that non-contiguous blocks of data have been received.

Congestion control in SACK, NewReno and Reno TCP is composed of three phases: slow start, congestion avoidance and fast retransmit/fast recovery. Three state variables, *cwnd* (congestion window), *rwnd* (receiver's advertised window) and *ssthresh* (slow start threshold), are maintained at the sender to deal with network congestion. In addition, SACK TCP has an extra variable called *pipe* at the sender that represents the estimated number of outstanding segments. SACK TCP also has a data structure called *scoreboard* at the sender side that keeps track of the contiguous data blocks that have arrived at the receiver. Retransmission timeout (RTO) is an important parameter in TCP congestion control. It has a minimum of one second and RFC 2988 [13] suggests that a maximum value may be placed on RTO. In our simulation, this maximum value is 64 seconds.

12.3 TCP Resilience Behavior

Understanding the resilience behavior of TCP is the first step to recommending TCP-oriented objectives. The general behavior of SACK, NewReno, and Reno TCP in the case of network failures is presented in this section. Some additional details of TCP resilience for different scenarios will be described in Section 12.4.

12.3.1 Simulation

In our research, we studied the behavior of TCP in the case of single TCP flow and multiple TCP flows. In the case of single TCP flow, a client and server are connected across a continental-scale simulation network. Each node is connected to a local router via a high-speed LAN link. The local routers are connected to the core network via access links. Three access link rates are commonly used in real-life systems: DS0 (64 Kbps), DS1 (1.544 Mbps) and OC-3c (155 Mbps). In terms of bandwidth, these digital access rates, DS0, DS1, and OC-3c, are comparable to Dial-Up, ADSL, and Fast Ethernet in our daily life, respectively. Based on the fact that servers are usually connected to the Internet via high-speed links while client-side access link rates vary a lot, our simulation fixes the server-side access at OC-3c and varies the client-side access link from DS0 to DS1 and OC-3c. The core network in our simulation has an NSFNET-like topology shown in Fig. 12.1. Core routers (Cisco 12008) are connected via OC-192 (10 Gbps), which is common in backbone networks nowadays. The client resides in Palo Alto and the server is located at Princeton.

As shown in Fig. 12.1, a packet discarder model, used to simulate outages, is in the middle of the link connecting Salt Lake City to Palo Alto. We can specify

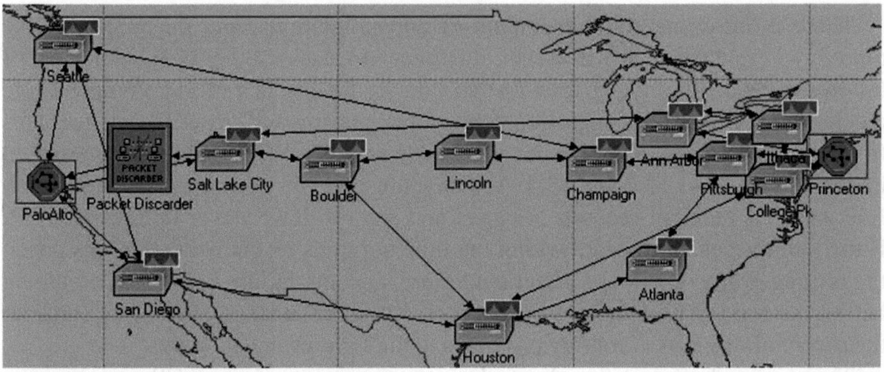

Fig. 12.1 Experimental network topology

either the number of packets to be dropped or a certain time period during which all packets are dropped. Our experiments simulate a unidirectional failure of packets going from Salt Lake City to Palo Alto (i.e. in the server-to-client path). Packets traveling the other way get to their destination safely. A unidirectional failure would be unusual in a transport network. However, this assumption was made by many network researchers to reflect the reality of today's Internet: routes for IP packets are often asymmetric [14]. Thus a failure in the underlying network will often only affect a session in one direction.

In the case of multiple TCP flows, there are eight TCP sessions altogether in the NSFNET-like simulation network. The eight clients are all attached to the router at Palo Alto via DS1 (Comparable to ADSL). OC-3c is used to connect the eight servers with the routers at Boulder, Lincoln, Champaign, Pittsburgh, Princeton, College Park, Ann Arbor, and Houston, respectively. Figure 12.2 illustrates the details. The dashed lines in Fig. 12.2 indicate that these lines represent multiple hops rather than a single link.

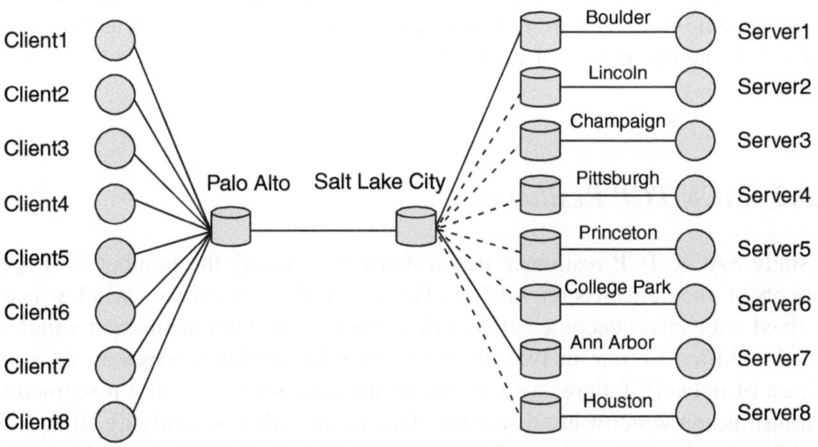

Fig. 12.2 The case of multiple TCP flows

There is only one routing domain in our simulations, and the NSFNET-like topology is relatively old. However, this paper focuses on the TCP-layer view of failures. That is, this paper tries to find out, in the absence of any compensating mechanisms, how TCP congestion control mechanisms will react to outages. This first-step experiment generated many valuable results, some of which are presented in detail in Sections 12.3 and 12.4. Actually, it usually takes routing protocols (both IGP and EGP) tens of seconds to detect and react to lower-layer failures [15, 16]. If the failures can be restored within the time horizons recommended in this paper, the routing protocol will not detect the failure, and any failure will be restored long before the routing protocol could converge. For these reasons, we do not consider the potential reaction of routing protocols to the failures under study.

The receive buffer at the client plays an important role in TCP performance. *rwnd* is actually a parameter indicating the available space in the receive buffer. Without failures, we can assume *rwnd* is equal to the receive buffer for simplicity. During a TCP session, the sending TCP continuously compares the outstanding unacknowledged traffic with *cwnd* and *rwnd*. Whenever the outstanding traffic is less than the smaller of these two variables by at least one SMSS (sender maximum segment size), the sender will send out some segments if there are any waiting to be sent. Generally the receive buffer size (*rbuff*) is set as:

$$rbuff = \text{bandwidth} * \text{round-trip-time} = r^{*}\tau \qquad (12.1)$$

where r stands for bandwidth and τ is RTT. This is commonly called the bandwidth-delay product [9].

The TCP session in our simulation must be long enough to test scenarios with varying failure durations. We chose FTP as the application-layer protocol and made the transmitted file large enough to fulfill this requirement. For DS0, DS1 and OC-3c client-side access links, we used 5 MB, 10 MB and 20 MB files, respectively. In reality, the duration of TCP flows covers a very large range. However, as P2P applications become more and more popular, long TCP flows are expected to account for a very large percentage of the total traffic. This paper focuses on long-running TCP flows.

12.3.2 SACK TCP Resilience

We study SACK TCP resilience performance by varying the number of dropped segments in one transmission window. The size of the transmission window is usually fixed once *cwnd* becomes greater than *rwnd*. And the transmission window is normally shifted by one or two segments after the sender receives an ACK. For the case of network failure, we considered the case where the first n segments in the transmission window are discarded. This is equivalent to studying all the other possible cases (loss of segments 2 through n + 1, 3 through n + 2, etc.) because the reaction of TCP is invariant under such shifts.

In our research, we use Transfer Time Increase (TTI) to quantify the impact of a network failure:

$$\text{TTI} = \text{ATT} - \text{NTT} \qquad (12.2)$$

where ATT stands for Actual Transfer Time in the case of network failure, and NTT means Normal Transfer Time in the case of no network failure.

We use a scenario with DS1 (Comparable to ADSL in our daily life) client access and 32 KB receive buffer as a typical example to illustrate the general behavior of TCP in the case of network failures. In our experiment, SMSS (Sender Maximum Segment Size) was set to 1460 bytes. The values of *cwnd*, *rwnd*, and *ssthresh* were initialized to 1460 bytes, 32 KB, and 64 KB, respectively. By 32 KB or 64 KB, we mean a multiple of SMSS that is just above 32 KB or 64KB. For example, in our model, SMSS is 1460 bytes, so by 32 KB, we mean $1460 * 23 = 33580$ bytes. For clarity, we consider 1460 byte segments as the data units. Thus, *cwnd*, *rwnd*, and *ssthresh* were set to 1, 23, and 45, respectively.

In our example, a long SACK TCP session starts at time 0 and the Packet Discarder begins to drop packets at 30 seconds, as shown in Fig. 12.3. The four curves in Fig. 12.3 illustrate the changes in the sender's congestion window over time in four different cases. Before the failure starting point (marked by "X"), the four curves overlap each other because during that period they describe essentially the same conditions. After point "X", they split into four different curves. These four curves correspond to four resilience cases.

Case 1: No Dropped Segments

The 0 drop curve corresponds to the scenario that no segment is discarded. Thus SACK goes through the normal state changes in this case. That is, at the beginning,

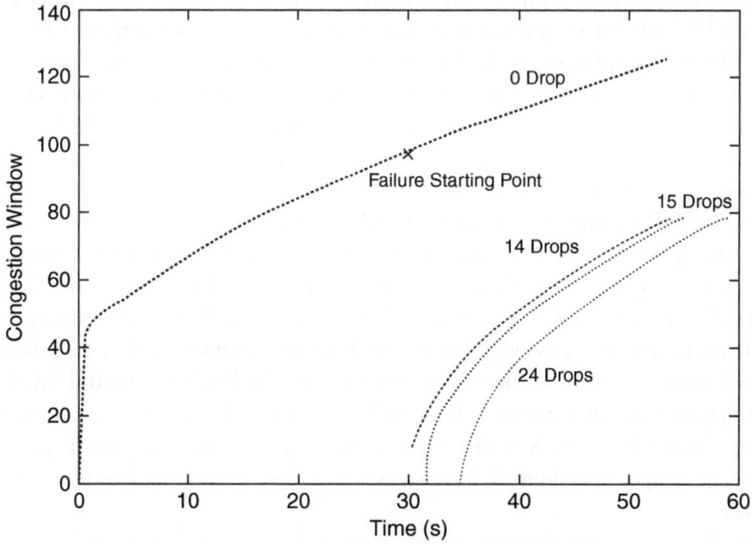

Fig. 12.3 Congestion window vs. transfer time (SACK TCP DS1-32K case)

the congestion window is initially set to one and TCP is in slow start. *cwnd* increases exponentially as the sender receives acknowledgements until *cwnd* equals *ssthresh* (initially *ssthresh* is 45). Then TCP transitions into congestion avoidance during which *cwnd* increases by one every RTT. The turning point on the curve marks the start of this slow increase period. Since *cwnd* increases much faster before the turning points, the points before the turning point are spread much more sparsely compared with the points after the turning point. Without a network failure, TCP stays in congestion avoidance until the file is completely transferred and the TCP session is terminated.

Case 2: Dropped Segments, Duplicate ACKs, and No Timeout

Now we consider the cases in which some segments are lost during a failure. In this scenario, although the first several segments in the transmission window are discarded, the rest of the segments will arrive at the client and trigger duplicate ACKs. After the sender receives three duplicate ACKs, TCP transitions into fast retransmit/fast recovery. It sets *pipe* to the number of outstanding segments and retransmits the earliest unacknowledged segment. In normal conditions, the number of outstanding segments should be equal to *rwnd*. But when fast retransmit/fast recovery occurs, the sender assumes that one segment has been dropped and hence sets *pipe* to (*rwnd* − 1). In this case, *pipe* is set to $(23 - 1) = 22$. The sender also sets *ssthresh* to *rwnd*/2 and sets *cwnd* to (*ssthresh* + 3) [10]. Given the method used to calculate *ssthresh*, we have the following relation between *cwnd* and *pipe*:

$$cwnd = \lfloor rwnd/2 \rfloor + 3 \tag{12.3}$$

During fast recovery, *pipe* is increased by one when the sender either retransmits an old segment or transmits a new segment, and it is decreased by one for each additional duplicate ACK. For each partial ACK, *pipe* is decreased by two rather than one because each partial ACK in fact represents two segments that have left the transmission link: the original segment that is assumed to be lost and the retransmitted segment. When *pipe* becomes less than *cwnd*, the sender will check the *scoreboard* and either retransmit the earliest unacknowledged segment or transmit a new segment when there are no unacknowledged segments.

We use n_D to denote the number of duplicate ACKs that have been received by the sender by the time that *pipe* has just become less than *cwnd*. Note that after TCP transitions into fast retransmit/fast recovery, TCP immediately retransmits the earliest unacknowledged segment, thus *pipe* is increased by one due to this retransmission. Also, this retransmission will finally lead to a partial ACK if the ACK triggered by the retransmission will not take TCP out of fast retransmit/fast recovery. The partial ACK will decrease *pipe* by two. As mentioned previously, *pipe* is set to (*rwnd*-1) when TCP transitions into fast retransmit/fast recovery. Thus we have:

$$((rwnd - 1) + 1 - 2) - n_D = cwnd - 1 \tag{12.4}$$

So:

$$n_D = rwnd - cwnd - 1 \tag{12.5}$$

From Eq. (12.3) and (12.5), we have:

$$n_D = rwnd - (\lfloor rwnd/2 \rfloor + 3) - 1 = rwnd - \lfloor rwnd/2 \rfloor - 4 \tag{12.6}$$

If we use $n_{C,S}$ to denote the critical number of lost segments in this case, we arrive at:

$$n_{C,S} = rwnd = n_D = \lfloor rwnd/2 \rfloor + 4 \tag{12.7}$$

In the normal state the receiver only sends out an ACK for every second full-sized segment, or within 200 ms of the arrival of the first unacknowledged segment. Also, out-of-order segments should be acknowledged immediately. Thus when the first out-of-order segment in the window arrives, if there is no unacknowledged segment at the receiver, this segment will trigger the receiver to send out a duplicate ACK. We call this case Type I Failure. If instead the first out-of-order segment arrives within 200 msec of an unacknowledged segment, the receiver will not send out a duplicate ACK. The receiver will only transmit an acknowledgement of the previously-unacknowledged segment. Each segment following the first out-of-order segment results in a duplicate ACK. We call this case Type II Failure. Figure 12.4

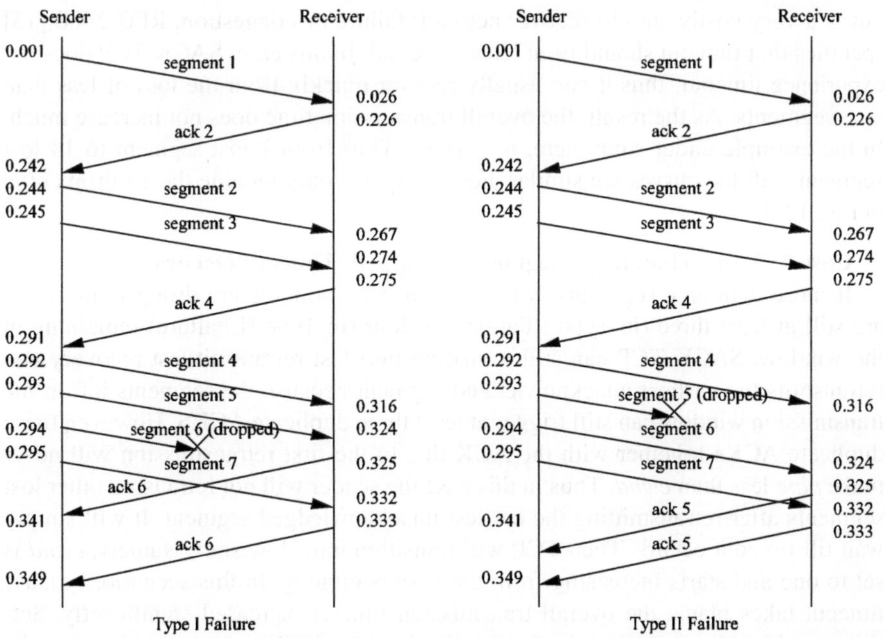

Fig. 12.4 Data flow of Type I Failure and Type II Failure (in seconds)

presents the initial data flow of two failure scenarios and illustrates the difference between Type I Failure and Type II Failure. The case where the sixth segment is lost, and that where the fifth segment is lost correspond to Type I and Type II Failure, respectively. Equation (12.7) above applies to Type I Failure. However, for Type II Failure, Equation (12.7) must be modified slightly, decreasing $n_{C,S}$ by one to account for the segment triggering the ACK for the previously-unacknowledged segment. Thus, we have:

$$n_{C,S} = \begin{cases} \lfloor rwnd/2 \rfloor + 4 \text{ in Type I Failure} \\ \lfloor rwnd/2 \rfloor + 3 \text{ in Type II Failure} \end{cases} \quad (12.8)$$

Note that the current example illustrates a Type II failure so that $n_{C,S}$ is $23/2 + 3 = 14$.

If fewer than $n_{C,S}$ segments within the transmission window are lost, there are still many segments arrive at the receiver, triggering a large quantity of duplicate ACKs. These duplicate ACKs, together with the possible partial ACK due to the first retransmission when TCP transitions into fast retransmit/fast recovery, are enough for the sender to make *pipe* less than *cwnd*. Then the sender can retransmit other lost segments after retransmitting the earliest unacknowledged segment when it first switches into fast retransmit/fast recovery. The sender keeps sending segments until a non-duplicate ACK arrives acknowledging all data that was outstanding when fast retransmit/fast recovery was entered. Then TCP exits fast retransmit/fast recovery, switches into congestion avoidance and gets back to its normal state. For TCP, timeout is a very costly step to recover network failure or congestion. RFC 2988 [13] specifies that timeout should be at least 1 second. In this case, SACK TCP does not experience timeout, thus it can usually recover quickly from the loss of less than $n_{C,S}$ segments. As the result, the overall transmission time does not increase much. In the example under study here, $n_{C,S}$ is 14. Thus from 1 lost segment to 14 lost segments, all the curves are similar. For clarity, we only include the 14-drop curve in Fig. 12.3.

Case 3: More Than $n_{C,S}$ Segments Dropped, Timeout Occurs

If more than $n_{C,S}$ segments in the transmission window are dropped and there are still at least three (in Type I Failure) or four (in Type II Failure) remaining in the window, SACK TCP can still transition into fast retransmit/fast recovery and retransmits the earliest unacknowledged segment because the segments left in the transmission window can still trigger at least three duplicate ACKS. However, these duplicate ACKs together with the ACK due to the first retransmission will never make *pipe* less than *cwnd*. Thus in this case the sender will not retransmit other lost segments after retransmitting the earliest unacknowledged segment. It will simply wait till timeout occurs. Then TCP will transition into slow start. Namely, *cwnd* is set to one and starts increasing from the very beginning. In this scenario, because timeout takes place, the overall transmission time is increased significantly. Setting *cwnd* back to one after the failure also impairs TCP performance because the network bandwidth is not fully utilized when *cwnd* is very small. This also has an impact on the overall transmission time. In short, SACK TCP suffers much in this scenario.

If less than three (in Type I Failure) or four (in Type II failure) segments remain in the transmission window, fast retransmit/fast recovery will not occur because there will be less than three duplicate ACKs. This also leads to timeout. When the retransmission timer expires TCP will transition into slow start. Although TCP does not experience fast retransmit/fast recovery in this scenario, we still categorize it into Case 3 because timeout is the main factor, and thus there is not a significant difference between these two scenarios in terms of TTI. Hence, from 15 lost segments to 23 lost segments, all the curves are similar to the 15-drop curve. We only include the 15-drop curve in Fig. 12.3 for clarity.

Case 4: Retransmitted Segment Also Dropped

If the network failure lasts long enough so that the segment retransmitted due to timeout is also dropped, things change again. This is because when the retransmitted segment is sent out, the retransmission timer has been doubled. If the retransmission fails, the sender will wait for twice the previous RTO before timing out and retransmitting the earliest unacknowledged segment again. Waiting for twice the previous RTO increases TTI significantly. This corresponds to the 24-drop curve in Fig. 12.3. If the repeated retransmission does not succeed, the sender has to wait for four times the original RTO to retransmit a third time. This process goes on until TCP gives up this connection. For clarity, we did not include the curves of 25, 26, etc. dropped segments.

The segments discarded after the first 23 segments are actually out of the current transmission window. However, since they are not normal segments but some retransmitted segments due to timeout, and the loss of them has a significant impact on SACK TCP resilience, we include the scenarios of discarding these segments in this paper for the purpose of completeness.

Figure 12.5 presents TTI vs. the number of dropped segments. From 1 drop to 14 drops, TTI does not increase much. When there are 15 drops, TTI increases dramatically. 24 drops leads to another significant increment. The two serious changes in TTI correspond to the two timeouts.

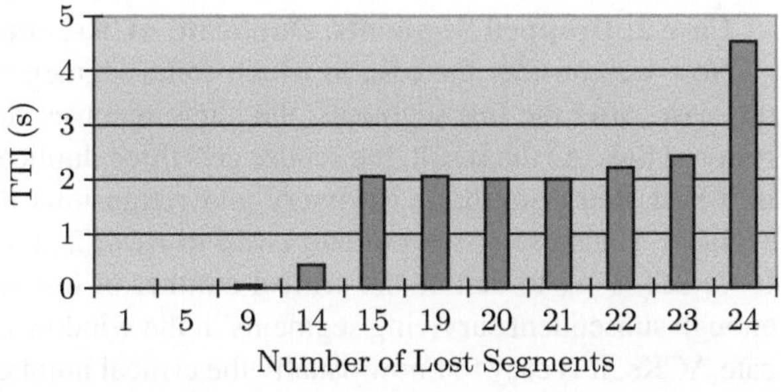

Fig. 12.5 TTI vs. Number of Lost Segments (SACK TCP DS1-32K case)

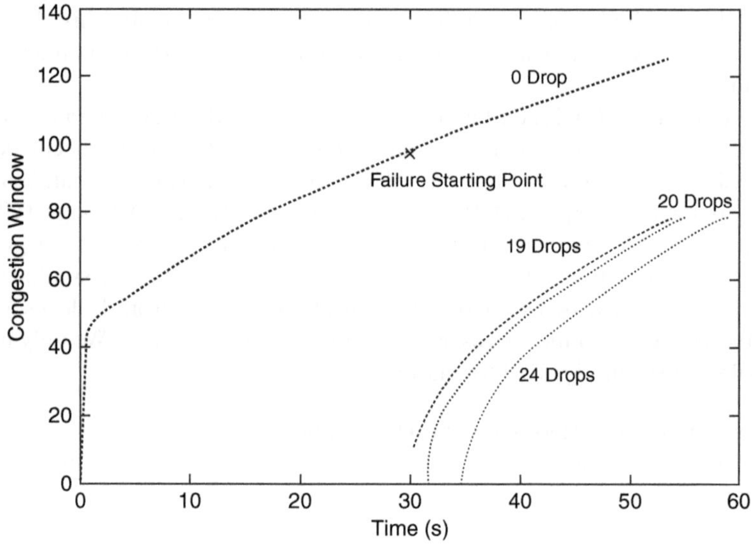

Fig. 12.6 Congestion Window vs. Transfer time (NewReno DS1-32K case)

12.3.3 NewReno TCP Resilience

For NewReno TCP, a similar experimental setup is used, but different experimental results are obtained. As shown in Fig. 12.6, a long NewReno TCP session also starts at time 0 and the Packet Discarder begins to drop packets at 30 s. The four curves in Fig. 12.6 illustrate the changes in the sender's congestion window over time in four different cases. Again, before the failure starting point (marked by "X"), the four curves overlap each other; after point "X", they transition into four different curves. These four curves also correspond to four resilience cases.

Case 1: No Dropped Segments
The curve labeled "0 Drop" is the same one illustrated in Section 12.3.2. Without network failures SACK and NewReno TCP behave in the same way.

Case 2: Dropped Segments, Duplicate ACKs, and No Timeout
Now we consider the case in which some segments are lost during a failure. In this case, after the lost segments, the client receives subsequent segments over the restored link. As the result, the sender gets three duplicate ACKs. Then it transitions into Fast Retransmit/Fast Recovery and retransmits the earliest unacknowledged segment. It also sets *ssthresh* and *cwnd* to *rwnd*/2 and $(rwnd/2 + 3)$, respectively. If we use $n_{C,NR}$ to denote the critical number of lost segments when there are just enough subsequent surviving segments in the window of data to trigger three duplicate ACKs, it is easy to know usually the critical number is equal to (*rwnd* - 3). Due to the TCP acknowledging mechanism illustrated in Section 12.3.2, we can get the

following formula:

$$n_{C,NR} = \begin{cases} rwnd - 3 \text{ in Type I Failure} \\ rwnd - 4 \text{ in Type II Failure} \end{cases} \tag{12.9}$$

Note that the NewReno DS1-32K example also illustrates a Type II failure, thus $n_{C,NR}$ is $23 - 4 = 19$.

If less than $n_{C,NR}$ segments in a window of data are lost, enough surviving segments can arrive at the receiver and trigger enough duplicate ACKs to make TCP transition into Fast Retransmit/Fast Recovery. In this case, the earliest unacknowledged segment is retransmitted and the retransmission leads to a partial ACK. The partial ACK will then make the sender retransmit the earliest unacknowledged segment at that moment. This retransmitted segment will result in another partial ACK, and thus leads to another retransmission. This process goes on until a non-duplicate ACK arrives acknowledging all data that was outstanding when TCP transitioned into Fast Retransmit/Fast Recovery, then TCP switches into Congestion Avoidance by setting *cwnd* back to *ssthresh*. We should note that each time the sender receives a partial ACK, it does one retransmission and thus recovers one lost segment. Namely, it takes NewReno TCP a whole RTT to recover one lost segment. Thus in a sense, RTT determines the final TTI value. If RTT is comparatively long, TTI increases dramatically with the number of lost segments; otherwise, TTI almost remains unchanged. In the NewReno DS1_32K case, RTT is relatively small, thus TTI does not increase much. In this case $n_{C,NR}$ is 19, so from 1 lost segment to 19 lost segments, all the curves are similar. For clarity, we only include the 19-drop curve in Fig. 12.6.

SACK TCP has a different mechanism to deal with partial ACKs. In Section 12.3.2, we have mentioned that *pipe* is decremented by one for each additional duplicate ACK, but it is decreased by two rather than one for each partial ACK. This additional decrease in *pipe* results in a faster recovery process: one partial ACK leads to two retransmissions. The two retransmissions will trigger another two partial ACKs and eventually will lead to another four retransmissions. This process goes on until a non-duplicate ACK arrives acknowledging all data that was outstanding when TCP transitioned into Fast Retransmit/Fast Recovery. Hence, within one RTT, usually many more lost segments can be recovered with SACK TCP than with NewReno TCP. This is why with SACK TCP, TTI does not increase much when less than $n_{C,S}$ segments within one window are lost, regardless of the length of RTT. In contrast, the TTI of NewReno is influenced by RTT in this situation.

Case 3: More Than $n_{C,NR}$ Segments Dropped, Timeout Occurs

On the other hand, if more than $n_{C,NR}$ segments in a window of data are dropped, Fast Retransmit/Fast Recovery will not occur because there will not be enough duplicate ACKs. This leads to timeout. When the retransmission timer expires TCP will transition into Slow Start and retransmit the first lost segment. In this scenario, timeout plays the major role in terms of TTI, and thus the overall transfer time does not increase much with the number of lost segments. In the NewReno DS1_32K

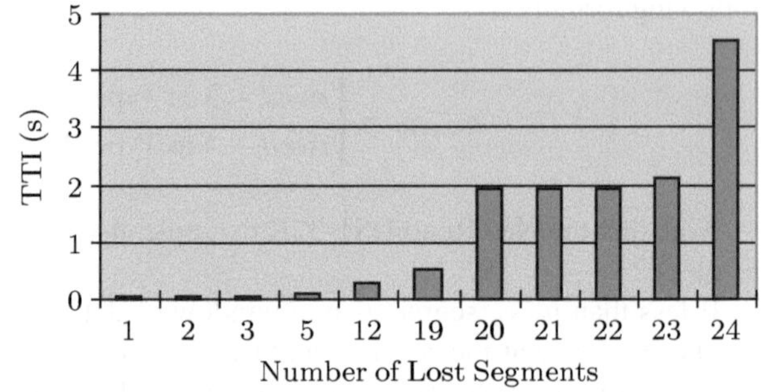

Fig. 12.7 TTI vs. No. of Lost Segments (NewReno DS1-32K case)

case, from 20 lost segments to 23 lost segments, all the curves are similar. We only include the 20-drop curve in Fig. 12.6 for clarity.

Case 4: Retransmitted Segment Also Dropped

If the network failure lasts long enough so that the segment retransmitted due to timeout is also dropped, NewReno TCP will experience the same doubling calculation that is illustrated in Section 12.3.2. A similar 24-drop curve is included in Fig. 12.6. For clarity, we did not include the curves of 25, 26, etc. dropped segments. Figure 12.7 illustrates the overall trend by presenting TTI changes vs. the number of dropped segments.

12.3.4 Reno TCP Resilience

For Reno TCP, again, we use a similar experimental setup. As shown in Fig. 12.8, a long Reno TCP session also starts at time 0 and the Packet Discarder begins to drop packets at 30 s. Before the failure starting point (marked by "X"), the four

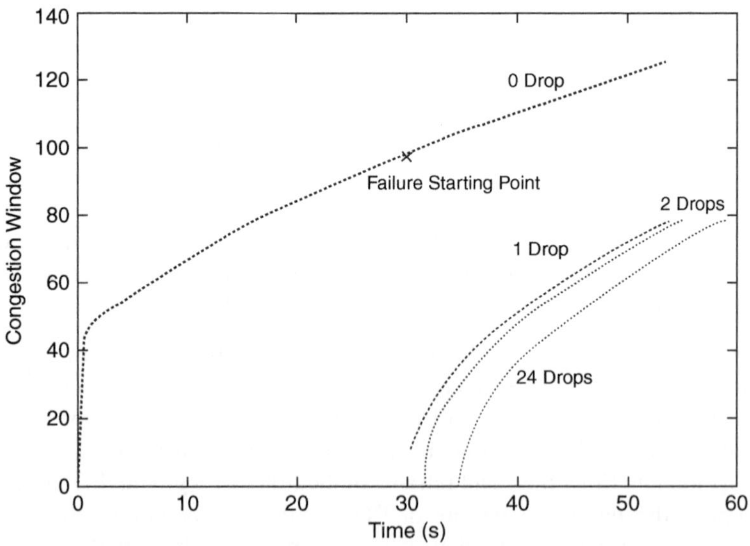

Fig. 12.8 Congestion Window vs. Transfer Time (Reno DS1-32K case)

curves overlap; after point "X", they split into four different curves, corresponding to four cases.

Case 1: No Dropped Segments

Similarly, the curve labeled "0 Drop" is the same one presented in Section 12.3.2. Without network failures SACK and Reno TCP behave in the same fashion.

Case 2: One Dropped Segments, Duplicate ACKs, and No Timeout

Now we consider the case in which one segment is lost during a failure. As the result of the client receiving subsequent segments, TCP transitions into Fast Retransmit/Fast Recovery after the sender gets three duplicate acknowledgements. The sender retransmits the earliest unacknowledged segment, sets *ssthresh* to *rwnd*/2 and sets *cwnd* to $(rwnd/2+3)$. For each additional duplicate ACK, *cwnd* increases by one. This process goes on until the retransmitted segment reaches the receiver and a new ACK acknowledgement for all outstanding segments is received by the sender. At that point *cwnd* is set back to the current *ssthresh*, and Congestion Avoidance starts again because *cwnd* is now equal to *ssthresh*. In short, Reno TCP can usually recover effectively from the loss of one segment. This is illustrated by the 1-drop curve in Fig. 12.8. In this scenario, the overall transfer time does not increase. We use $n_{C,R}$ to denote the critical number in this case. Obviously, $n_{C,R}$ is always equal to 1.

Case 3: More Than One Segment Dropped, Timeout Occurs

Losing two segments makes a difference. Before the ACK for the first retransmission is received by the sender, the recovery process is similar to that when only one segment is lost. As before, the ACK for the first retransmission only acknowledges the first lost segment. This segment has been retransmitted due to Fast Retransmit/Fast Recovery, while the second lost segment has not been retransmitted yet. The Fast Retransmit/Fast Recovery algorithm in Reno TCP assumes that only one segment has been lost, so the sender does not immediately retransmit the second lost segment. Because two segments have been lost, TCP will eventually time out, and switch into Slow Start. TCP will have to retransmit the earliest segment that has not been acknowledged, which in this case is the second lost segment. This timeout results in the large gap between the 1-drop and 2-drop curves in Fig. 12.8.

In the case of more than two lost segments, if there are still three or more non-retransmitted segments following the lost segments that arrive at the receiver, enough duplicate ACKs will reach the sender to trigger Fast Retransmit/Fast Recovery. In this scenario, losing more than two segments leads to the same recovery process as losing two segments. On the other hand, if less than three segments follow the lost segments, Fast Retransmit/Fast Recovery will not occur because there will not be enough duplicate ACKs. When the retransmission timer expires, TCP transitions into Slow Start and retransmits the first lost segment. Although TCP experiences different transitions than in the case of two lost segments, total transfer time does not increase dramatically because timeout is the main factor. Hence, from 3 lost segments to 23 lost segments, all the curves are similar to the 2-drop curve in Fig. 12.8.

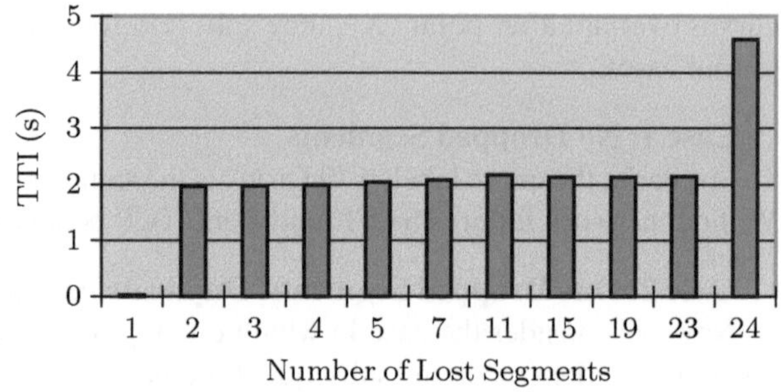

Fig. 12.9 TTI vs. Number of Lost Segments (Reno DS1-32K case)

Case 4: Retransmitted Segment Also Dropped

If the network failure lasts long enough so that the segment retransmitted due to timeout is also dropped, things change again. This is because when the retransmitted segment is sent out, the retransmission timer has been doubled. If the retransmission fails, the sender will wait for twice the previous RTO before timing out and retransmitting the earliest unacknowledged segment again. This corresponds to the 24-drop curve in Fig. 12.8. If the repeated retransmission does not succeed, the sender has to wait for four times the original RTO to retransmit a third time. This process goes on until TCP gives up this connection. For clarity, we did not include the curves of 25, 26, etc. segments dropped, but it is not difficult to imagine what they should look like in Fig. 12.8. We can observe the Reno TCP general behavior in terms of TTI in Fig. 12.9.

12.4 TCP-Oriented Restoration Objectives

After understanding the resilience behavior of TCP, we propose some restoration objectives based on TCP resilience behavior in this section. Restoration objectives for each of SACK, NewReno and Reno TCP are first proposed in Sections 12.4.1, 12.4.2, and 12.4.3. Then the experimental results in the case of multiple TCP flows are presented in Section 12.4.4 to show the recommended objectives work well for SACK, NewReno and Reno TCP in both single-TCP and multiple-TCP scenarios. Finally, the overall objectives are proposed in Section 12.4.5.

12.4.1 Restoration Objectives for SACK TCP

The general behavior of SACK, NewReno and Reno TCP was presented in Section 12.4. Some more simulation scenarios in the case of a single TCP session are described in Sections 12.4.1, 12.4.2, and 12.4.3 in order to propose appropriate restoration objectives for SONET/SDH networks. These simulations differ by the size of *rwnd*. As mentioned previously, the bandwidth-delay product, $r\tau$, is commonly used to size *rwnd*. In our simulations, the RTT for DS0, DS1 and OC-3c access is 210 ms, 41 ms and 26 ms respectively, so $r\tau$ has values of 1680, 7913 and

505440 bytes respectively. For each access link rate, we experimented with 8 different *rwnd* sizes, from 8 KB to 1024 KB. By 8 KB, we mean a multiple of SMSS that is just above 8 KB. For example, in our simulation, SMSS is 1460 bytes, so by 8 KB, we mean 1460*6 = 8760 bytes.

We have demonstrated that losing less than $n_{C,S}$ segments typically does not increase SACK TCP transfer time significantly. Losing ($n_{C,S} + 1$) segments makes a difference, and subsequent losses have little impact until it comes to the loss of the retransmitted copy. Generally speaking, this applies to all SACK TCP cases with different access link rates and varied *rwnd* sizes. The detailed resilience performance of SACK TCP in terms of "TTI vs. Outage Duration" is presented in Figs. 12.10, 12.11 and 12.12.

To link the number of lost segments to outage duration, we define SACK Level-1 Restoration Requirement ($T_{1,S}$) as the period from the moment that network failure occurs to the moment just before the segment following the dropped $n_{C,S}$ segments arrives. We also define SACK Level-1 Restoration Objective ($\tau_{1,S}$) as the period from the moment that the first dropped segment arrives at the failure point to the moment just before the segment following the dropped $n_{C,S}$ segments arrives. The subtle difference between these two concepts can be illustrated using the sample segment arrival pattern in Fig. 12.13. In this example, a network failure occurs at 0.5 ms. The last segment that leaves the failure point before the failure occurs arrives at the failure point at 0 msec. The first segment that is dropped due to this failure arrives at the failure point at 1 ms. After this, subsequent segments arrive at the failure point every 1 ms. Furthermore, we assume that $n_{C,S}$ is equal to 6. Apparently, in order to avoid losing more than $n_{C,S}$ segments, the network should be restored

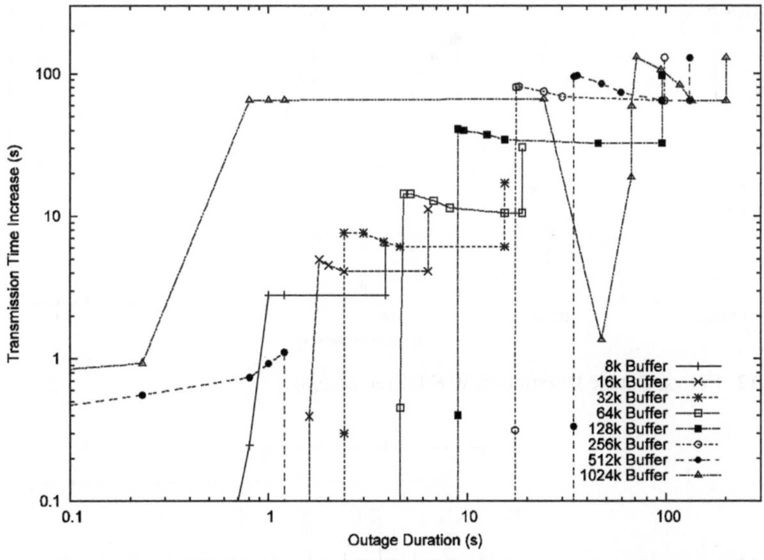

Fig. 12.10 TTI vs. Outage Duration (SACK DS0 access)

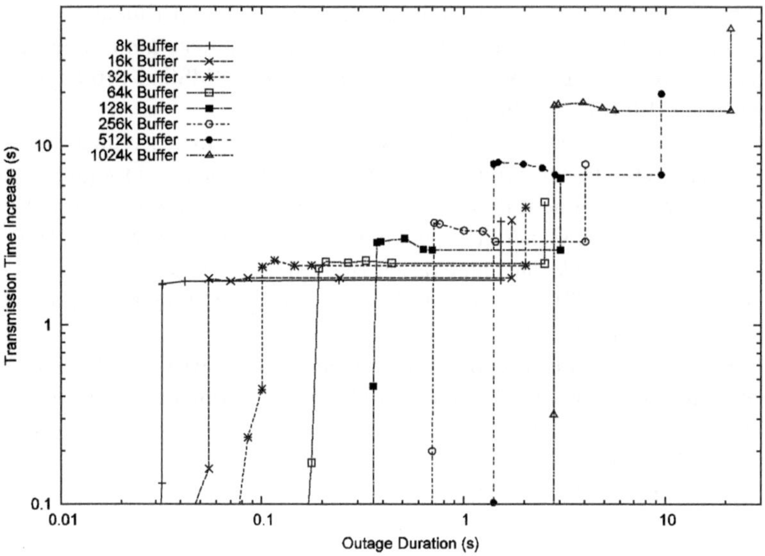

Fig. 12.11 TTI vs. Outage Duration (SACK DS1 access)

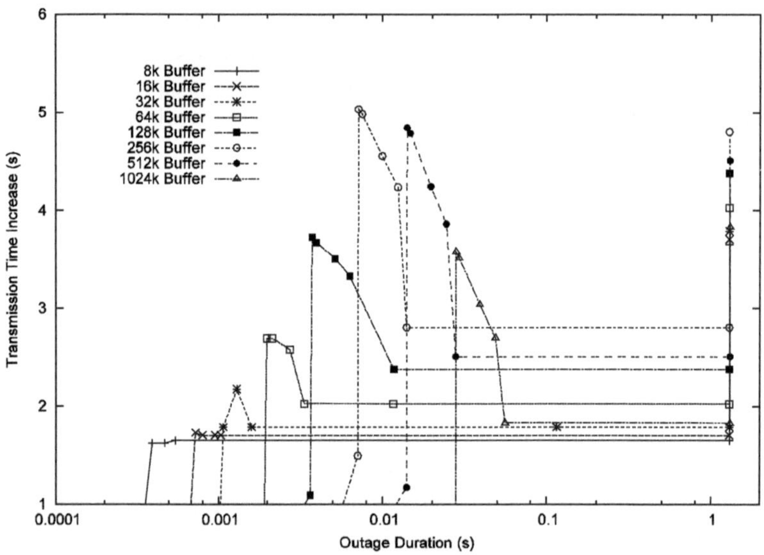

Fig. 12.12 TTI vs. Outage Duration (SACK OC-3c access)

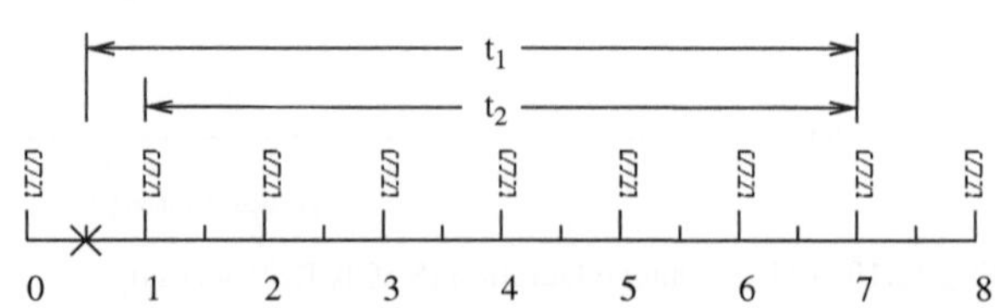

Fig. 12.13 Sample segment arrival pattern ($n_{C,S} = 6$)

within t_1. That is, network restoration should be completed within 6.5 ms of the moment that the network failure takes place. This corresponds to $T_{1,S}$. In the case that a failure occurs between 0 msec and 1 msec, the failure could take place at many different moments, leading to varied values of $T_{1,S}$. For instance, the failures at 0.2 msec and 0.7 msec will result in the values of 6.8 msec and 6.3 ms, respectively. In the extreme, when the failure occurs at the moment very close to 1 msec, the point of time just before the first segment that is dropped by the failure arrives, this failure will lead to the minimum value of $T_{1,S}$, 6 msec. This minimum value is indicated by t_2 in Fig. 12.13 and it corresponds to $\tau_{1,S}$. From the perspective of SONET/SDH operators, $\tau_{1,S}$, instead of $T_{1,S}$, should be used as a restoration objective. This is because $T_{1,S}$ is a specific restoration requirement for a particular scenario. It works perfectly for the specific scenario, but it might not work in other scenarios. However, restoring the network within $\tau_{1,S}$ guarantees that no more than $n_{C,S}$ segments will be lost. For example, if 6.5 ms (one of the varied values of $T_{1,S}$) is used as the restoration objective, then it is possible that the segment arrives at 7 msec will be dropped in the scenario that the failure occurs at 0.7 msec. That is, more than $n_{C,S}$ segments are lost due to the failure and the performance of SACK TCP will be seriously degraded. However, if 6 msec (Namely, $\tau_{1,S}$) is chosen as the restoration objective, then no matter when the failure occurs, at most $n_{C,S}$ segments could be dropped.

Apparently, $\tau_{1,S}$ is the time during which $(rwnd/2+4)$ or $(rwnd/2+3)$ segments pass the failure point. So we have:

$$\tau_{1,s} = \begin{cases} ((\lfloor rwnd/2 \rfloor + 4)^{*}PS)/BW & \text{in Type I Failure} \\ ((\lfloor rwnd/2 \rfloor + 3)^{*}PS)/BW & \text{in Type II Failure} \end{cases} \tag{12.10}$$

where PS stands for packet size and BW is the bottleneck bandwidth experienced by the TCP session being protected. In our simulations, PS is equal to 1500 bytes. This is because from the viewpoint of SONET/SDH networks, each packet is composed of three parts: the 1460-byte TCP payload, the 20-byte IP header, and the 20-byte TCP header. Actually, 1500 bytes is the maximum payload size of an Ethernet frame. The bottleneck bandwidth in our experiments is mostly decided by the client access rate. So for the DS1-32K example presented previously, BW is equal to 1.544 Mbps (DS1). Note that the example SACK TCP session experienced a Type II failure and rwnd equals 23, thus in this scenario $\tau_{1,S} = ((23/2 + 3)^{*}(1500^{*}8))/(1.544^{*}10^6) = 108.81$ msec. Obviously, $\tau_{1,S}$ increases with *rwnd*. If *rwnd* is large enough so that we can approximate $\tau_{1,S}$ as follows:

$$\tau_{1,s} = ((\lfloor rwnd/2 \rfloor)^{*}PS)/BW \tag{12.11}$$

then $\tau_{1,S}$ approximately doubles as *rwnd* doubles. This is illustrated in Figs. 12.10, 12.11 and 12.12

We define SACK Level-2 Restoration Requirement ($T_{2,S}$) as the period from the moment network failure occurs to the moment just before the copy retransmitted due to timeout arrives. And we define SACK Level-2 Restoration Objective ($\tau_{2,S}$) as the period from the moment that the first dropped segment arrives at the failure point to the moment just before the copy retransmitted due to timeout arrives. Similarly, $\tau_{2,S}$, instead of $T_{2,S}$, should be used as a restoration objective. $\tau_{2,S}$ is not as straightforward because it is mainly related to RTO, and RTO is influenced by many factors [9]. RTO usually increases with *rwnd*, and has a minimum value of 1 s. Thus, when RTO is greater than 1 s, $\tau_{2,S}$ increases with *rwnd*. This is illustrated in Figs. 12.10 and 12.11. When RTO is at its minimum of 1 s, $\tau_{2,S}$ does not change much and is independent of *rwnd*. This can be observed in Fig. 12.12. In any case, $\tau_{2,S}$ is always greater than 1 second.

For SONET/SDH operators, either $\tau_{1,S}$ or $\tau_{2,S}$ can be chosen as a restoration objective when SACK TCP is the transport layer protocol. If the restoration can be finished within $\tau_{1,S}$, the overall transfer time will not increase much in the case of network failures. If the restoration time is in the range of $\tau_{1,S}$ to $\tau_{2,S}$, the overall transfer time is increased but it is guaranteed that the TTI is around a fixed value. If possible, $\tau_{1,S}$ should be adopted as the restoration objective because it leads to better resilience than does $\tau_{2,S}$. Other thresholds can be defined on the basis of a third timeout and so on. However, we know that $\tau_{2,S}$ is certainly greater than 1 s. This is already much larger than the de facto target of 50 msec.

According to Figs. 12.10, 12.11 and 12.12, $\tau_{1,S}$ should be adopted as the restoration objective for the scenarios with DS0 and DS1 access. This is because in these cases, $\tau_{1,S}$ is mostly greater than 50 ms, a feasible objective that has been implemented in SONET/SDH networks. For OC-3c access, $\tau_{1,S}$ is mostly less than 10 ms and thus too small to be realistically attainable. In this case, $\tau_{2,S}$ should be chosen as the objective.

There are some exceptions to these typical cases. We should take these exceptions into consideration when we plan to adopt $\tau_{1,S}$ or $\tau_{2,S}$ as the restoration objective for SONET/SDH networks. First, the 512 KB and 1024 KB curve in Fig. 12.10 illustrate situations in which very large receive buffers lead to a calculated value of RTO that is greater than the TCP-defined maximum of 64 s. This puts TCP into Slow Start many times unnecessarily and dramatically changes the normal recovery process. Thus the 512 KB and 1024 KB curve are very irregular. Secondly, in Fig. 12.12, we note that a 1024 KB buffer mostly leads to a shorter TTI than does a 512 KB buffer. This is exceptional because generally TTI increases with *rwnd*. However, the bandwidth-delay product for OC-3c access is 505440 B, and after the failure *ssthresh* is set to half the current flight size, which is around 256 KB in the case of a 512 KB buffer. Setting *ssthresh* to a value less than $r\tau$ hurts link utilization and leads to a longer TTI. Thirdly, in Fig. 12.12, we observe that when outage duration is between $\tau_{1,S}$ and $\tau_{2,S}$, TTI decreases dramatically with outage duration. This is again the result of the large value of $r\tau$. When outage duration is in this range, the sender times out and finally gets into Congestion Avoidance. In the case of OC-3c access, *cwnd* increases with the number of lost segments (corresponding to longer outage duration) when TCP transitions into Congestion Avoidance. In this scenario, the network connection is not fully utilized after the failure because *cwnd* is always less than $r\tau$, so a larger *cwnd* due to longer failure time leads to shorter TTI. Fourthly, in Fig. 12.12, after $\tau_{1,S}$, TTI increases as the *rwnd* increases from 8 KB to 256 KB and it decreases as the *rwnd* increases from 256–512 KB. We know that in the case of OC-3c access $r\tau$ is 505440 bytes. Hence, the curves for 8–256 KB are for receiver window sizes of less than $r\tau$ and those for 512–1024 KB are for sizes greater than $r\tau$. A value for *rwnd* less than $r\tau$ leads to poorer link utilization and so to larger NTT [9]. NTT is the baseline value used to calculate TTI in Equation (12.2). Thus, we have two different classes in terms of TTI, above and below $r\tau$. They are essentially not comparable.

12.4.2 Restoration Objectives for NewReno TCP

We have illustrated that for NewReno TCP, when less than $n_{C,NR}$ segments are lost, TTI is affected by RTT. Losing ($n_{C,NR} + 1$) segments makes a difference, and subsequent losses have little impact until it comes to the loss of the retransmitted copy. Generally, this applies to all NewReno TCP cases with different access link rates and varied *rwnd* sizes. The detailed resilience performance in terms of "TTI vs. Outage Duration" is presented in Figs. 12.14, 12.15 and 12.16.

We define NewReno Level-1 Restoration Requirement ($T_{1,NR}$) as the period from the moment that network failure occurs to the moment just before the segment following the dropped $n_{C,NR}$ segments arrives. We also define NewReno Level-1 Restoration Objective ($\tau_{1,NR}$) as the period from the moment that the first dropped segment arrives at the failure point to the moment just before the segment following the dropped $n_{C,NR}$ segments arrives. Similarly, $\tau_{1,NR}$, instead of $T_{1,NR}$, should

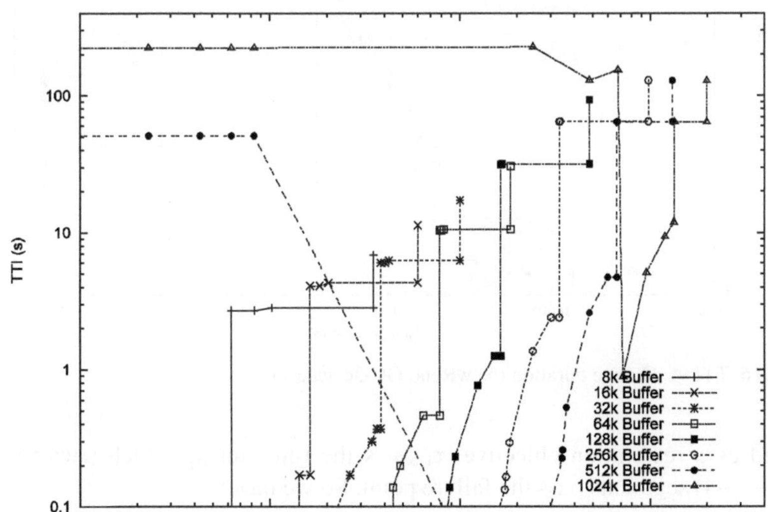

Fig. 12.14 TTI vs. Outage duration (NewReno DS0 access)

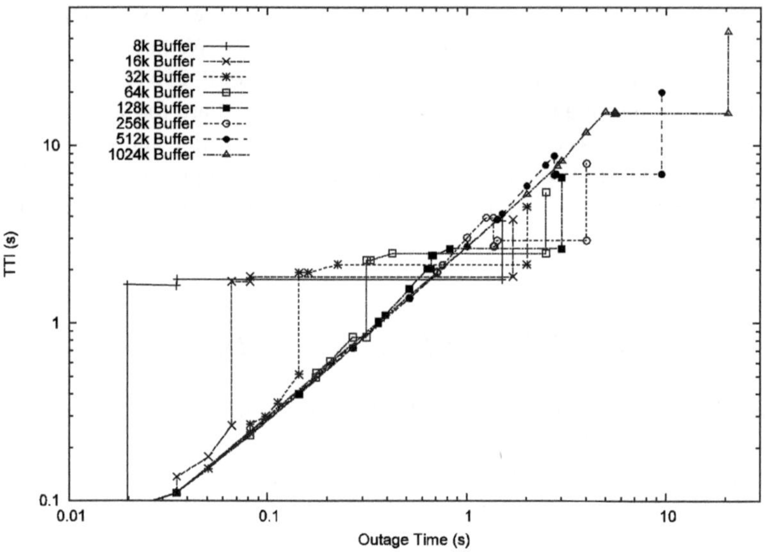

Fig. 12.15 TTI vs. Outage duration (NewReno DS1 access)

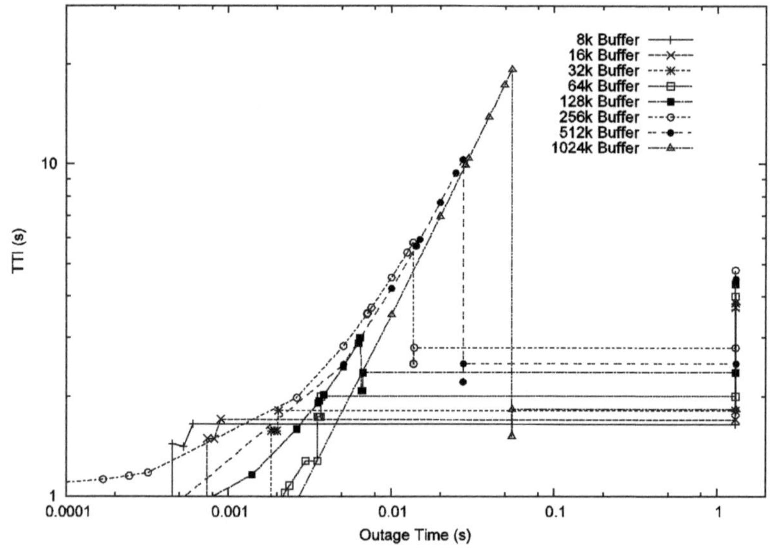

Fig. 12.16 TTI vs. Outage duration (NewReno OC-3c access)

be used as a restoration objective. $\tau_{1,\mathrm{NR}}$ is the time during which (*rwnd* - 3) or (*rwnd* − −4) segments pass the failure point. So we have:

$$\tau_{1,\mathrm{NR}} = \begin{cases} ((rwnd-3)^{*}PS)/BW & \text{in Type I Failure} \\ ((rwnd-4)^{*}PS)/BW & \text{in Type II Failure} \end{cases} \qquad (12.12)$$

where PS stands for packet size and BW is the bottleneck bandwidth experienced by the TCP session being protected. Note that the example NewReno TCP session experienced a Type II failure, *rwnd* equals 23, PS is equal to 1500 Bytes, and BW equals 1.544 Mbps. Thus in this scenario $\tau_{1,NR} = ((23-4)^*(1500^*8))/(1.544^*10^6) =$ 147.67 msec.

Apparently, $\tau_{1,NR}$ increases with *rwnd*. If *rwnd* is large enough so that we can approximate $\tau_{1,NR}$ as follows:

$$\tau_{1,NR} = (rwnd^*PS)/BW \tag{12.13}$$

then $\tau_{1,NR}$ approximately doubles as *rwnd* doubles. This is illustrated in Figs. 12.14, 12.15 and 12.16. From Equations (12.11) and (12.13), we conclude that $\tau_{1,NR}$ is approximately twice as large as $\tau_{1,S}$ when *rwnd* is large.

We define NewReno Level-2 Restoration Requirement ($T_{2,NR}$) as the period from the moment network failure occurs to the moment just before the copy retransmitted due to timeout arrives. And we define NewReno Level-2 Restoration Objective ($\tau_{2,NR}$) as the period from the moment that the first dropped segment arrives at the failure point to the moment just before the copy retransmitted due to timeout arrives. Again, $\tau_{2,NR}$, instead of $T_{2,NR}$, should be used as a restoration objective. $\tau_{2,NR}$ is essentially the same as $\tau_{2,S}$. So all conclusions about $\tau_{2,S}$ also apply to $\tau_{2,NR}$.

Either $\tau_{1,NR}$ or $\tau_{2,NR}$ can be chosen as a restoration objective when NewReno TCP is the transport layer protocol. Generally, $\tau_{1,NR}$ leads to shorter TTI than does $\tau_{2,NR}$. But with NewReno TCP, if the restoration can be finished within $\tau_{1,NR}$, the overall transfer time is influenced by RTT. If RTT is relatively small, the overall transfer time does not change much as restoration time increases; otherwise, TTI increases with restoration time. In Figs. 12.14, 12.15 and 12.16, we observe that, when restoration time is less than $\tau_{1,NR}$, TTI does not change much in the DS0 scenario, but RTT plays a role in the DS1 scenario and TTI increases dramatically with outage duration in the OC-3c scenario. It is interesting that in the OC-3c scenario, for large receive buffers, restoration times longer than $\tau_{1,NR}$ lead to better resilience (i.e. decreased TTI).

Generally speaking, $\tau_{1,NR}$ should be adopted as the restoration objective for the scenarios with DS0 or DS1 access. This is because in these cases, $\tau_{1,NR}$ is mostly at least 50 msec. For OC-3c access, $\tau_{1,NR}$ is mostly less than 35 msec and thus too small to be realistically attainable. In this case, $\tau_{2,NR}$ should be chosen as the objective.

There are also some exceptions for NewReno TCP. We should be careful about the exceptions when we plan to adopt $\tau_{1,NR}$ or $\tau_{2,NR}$ as the restoration objective for SONET/SDH networks. The exceptions due to large *rwnd*, *ssthresh* halving and insufficient $r\tau$ presented in Section 12.4.1 also apply to NewReno TCP and they can be observed in Figs. 12.14, 12.15 and 12.16. The exception with SACK TCP that TTI decreases with outage duration when restoration is finished between $\tau_{1,S}$ and $\tau_{2,S}$ in the OC-3c access case does not occur in NewReno TCP because with NewReno TCP, after $\tau_{1,NR}$ there are only 3 or 4 segments left in the window of data, these segments do not make a significant change to TTI.

12.4.3 Restoration Objectives for Reno TCP

Previously, we demonstrated that losing one segment typically does not change Reno TCP performance dramatically. However, two losses make a difference, and subsequent losses have little impact until it comes to the loss of the retransmitted copy due to timeout. Generally, this applies to all Reno TCP cases with different access link rates and varied *rwnd* sizes. Considering the absolute time period rather than the number of lost segments, we can find the relationship between TTI and network failure duration. The detailed resilience performance in terms of "TTI vs. Outage Duration" is presented in Figs. 12.17, 12.18 and 12.19.

We define Reno Level-1 Restoration Requirement ($T_{1,R}$) as the period from the moment that network failure occurs to the moment just before the segment following the first dropped segment arrives. We also define Reno Level-1 Restoration Objective ($\tau_{1,R}$) as the period from the moment that the first dropped segment arrives at the failure point to the moment just before the segment following the first dropped segment arrives. Similarly, $\tau_{1,R}$, instead of $T_{1,R}$, should be used as a restoration objective.

Essential, $\tau_{1,R}$ is the maximum failure period that allows at most one segment to be dropped. Since two losses make a difference for Reno TCP, segment arrival pattern at the failure point determines the value of $\tau_{1,R}$. The mechanism of generating acknowledgements is related to the segment arrival pattern. RFC 2581 [8] suggests that an ACK should be generated for at least every second segment or within 500 msec of the arrival of the first unacknowledged segment (typically it is 200 ms instead). Based on this mechanism, for DS0 access, an ACK usually acknowledges just one segment and thus triggers transmitting only one segment; for DS1 access

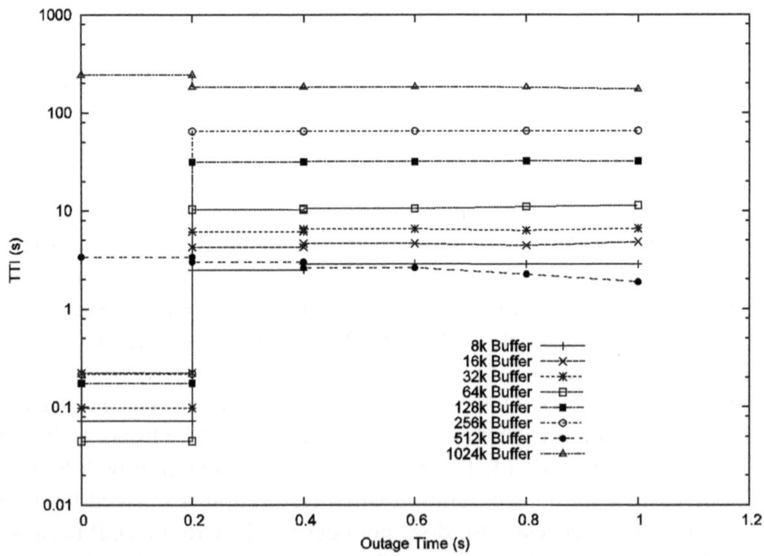

Fig. 12.17 TTI vs. Outage duration (Reno DS0 access)

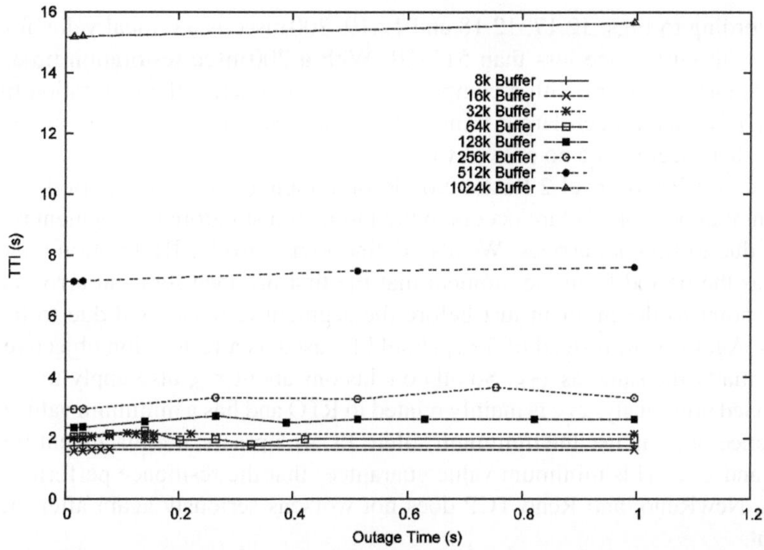

Fig. 12.18 TTI vs. Outage duration (Reno DS1 access)

and OC-3c access, the link is so fast that it is usually two consecutive segments that result in an ACK, which makes the sender transmit segments in pairs. As a result, for DS0 access, there is usually a 200 ms gap between segments, resulting in a 200 ms $\tau_{1,R}$. For DS1 access and OC-3c access (with buffer sizes of 512 K and 1024 K), segments cluster in pairs. For a pair of segments in this case, there is a near-zero gap between them, leading to a near-zero $\tau_{1,R}$.

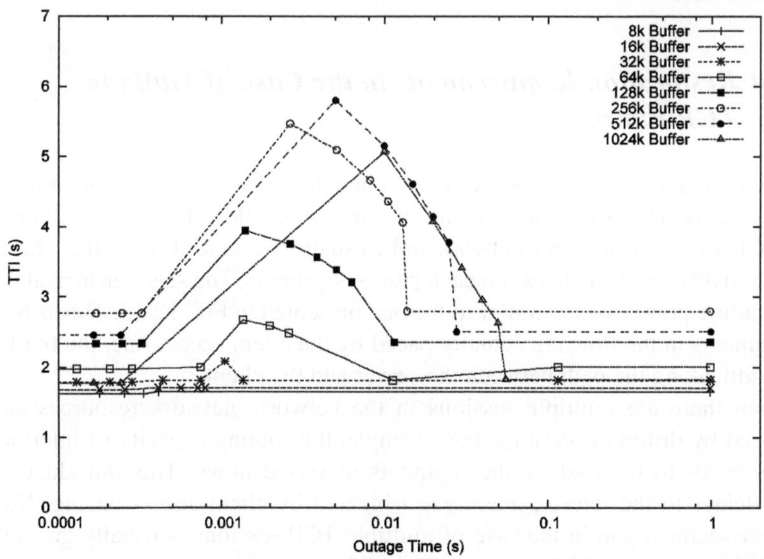

Fig. 12.19 TTI vs. Outage duration (Reno OC-3c access)

According to Figs. 12.17, 12.18 and 12.19, 200 msec is a critical value for DS0 access with buffer size less than 512 KB. With a 200 msec restoration time, TCP can guarantee recovery with a comparatively low impact; if the restoration time is longer, TCP has a much larger transfer time increase. For DS1 access and OC-3c access, there seem to be no critical value.

We define Reno Level-2 Restoration Requirement ($T_{2,R}$) as the period from the moment that network failure occurs to the moment just before the segment retransmitted due to timeout arrives. We also define Reno Level-2 Restoration Objective ($\tau_{2,R}$) as the period from the moment that the first dropped segment arrives at the failure point to the moment just before the segment retransmitted due to timeout arrives. Again, $\tau_{2,R}$, instead of $T_{2,R}$, should be used as a restoration objective. $\tau_{2,R}$ is essentially the same as $\tau_{2,S}$. So all conclusions about $\tau_{2,S}$ also apply to $\tau_{2,R}$. As mentioned previously, $\tau_{2,S}$ is mainly related to RTO and has a minimum value of 1 s. In practice, we can use the minimum value, 1 s, as a practical replacement for $\tau_{2,S}$, $\tau_{2,NR}$, and $\tau_{2,R}$. This minimum value guarantees that the resilience performance of SACK, NewReno, and Reno TCP does not worsens seriously again after the first timeout.

Either $\tau_{1,R}$ or $\tau_{2,R}$ can be chosen as a restoration objective when Reno TCP is the transport layer protocol. However, we should note that $\tau_{2,R}$ is more useful since $\tau_{1,R}$ is usually a near-zero value except for the scenarios in which the client access rate is very low, such as the DS0 cases in our research.

There are some exceptions for Reno TCP, too. Again, we should take exceptions into account when we try to adopt $\tau_{1,R}$ or $\tau_{2,R}$ as the restoration objective. The exceptions due to large receive buffers, *ssthresh* halving and insufficient $r\tau$ presented in Section 12.4.1 also apply to Reno TCP and they can be observed in Figs. 12.17, 12.18 and 12.19.

12.4.4 Restoration Requirements in the Case of Multiple TCP Flows

In the case that there is only one TCP session in the network, there is no other traffic competing for network resources with the single TCP flow. In our simulations with one TCP flow, on the core network link, usually we can observe that there is a comparatively fixed gap between each pair of segments. The segment arrival pattern at the failure point is very similar to the one presented in Fig. 12.13. This is because the segments in the flow are actually paced by the client access rate, the bottleneck bandwidth along the path between the server and the client.

When there are multiple sessions in the network, network resources have to be shared by different sessions. For example, the routing capacity of intermediate routers needs to be used by the segments in varied flows. This introduces additional delays to the inter-segment gap triggered by client access pacing. Namely, the inter-segment gap in the case of multiple TCP sessions is usually greater than the gap in the single-TCP scenario. Thus, if a restoration objective can guarantee that no more than $n_{C,S}$ (or $n_{C,NR}$, or $n_{C,R}$) segments can be dropped in single-TCP case, then in the scenario of multiple TCP sessions, with the same restoration objective, at the most only $n_{C,S}$ (or $n_{C,NR}$, or $n_{C,R}$) segments could be lost. That is, the restoration objectives presented previously for the single-TCP cases also work for the scenarios of multiple TCP sessions. Actually, the objectives in Sections 12.4.1, 12.4.2 and 12.4.3 are the lower bounds for all different scenarios. Both the Level-1 and Level-2 objectives are the lower bounds. However, as mentioned previously, all Level-2 objectives are influenced by many factors and always greater than 1 second. In practice, if a Level-2 objective is a proper restoration objective, we can simply use 1 s as the restoration objective. Apparently, this uniform 1-s objective works for both single-TCP and multiple-TCP scenarios due to the standard retransmission timer calculation algorithm [13]. In this section, we focus on comparing the specific Level-1 Restoration Requirements for various TCP flows in different simulation scenarios with the corresponding Level-1 Restoration Objectives in order to demonstrate that the restoration objectives presented in Sections 12.4.1, 12.4.2, and 12.4.3 also work well in the case of multiple TCP sessions.

In our multiple-TCP simulations, the client access rate is fixed at DS1 and *rwnd* is always equal to 32 KB. This was set on purpose so that the experimental results could be compared with the results from the sample single-TCP simulations presented in Section 12.3. Also, in each simulation run, all different TCP flows employ the same TCP version. That is, at any time, there are only eight SACK (or NewReno, or Reno TCP) flows in the continental-scale network. We believe that the TCP protocol itself is complex enough that it is necessary to first understand how TCP behaves in this baseline scenario before exploring the impact of additional variables.

Table 12.1 summarizes the detailed Level-1 Restoration Requirements for different TCP flows when the failure occurs at 30 s. When SACK TCP is employed, different flows in the network require different specific recovery periods in order to restore the transmission efficiently. For example, for the flow from Boulder to Palo Also and the flow from Lincoln to Palo Alto, the required recovery periods are 119.07 ms and 109.29 msec, respectively. The minimum recovery period among these varied $T_{1,S}$ is 110.05 ms. Namely, in the case that a failure occurs at 30 s,

Table 12.1 Level-1 Restoration requirements for different TCP Flows (In the case that the failure starts at 30 ms, client access is ADSL or DS1, *rwnd* is equal to 32 KB)

	$T_{1,S}$ (SACK TCP)	$T_{1,NR}$ (NewReno TCP)	$T_{1,R}$ (Reno TCP)
Boulder => Palo Alto	119.07 ms	158.03 ms	10.34 ms
Lincoln => Palo Alto	109.29 ms	148.26 ms	0.56 ms
Champaign => Palo Alto	111.86 ms	150.82 ms	3.13 ms
Pittsburgh => Palo Alto	122.83 ms	161.80 ms	14.10 ms
Princeton => Palo Alto	116.71 ms	155.68 ms	7.98 ms
College Pk => Palo Alto	122.62 ms	161.59 ms	13.90 ms
Ann Arbor => Palo Alto	116.94 ms	155.91 ms	8.21 ms
Houston => Palo Alto	110.05 ms	149.02 ms	1.32 ms
Minimum	109.29 ms	148.26 ms	0.56 ms
Average	116.17 ms	155.14 ms	7.44 ms
Variance	28.37	28.38	28.39

Table 12.2 Level-1 restoration requirements for different TCP flows (In the case that the failure starts at 20 ms, client access is ADSL or DS1, *rwnd* is equal to 32 KB)

	$T_{1,S}$ (SACK TCP)	$T_{1,NR}$ (NewReno TCP)	$T_{1,R}$ (Reno TCP)
Boulder => Palo Alto	124.24 ms	163.22 ms	15.52 ms
Lincoln => Palo Alto	114.47 ms	153.44 ms	5.74 ms
Champaign => Palo Alto	117.04 ms	156.01 ms	8.31 ms
Pittsburgh => Palo Alto	112.47 ms	151.44 ms	3.74 ms
Princeton => Palo Alto	121.89 ms	160.86 ms	13.16 ms
College Pk => Palo Alto	112.26 ms	151.23 ms	3.53 ms
Ann Arbor => Palo Alto	122.12 ms	161.09 ms	13.39 ms
Houston => Palo Alto	115.23 ms	154.20 ms	6.42 ms
Minimum	112.26 ms	151.23 ms	3.53 ms
Average	117.47 ms	156.44 ms	8.73 ms
Variance	21.91	21.92	21.98

as long as the network can be restored within 110.05 msec, all the eight flows in the simulation will not be seriously impacted. As mentioned previously, in the scenario of 32 KB and DS1 access, $\tau_{1,S}$ is equal to 108.81 ms. So if $\tau_{1,S}$ is used as the restoration objective, no flow in the simulation will be affected seriously by the network failure. This shows that $\tau_{1,S}$ is actually the lower bound. When NewReno is employed, the minimum $T_{1,NR}$ is 148.26 ms when the failure takes place at 30 s. This is also greater than $\tau_{1,NR}$ in the case of 32 KB and DS access, 147.67 msec. When Reno is employed, the minimum is 0.56 ms. The near-zero $\tau_{1,R}$ the case of 32 KB and DS access is less than this very small minimum. The average and variance of different Level-1 Restoration Requirements for SACK, NewReno, and Reno are also included in Table 12.1.

The detailed Level-1 Restoration Requirements for SACK, NewReno, and Reno in the scenarios that the failure occurs at 20 s and 40 s are included in Tables 12.2 and 12.3, respectively. The results also show that the restoration objectives proposed in Sections 12.4.1, 12.4.2, and 12.4.3 are the lower bounds that work in both single-TCP and multiple-TCP cases.

Table 12.3 Level-1 restoration requirements for different TCP flows (In the case that the failure starts at 40 ms, client access is ADSL or DS1, *rwnd* is equal to 32 KB)

	$T_{1,S}$ (SACK TCP)	$T_{1,NR}$ (NewReno TCP)	$T_{1,R}$ (Reno TCP)
Boulder => Palo Alto	113.89 ms	152.85 ms	5.16 ms
Lincoln => Palo Alto	119.65 ms	158.62 ms	10.92 ms
Champaign => Palo Alto	122.22 ms	161.19 ms	13.49 ms
Pittsburgh => Palo Alto	117.65 ms	156.62 ms	8.92 ms
Princeton => Palo Alto	111.53 ms	150.50 ms	2.30 ms
College Pk => Palo Alto	117.44 ms	156.41 ms	8.71 ms
Ann Arbor => Palo Alto	111.76 ms	150.73 ms	3.03 ms
Houston => Palo Alto	120.41 ms	159.38 ms	11.68 ms
Minimum	111.53.ms	150.50 ms	3.03 ms
Average	116.82 ms	155.79 ms	8.03 ms
Variance	16.18	16.19	16.97

12.4.5 Restoration Objective Recommendation

Based on the experimental results illustrated in Figs. 12.10–12.12, 12.14–12.19, our recommendation on restoration objectives for SONET/SDH networks carrying TCP traffic are summarized in Table 12.4. Specifically:

- For low-rate access (e.g. Dial-Up and DS0), we recommend $\tau_{1,S}$ or $\tau_{1,NR}$ to be the restoration objective if SACK or NewReno TCP is used. This is because in this situation $\tau_{1,S}$ or $\tau_{1,NR}$ is the threshold after which TTI increases markedly. If Reno TCP is the transport layer protocol, 200 ms is recommended.
- For medium-rate access (e.g. ADSL and DS1), if SACK or NewReno TCP is the transport layer protocol, then $\tau_{1,S}$ or $\tau_{1,NR}$ should be chosen as the restoration objective for the same reason mentioned previously. If Reno TCP is used, $\tau_{2,R}$ should be adopted as the restoration objective because $\tau_{1,R}$ is a near-zero value in this case.
- For high-rate access (e.g. Fast Ethernet and OC-3c), $\tau_{1,S}$, $\tau_{1,NR}$ and $\tau_{1,R}$ are all too small to be realistically attainable. Thus, $\tau_{2,S}$, $\tau_{2,NR}$ and $\tau_{2,R}$ should be chosen as the restoration objectives for SACK, NewReno and Reno TCP, respectively.

Table 12.4 provides a clear guideline for adopting a TCP-oriented restoration objective for a certain scenario. According to the guideline, once the information about the type of TCP variant, *rwnd*, PS and BW is available, we can easily find the corresponding restoration objective. And this objective can be used as the restoration target for SONET/SDH networks carrying TCP traffic of the corresponding type.

Table 12.4 can be used for a certain scenario. In our research, we would also like to find the restoration objectives for a set of typical scenarios that co-exist today or will co-exist in the foreseeable future. Once the objective for each scenario in the set is available, we can find a uniform objective and make some quantitative recommendations on restoration objectives for current and future SONET/SDH networks that deal with the co-existence of these scenarios.

We used TCP variant type, *rwnd*, PS and BW to define the typical scenarios. Once these parameters are specified, we can use Table 12.4 to find the appropriate restoration objective for the specific scenario. In terms of TCP variant type, SACK, NewReno and Reno were all included in the typical scenarios since they

Table 12.4 TCP-Oriented restoration objectives

Access Rate	SACK-Oriented objectives	NewReno-Oriented objectives	Reno-Oriented objectives
Low-Rate Access (Such as Dial-Up and DS0)	$\tau_{1,S}$	$\tau_{1,NR}$	200 ms
Medium-Rate Access (Such as ADSL and DS1)	$\tau_{1,S}$	$\tau_{1,NR}$	$\tau_{2,R}$
High-Rate Access (Such as Fast Ethernet, ADSL2+, and OC-3c)	$\tau_{2,S}$	$\tau_{2,NR}$	$\tau_{2,R}$

Table 12.5 Level-1 restoration objectives (in the case of *rwnd* = 16 KB and Type II Failure)

Client access rate	$\tau_{1,S}$ (SACK TCP)	$\tau_{1,NR}$ (NewReno TCP)	$\tau_{1,R}$ (Reno TCP)
DS0: 64 Kbps	1687.50 ms	1500 ms	200 ms
DS1: 1.544 Mbps	69.95 ms	62.18 ms	approximately 0 msec
ADSL2+: 24 Mbps	4.50 ms	4.00 ms	approximately 0 msec
OC-3c: 155 Mbps	0.70 ms	0.62 ms	approximately 0 msec
GigabitEthernet:1 Gbps	0.11 ms	0.01 ms	approximately 0 msec

have been widely deployed [7] and will co-exist for the near future. In terms of *rwnd*, 16 KB, 32 KB and 64 KB were used to as the typical sizes for *rwnd*. Most commonly used operating systems use a range of 16–64 KB to size *rwnd*. Emphasis was put on 32 KB and 64 KB because receive window has been increasing as access rate increases. In terms of PS, we chose the payload size of an Ethernet frame, 1500 bytes, as the de facto PS due to the popularity of Ethernet. In terms of BW, aside from the three access rates used in our simulations, we adopted two new access technologies, ADSL2+ and Gigabit Ethernet, to reflect the foreseeable future.

These parameters led to a large set of scenarios. We divided these scenarios into three subsets by the size of *rwnd*. The first, second and third subset corresponds to the scenarios with 16 KB, 32 KB and 64 KB *rwnd*. For each certain scenario, there could be two restoration objectives, "level-1" or "level-2". If a "level-2" objective is recommended, we simply used 1 second as the practical replacement. The details of "level-1" objectives are illustrated as follows.

Table 12.5 summarizes "level-1" restoration objectives for the scenarios in the first subset. We assumed that all network failures that occur to the typical scenarios are Type II Failure. This does not have a serious impact on the experimental results because there is only one packet difference between Type I and Type II Failure. As shown in Table 12.5, for this subnet, the restoration objective decreases as access rate in increased for SACK TCP. The objective range is from 1687.50 ms for DS0 to 0.11 ms for Gigabit Ethernet. For NewReno TCP, the restoration objective decreases in a similar manner. For Reno TCP, $\tau_{1,R}$ is 200 msec for DS0 access and approximately 0 msec for all other accesses.

Table 12.6 contains "level-1" restoration objectives for the second subset. In this subset, for SACK TCP, the restoration objective decreases from 2625.00 ms to 0.17 ms as the access rate increases from DS0 to Gigabit Ethernet. For NewReno

Table 12.6 Level-1 restoration objectives (in the case of *rwnd* = 32 KB and Type II Failure)

Client access rate	$\tau_{1,S}$ (SACK TCP)	$\tau_{1,NR}$ (NewReno TCP)	$\tau_{1,R}$ (Reno TCP)
DS0: 64 Kbps	2625.00 ms	3562.5 ms	200 ms
DS1: 1.544 Mbps	108.81 ms	147.67 ms	approximately 0 msec
ADSL2+: 24 Mbps	7.00 ms	9.50 ms	approximately 0 msec
OC-3c: 155 Mbps	1.08 ms	1.47 ms	approximately 0 msec
GigabitEthernet:1 Gbps	0.17 ms	0.23 ms	approximately 0 msec

Table 12.7 Level-1 restoration objectives (in the case of *rwnd* = 64 KB and Type II Failure)

Client access rate	$\tau_{1,S}$ (SACK TCP)	$\tau_{1,NR}$ (NewReno TCP)	$\tau_{1,R}$ (Reno TCP)
DS0: 64 Kbps	4687.50 ms	7687.50 ms	200 ms
DS1: 1.544 Mbps	194.30 ms	318.65 ms	approximately 0 msec
ADSL2+: 24 Mbps	12.50 ms	20.50 ms	approximately 0 msec
OC-3c: 155 Mbps	1.94 ms	3.17 ms	approximately 0 msec
GigabitEthernet:1 Gbps	0.30 ms	0.49 ms	approximately 0 msec

TCP, the restoration objective decreases in a similar fashion. For Reno TCP, again, $\tau_{1,R}$ is 200 ms for DS0 access and approximately 0 ms for all other accesses.

Table 12.7 includes "level-1" restoration objectives for the third subset. For both SACK and NewReno TCP, the restoration objective decreases as the access rate is increased. The range of restoration objectives for SACK and NewReno TCP are 4687.50 to 0.30 ms and 7687.50 to 0.49 ms, respectively. For Reno TCP, $\tau_{1,R}$ is a near-zero value except for the DS0 access case in which $\tau_{1,R}$ is 200 ms.

We first present the final results for SACK TCP. According to Tables 12.5, 12.6 and 12.7, $\tau_{1,S}$ is always greater than 1500 msec for low-rate access, such as DS0. Thus 1500 ms can be chosen as the objective for SACK TCP with low-rate access. For medium-rate access, such as DS1, $\tau_{1,S}$ is around 70 ms when *rwnd* equals 16 KB and it is greater than 100 msec when *rwnd* is equal to 32 KB or 64 KB. Considering the migration of client access from low-rate access (such as ADSL) to high-rate technologies (such as ADSL2+) and receive window from 16 KB to 64 KB or higher, we believe that 16 KB receive window will be replaced by larger windows soon. Thus 100 msec is a proper objective for SACK TCP with medium-rate access. For high-rate access, including ADSL2+, OC-3c and Gigabit Ethernet, $\tau_{1,S}$ falls in the range from 0.11 to 0.30 ms. This is too short to be realistically attainable, thus $\tau_{2,S}$ should be used as the restoration objective for SACK TCP with high-rate access. For simplicity, we can adopt 1000 msec as the objective. According to Equation (12.10), we can easily conclude that the objectives of 1500 ms and 100 ms also apply to Dial-Up and ADSL, respectively. SACK-oriented restoration objectives for the typical scenarios under discussion are summarized in the second column in Table 12.8.

Table 12.8 TCP-oriented restoration objectives based on typical scenarios

Access Rate	SACK-oriented objectives	NewReno-oriented objectives	Reno-oriented objectives	Overall objectives (the minimum)
Low-Rate Access (Such as Dial-Up or DS0)	1500 ms	1500 ms	200 ms	200 ms
Medium-Rate Access (Such as ADSL or DS)	100 ms	100 ms	1000 ms	100 ms
High-Rate Access (Such as Fast Ethernet, ADSL2+, or OC-3c)	1000 ms	1000 ms	1000 ms	1000 ms

Similarly, 1500 ms, 100 ms and 1000 ms are recommended for NewReno with low-rate, medium-rate and high-rate access. The details are summarized in the third column in Table 12.8.

For Reno TCP, when the access rate is DS0 or Dial-Up, $\tau_{1,R}$ is 200 ms and it should be recommended as the restoration objective. For medium-rate and high-rate access, $\tau_{1,R}$ is too short to be adopted as an objective. As a result, 1000 ms, the practical replacement for $\tau_{2,R}$, is recommended. This is presented in the fourth column in Table 12.8.

In addition, we can define the overall objective for each type of access rate as the minimum of the three recommended objectives in the same row. The overall objective indicates that for the corresponding access rate, the impact of network failures will not be serious as long as the failures can be restored within the recommended period, no matter which type of TCP variant is involved. For low-rate, medium-rate and high-rate access, the overall objectives are 200, 100 and 1000 ms, respectively.

The minimum of the overall objectives, 100 ms, is the uniform restoration objective for SONET/SDH networks dealing with the set of typical scenarios under discussion. This 100 ms can be recommended as the restoration objective for SONET/SDH networks that deal with the co-existence of these scenarios. Note that this objective is already much looser than the original 50 msec requirement.

However, considering the migration of client access from low-rate ADSL to high-rate ADSL2+ and *rwnd* from 16 KB to 64 KB or even larger, we expect that the last row in Table 12.8 indicates the appropriate restoration objective for the typical scenarios in the near future. That is, 1000 ms, instead of 100 ms should be the restoration target for SONET/SDH networks designed for next-step TCP traffic.

12.5 Conclusions

SONET/SDH has been widely used to build Internet backbones. The restoration capability of SONET/SDH determines how efficiently the Internet can recover from network malfunctions. The default restoration objective in SONET/SDH is for restoration to occur in 50 milliseconds or less. This was derived from the requirements of conventional telephone traffic. Unfortunately this same standard has been forced onto the SONET/SDH transport systems supporting the Internet. Considering that TCP-based P2P file transfer has become the dominant application in the Internet, we studied the reaction of TCP to network failures in a continental-scale network in order to propose the proper restoration objectives for SONET/SDH networks carrying Internet traffic.

We studied the resilience performance of SACK, NewReno, and Reno TCP in the case of a single TCP session and multiple TCP flows. Our experimental results demonstrate that the traditional 50 ms recovery time is not suitable for Internet backbone links carrying P2P file transfer traffic. With SACK TCP, we found two restoration objectives, $\tau_{1,S}$ and $\tau_{2,S}$. $\tau_{1,S}$ is given by Equation (12.10), and $\tau_{2,S}$ is closely related to RTO. With NewReno TCP, we also found two restoration objectives, $\tau_{1,NR}$ and $\tau_{2,NR}$. $\tau_{1,NR}$ is given by Equation (12.12), and $\tau_{2,NR}$ is essentially the same as

$\tau_{2,S}$. $\tau_{1,NR}$ is approximately twice as large as $\tau_{1,S}$ when *rbuff* is large. For Reno TCP, two restoration objectives, $\tau_{1,R}$ and $\tau_{2,R}$, were defined too. $\tau_{1,R}$ is approximately equal to 0 in most scenarios except for low-rate access, such as DS0. $\tau_{2,R}$ is also the same as $\tau_{2,S}$.

Generally, for different scenarios, one of these restoration objectives should be adopted according to the following guideline, summarized in Table 12.4.

- For low-rate access, such as Dial-Up or DS0, we recommend $\tau_{1,S}$ or $\tau_{1,NR}$ to be the restoration objective if SACK or NewReno TCP is used. This is because in this situation $\tau_{1,S}$ or $\tau_{1,NR}$ is the threshold after which TTI increases markedly. In our simulation, when *rwnd* is equal to 32 KB and the access rate is Dial-Up or DS0, $\tau_{1,S}$ and $\tau_{1,NR}$ are 2625 ms and 3562.5 ms, respectively. If Reno TCP is the transport layer protocol, 200 msec is recommended.
- For medium-rate access, such as ADSL or DS1, if SACK or NewReno TCP is the transport layer protocol, then $\tau_{1,S}$ or $\tau_{1,NR}$ should be chosen as the restoration objective for the same reason mentioned previously. In our simulation, when *rwnd* is equal to 32 KB and the access rate is ADSL or DS1, $\tau_{1,S}$ and $\tau_{1,NR}$ are 108.81 msec and 147.67 msec, respectively. If Reno TCP is used, $\tau_{2,R}$ should be adopted as the restoration objective because $\tau_{1,R}$ is a near-zero value in this case. In practice, we can simply use 1 s to replace $\tau_{2,R}$.
- For high-rate access, such as Fast Ethernet, ADSL2+ or OC-3c, $\tau_{1,S}$, $\tau_{1,NR}$ and $\tau_{1,R}$ are all too small to be realistically attainable. Thus, $\tau_{2,S}$, $\tau_{2,NR}$ and $\tau_{2,R}$ should be chosen as the restoration objectives for SACK, NewReno, and Reno TCP, respectively. Namely, 1second should be used as the restoration objective no matter which version of TCP is used.

We studied a set of typical scenarios that co-exist today or will co-exist in the foreseeable future and expect that with the migration of client access from low-rate ADSL to high-rate Fast Ethernet or ADSL2+ and receive windows from 16 KB to 64 KB or even larger, a target of 1 second instead of 50 ms should be the restoration target for SONET/SDH networks carrying Internet traffic.

References

1. A. Antonopoulos, *Metrication and Performance Analysis on Resilience of Ring-Based Transport Network Solutions, GLOBECOM 1999*, Rio de Janeiro, Brazil, December 5–9, 1999.
2. *GR-499-CORE: Transport Systems Generic Requirements (TSGR): Common Requirements*, December, 1998.
3. *ITU-T G.841: Types and Characteristics of SDH Network Protection Architectures*, October, 1998.
4. G. Haßlinger, *ISP Platforms Under a Heavy Peer-to-Peer Workload*, in *LNCS 3485: Peer-to-Peer Systems and Applications*, p. 369–381, Springer, Berlin, Heidelberg, 2005.
5. *Managing Peer-To-Peer Traffic With Cisco Service Control Technology (CISCO White Paper)*, available from: http://www.cisco.com.
6. K. Tutschku, et al., *Traffic Characteristics and Performance Evaluation of Peer-to-Peer Systems*, in *LNCS 3485: Peer-to-Peer Systems and Applications*, pp. 383–397, 2005.

7. J. Padhye, et al., *On Inferring TCP Behavior, ACM SIGCOMM 2001*, San Diego, CA, USA, August 27–31, 2001.
8. *RFC 2581: TCP Congestion Control*, April, 1999.
9. W. R. Stevens, *TCP/IP Illustrated, Volume 1*. Addison Wesley, 2000.
10. *RFC 2018: TCP Selective Acknowledgement Options*, October, 1996.
11. K. Fall, et al. *Simulation-based Comparisons of Tahoe, Reno and SACK TCP*. Computer Communication Review, 26(3): pp. 5–21, July 1996.
12. *RFC 2582: The NewReno Modification to TCP's Fast Recovery Algorithm*, April, 1999.
13. *RFC 2988: Computing TCP's Retransmission Timer*, November, 2000.
14. *RFC 3168: The Addition of Explicit Congestion Notification (ECN) to IP*, September, 2001.
15. *RFC 2328: OSPF Version 2*, April, 1998.
16. *RFC 1267: Border Gateway Protocol 3*, 1991.

Index

MIX
Papier aus verantwortungsvollen Quellen
Paper from responsible sources
FSC® C105338

If you have any concerns about our products,
you can contact us on
ProductSafety@springernature.com

In case Publisher is established outside the EU,
the EU authorized representative is:
Springer Nature Customer Service Center GmbH
Europaplatz 3, 69115 Heidelberg, Germany

Printed by Libri Plureos GmbH
in Hamburg, Germany